PRICELESS FLORIDA

NATURAL ECOSYSTEMS AND NATIVE SPECIES

Ellie Whitney, Ph.D.

D. Bruce Means, Ph.D.

Anne Rudloe, Ph.D.

illustrated by
Eric Jadaszewski

PINEAPPLE PRESS, INC.
SARASOTA, FLORIDA

To Eldred Karker and David Ferrell, who understood instantly what this book was to be about, and to the memory of my husband, Jack Yaeger, whose love for wild Florida I now carry forward.

——Ellie Whitney

To all the animals, plants, and ecosystems that have gone extinct because of mankind, and to those we are now threatening.

——D. Bruce Means

To all the kids now playing on the beach who will continue this work in the next generation.

——Anne Rudloe

Inquiries should be addressed to:
Pineapple Press, Inc.
P.O. Box 3889
Sarasota, Florida 34230
www.pineapplepress.com

Library of Congress Cataloging-in-Publication Data
Whitney, Eleanor Noss.
    Priceless Florida : natural ecosystems and native species / by Ellie Whitney, D. Bruce Means, and Anne Rudloe ; illustrated by Eric Jadaszewsky. –1st ed.
        p. cm.
    Includes bibliographical references (p.   ) and indexes.
    ISBN 1-56164-309-2 (alk. paper) – ISBN 1-56164-308-4 (pbk. : alk. paper)
1. Biotic communities—Florida. 2. Natural history—Florida. I. Means, D. Bruce. II. Rudloe, Anne. III. Title.

    QH105.F6W46 2004
    577.8'2'09759—dc22

                                        2004046504

First Edition
10 9 8 7 6 5 4 3 2
Design by Peter Denes
Composition and layout by Ellie Whitney
Printed in China

Front Cover Photographs
Storm over tidal marsh, Cedar Key—Bruce Colin
Cardinal airplant (*Tillandsia fasciculata*)—Roger L. Hammer
Tiger swallowtail (*Pterouras glaucus*)—Peter Stiling
Pine Barrens treefrog (*Hyla andersonii*)—D. Bruce Means
Purple gallinule (*Porphyrula martinica*)—Florida Park Service
Round-tailed muskrat (*Neofiber alleni*)—Barry Mansell
Suwannee River cooter (*Pseudemys concinna suwanniensis*)— Doug Alderson
Coral reef off the Florida Keys—Thomas K. Gibson, Florida Keys National Marine Sanctuary

Spine Photographs
Scrub balm (*Dicerandra frutescens*)—Thomas Eisner
Barred owl (*Strix varia*)—Pam Sikes

Title Page Photograph
Atamasco lilies (*Zephyranthes atamasca*)—Lois Fletcher

Contents Page Photograph
Spotfin butterflyfish (*Chaetodon ocellatus*)—Florida Park Service

For photographer information, see page 402.

CONTENTS

ACKNOWLEDGMENTS

Florida native, summers: Ruby-throated hummingbird (*Archilocus colubris*).

The authors and publisher gratefully acknowledge the assistance of the Coastal Plains Institute in bringing this book to the public at a reduced price.

We are also grateful to many people for their contributions of expert knowledge and sound counsel: biologist Wilson Baker, naturalist Giff Beaton, oceanographer Henry Bittaker, wildlife biologist and writer Susan Cerulean, attorney Tonya Chavis, botanist Andre Clewell, hydrologist Clyde Conover, botanist Angus Gholson, botanist and naturalist Roger Hammer, photographer Ellen Hausel, meteorologist John Hope, the Reverend Cleve Horne, attorney Mallory Horne, environmental specialist Rosalyn Kilcollins, environmental educator Jim Lewis, geologists Harley Means and Tom Scott, paleontologist S. David Webb, and ichthyologist Jim Williams.

Three editors have contributed importantly to the quality of this book: Patricia L. Johnson, Dana Knighten, and Ken Scott. Without their input, this book would not have come to be.

We also have benefitted from the enthusiastic efforts of our able assistants: Brian Alexander, Bill Beers, Elizabeth Clark, Scott Hauge, Meredith Finley-Simonds, and Sabrina McGriff. Others have helped in numerous ways: Gray Bass, Dan Bracewell, James Burkhalter, Mary Cooke, Linda DeBruyne, Dan Dobbins, Linda Patton, Buddy Payton, and Charlene Smith. At Pineapple Press, June and David Cussen, Shé Heaton, and Alba Aragón have been wonderfully supportive. And special thanks go to the fine support staff at CompUSA—Matt Brown, Keith Geiman, Keith Hall, Chad McClemens, Joey Ricard, and Adam Thomas, and to Stephanie and Mark Weinberger of USA Photo.

The photographers who contributed images to these pages are listed on page 402. Every one of them has been most cooperative and we sincerely thank them all.

* * *

We also want to express gratitude individually:

Thank you, Council Raker. Your support has meant more to me than you can imagine. And thank you to my children, Lynn, Russ, Kara, and your families, and to my "honorary" children, Vicki and Pam, for your love and support. ——Ellie Whitney

Thanks to all those people who took my courses on the ecology of Florida over the past 22 years, and to my beloved sons, Harley and Ryan, and wife, Kathy Steinheimer, for enduring the passions—and the absences—of a naturalist father and husband. ——Bruce Means

I gratefully acknowledge the support of Bill Pinschmidt at Mary Washington College, Bill Herrnkind at Florida State University, the late Nixon Griffiths of the New York Zoological Society, and my husband Jack Rudloe, all of whom have helped me to discover the beauty of the living world. ——Anne Rudloe

FOREWORD

Priceless Florida is one of the most important books ever crafted about Florida's natural history. The rich heritage and biological diversity of Florida are skillfully presented for both laypersons and seasoned scientists to learn from and enjoy. The scientific and educational lessons embodied within these pages are simply astounding. Collected within this marvelous book are detailed natural histories of the native species and communities that comprise Florida. The ecological components and interactions are adeptly related and supported by the kinds of observations that can only be gained through years of experience in the swamps, pinelands, marshes, rivers, and bays of our incredible state.

Florida native: Sand fiddler crab (*Uca pugilator*). Huge populations of this burrowing crab work the soils of tidal marshes and swamps along both coasts of Florida. Chapter 16 gives some attention to the fiddler crab.

The text is organized clearly and logically, with chapters on the various communities grouped by broad ecological type in the manner in which The Nature Conservancy does its conservation work. Sufficient background is also provided to give even the most unfamiliar reader an understanding of the exciting worlds of botany, zoology, marine science, ecology, hydrology, and geomorphology. The book is beautifully illustrated, with myriad diagrams, flow charts, maps, and photographs.

Florida's inhabitants, human and otherwise, depend on the state's fragile environment for survival, but the challenges ahead are enormous. Rapid growth drives the need to secure a system of parks, forests, and wildlife management lands. Without a core system of conservation lands to sustain the ecosystem functions and services upon which all life depends, the quality of life for every Floridian will suffer. Humans also need open space for recreation, water supply, rejuvenation of spirit, and reconnection with nature.

I have no doubt that the readers of this book will become ardent proponents of this message, and most importantly, will love wild Florida even more.

—Victoria J. Tschinkel, State Director
The Nature Conservancy's Florida Chapter

PREFACE

Florida's landscape is much more varied than most people realize, and it supports a surprisingly great variety of ecosystems—several dozen, in fact. And in each ecosystem, a tremendous number of plant and animal species thrive and interact in webs of life that hardly anyone knows anything about.

We think that Florida's ecosystems should be more familiar to most Floridians than they are—for many reasons, and not least because they are so fascinating. They are extraordinarily complex, intricate, ancient, and full of secrets, only a few of which have been revealed to us. The more we learn about them, the greater our pleasure, sense of wonder, and feeling of connectedness. Besides, they are ours, both to enjoy and to protect.

This book is primarily about *healthy* ecosystems. Many books present short descriptions of the natural world and then devote many pages to the damage it has sustained. Those books serve a useful purpose, but in this one, *natural* ecosystems are given all the space available. Similarly, *native* species are in the spotlight throughout. They, too, are fascinating, because they are adapted so beautifully to Florida environments. What must an animal do to live all its life in a cave? What does it eat, how does it "see" in the dark? What sorts of animals live in the bark of a tree? What do they do to the tree, and to each other? How in the world do seaoats survive almost continual onslaughts of salt spray, and even thrive on salt? And how are the plants and animals of an ecosystem woven together in the web of life?

Florida native: Holywood lignumvitae (*Guajacum sanctum*). This extremely rare flower grows in the Florida Keys.

Like the ecosystems, Florida's native species, also, are far too little known—and there are so many. This book, in all its photographs and lists, names hundreds of Florida's natives, but those named represent fewer than five percent of the total that are out there. Each is a masterpiece of biological adaptation evolved over eons of time. Each is worthy of study, but most have not been described or even discovered or named. As a consequence, it is far beyond the scope of a book of this size to present a complete inventory of the native plants and animals in every Florida ecosystem. Rather, one or more groups are singled out for special notice in each chapter. The chapter on lakes offers some detail on dragonflies; the chapter on coastal uplands devotes a few pages to butterflies and migratory birds. Small fishes receive special attention in the chapter on rivers, and mollusks are given a close look in the chapter on estuaries. This mode of

PKS

Florida native: Green anole (*Anolis carolinensis*). The green anole occurs all over the southeastern United States and is the only anole native to Florida.

treatment is as fair as we can make it; even so, many groups have been left out altogether. Lists of some of the plant and animal species in each ecosystem accompany the text to remind the reader of their diversity.

About our approach: the book is written at several levels. For those who are unfamiliar with the geology, geography, or other background necessary to understand Florida's ecosystems, basic terms are presented in definition boxes like the one shown here in the margin, and other essentials are covered in special "background" boxes in the text. For those who are familiar with the terms and topics, the definitions and background boxes are easy to skip.

By the same token, most readers doubtless are more comfortable with the common names of species such as the ghost orchid, while other readers prefer the precision of the scientific names and want the flower identified as *Dendrophylax lindenii*. To accommodate both preferences, all species are identified by their common names, but those depicted in photos are identified both ways, and the Index to Species presents the scientific names for all species mentioned in the book.

Of the many ways in which we could have organized the book, we chose the system developed by The Nature Conservancy for the ecosystems of Florida. Beginning in 1981, The Nature Conservancy helped Florida establish the Florida Natural Areas Inventory (FNAI) to identify the state's natural communities, to single out noteworthy examples of each, and to locate populations of rare and endangered plant and animal species. FNAI identified 69 ecosystems in Florida. FNAI's classification scheme, shown in the appendix, provides the framework for this book.

As for the order of presentation, Chapter 1 presents an overview and makes clear what is meant by "natural" ecosystems and "native" species. Chapter 2 displays the climate, land, soil, and water to which every creature native to Florida is adapted. The next 19 chapters display the ecosystems themselves: first, the interior uplands, wetlands, and water bodies; then the coastal and offshore ecosystems. Finally, to put Florida's ecosystems in

An **ecosystem** is any assemblage of living things, together with the physical environment in which they occur. It may be of any size. A teaspoon of soil is an ecosystem; so is the whole planet Earth.

This book is devoted to Florida's natural ecosystems. A **natural ecosystem** (or **natural community**):

occurs naturally on the landscape wherever certain physical conditions occur.

is a distinct assemblage of populations of plants, animals, and other living things that are naturally associated with each other.

Florida's natural ecosystems include forests, prairies, cave communities, swamps, marshes, bogs, streams, ponds, estuaries, seagrass beds, coral reefs, and many more.

TN/NOAA/NURP

Native to Florida's offshore waters: A polychaete worm in the tropical Atlantic Ocean. Polychaetes are a major group of seafloor animals and are given attention in Chapter 20.

perspective, Chapter 22 tells the amazing story of how they came to be here, a history that can never be repeated, and Chapter 23 returns to the theme of their value.

We offer two cautions to readers. Some of the photo arrays on these pages show closely related plants or animals, but not in enough detail to enable readers to identify them as to species. The photos serve the same purpose as the lists—to give some sense of each community's diversity. Also, despite our efforts, we know there must be errors in the chapters. We will appreciate the reader taking the time to inform us of them.

We hope this book will lead to a better understanding of Florida's ecosystems than most people enjoy at present. Ecology is usually not taught in school and natural Florida is remote from many people's lives today. Most of us see more cars, roads, and shopping malls than forests, swamps, and lakes. Some 90-plus percent of Floridians live in urban areas: Florida is the nation's sixth most-urban state. Frog calls and bird songs are less and less frequently heard. But the natural ecosystems and their resident plants and animals are still out there, beyond the city limits, and they are all the more precious because so much reduced in area and number.

Only when one understands a thing, can one come to love and cherish it. We hope this book leads to greater understanding of natural Florida.

PART ONE

FLORIDA'S NATURAL

RICHES

Florida's natural riches include its native plants, animals, and other living things, its mild climate, its many landscape features and soils, and its numerous and diverse water bodies. Together, these living and nonliving elements form the state's diverse ecosystems. Each ecosystem is a distinct assortment of living things that assembles naturally wherever certain physical conditions occur. Chapter One introduces the ecosystems and the native species in them. Chapter Two describes the landscape context within which the ecosystems have developed.

Aquifer water seen from inside an aquatic cave, looking skyward.

1

CHAPTER ONE

NATURAL ECOSYSTEMS AND NATIVE SPECIES

Florida's landscape includes upwards of 50,000 square miles of diverse natural areas: forests, flatwoods, prairies, swamps, marshes, and waterways. It has 7,800 lakes, 1,700 rivers and streams, and more than 600 clearwater springs. It has a 1,300-mile shoreline of beaches, tidal marshes, and swamps, and its estuaries, seagrass beds, and coral reefs swarm with marine life. This book invites you to explore Florida's lands and waters and become acquainted with the natural ecosystems that assembled themselves here long before human beings ever arrived.

The term *natural*, as used here, is not meant to convey superiority—rather, it simply describes ecosystems that are relatively undisturbed by human influence, ecosystems that still function as they did in pre-Columbian times. This demarcation line is somewhat arbitrary. Even before Columbus, the native people who inhabited Florida purposefully altered the land somewhat by clearing, burning, and planting, but not nearly so drastically as the Europeans and Africans who succeeded them. *Natural* is therefore an approximate term that represents a range of values. Some natural ecosystems are nearly pristine and some resemble Florida's original ecosystems sufficiently to be considered natural for the purposes of this book.[1]

The term *natural ecosystem* has the same meaning as *natural community*, a phrase coined by the Florida Natural Areas Inventory (FNAI). FNAI keeps track of Florida's environmentally important lands and waters following guidelines set by The Nature Conservancy in 1981. FNAI's definition of a natural community appears in the margin.[2]

Importantly, natural ecosystems are all composed of populations of *native* plants, animals, and other living things. Like *natural*, *native* means that the population has existed here for a long time and is adapted to local conditions, including the presence of other native species. Many native species populations are well integrated into natural communities and contribute services of ecological value. Provided that its environment does not

Florida native: Round-tailed muskrat (*Neofiber alleni*). This rare animal is found naturally only in Florida and southeastern Georgia. It occurs in colonies in marshes.

A **natural ecosystem**, or **natural community**, is defined as follows:

It is a ***distinct assemblage of populations*** of living things that inhabits a natural area of land or water.

Its resident populations are ***naturally associated with each other***.

It occurs naturally ***on the landscape*** wherever certain physical conditions occur.

OPPOSITE: Natural ecosystems on Florida's landscape. A meandering stream flows to the Gulf of Mexico through several natural ecosystems on the Big Bend coast. Shown are pine-palmetto hammocks, tidal marshes, the stream with several smaller tributaries flowing into it, and at the shore, the stream's estuary.

Native shrub: Sweet pinxter azalea (*Rhododendron canescens*).
This shrub grows in acid sandy
soils along stream and swamp
margins in north Florida.

Native weed: Lyreleaf sage (*Salvia
lyrata*). This native plant grows
freely over most of Florida and is
especially common on disturbed
sites, but readily gives way to other
native plants.

Native weed: Bushy bluestem
(*Andropogon glomeratus*). This
plant, earlier called broom sedge,
readily colonizes disturbed sites,
then gives way to other native plant
populations.

Introduced exotic plant: Sa-
cred lotus (*Nelumbo nucifera*).
This lotus, an import from the
Old World, occasionally escapes into the wild and grows
in wetlands.

change greatly, a native species population is also likely to be capable of
perpetuating itself without elaborate management efforts. Native plants are
offered as examples here, but the same generalizations apply to native animals and other living things.

Some native plants have narrow habitat requirements: they grow only
on well-drained, sandy soil, or in acid wetlands, or in salt water, and they
can reproduce only in these settings. Some are widespread across the Southeast, the continent, the tropics, or the world. Others are completely restricted
to Florida (see "Endemic Species," later in this chapter).

Some natives are "weeds," that is, plants that move in readily and grow
rapidly wherever space becomes available. Weeds appear promptly where
erosion occurs, where trees fall, fields are plowed, forests are cut, or land is
cleared for roads or construction. Native weeds are important parts of many
ecosystems in transition times. They are the first to move in, to provide
habitat and plant food for insects and other small organisms, and to stabilize the soil while slower-growing native plants are getting established. Native weeds are normally somewhat controlled by their own natural enemies
and are succeeded by other native plants over time.

In contrast to natives, other species have been introduced into Florida
from other parts of the world where they are native. Some introduced (or
"exotic") species are relatively benign in local settings such as parks and
gardens, where they are used as ornamental plants. Either they do not reproduce at all, or they reproduce at a modest rate and it is easy to keep them
from spreading into natural areas. Some exotic species, however, become
invasive; that is, they tend to multiply out of control, taking over the territories of native species and disrupting their webs of relationships. Invasive
exotics are so widespread, and can be so destructive to Florida's natural

Invasive exotic pest plant: Cypressvine *(Ipomoea quamoclit)*. This vine is native to Mexico but invades disturbed sites all over the southern United States and throughout the tropics.

Invasive exotic pest plant (aquatic weed): Common water-hyacinth *(Eichhornia crassipes)*. This noxious weed has one of the fastest growth rates of any known plant. It overwhelms water bodies and displaces native aquatic plants.

Invasive exotic pest plant: Chinese tallowtree *(Sapium sebiferum)*. This tree invades natural areas, reproduces prolifically, crowds out native trees, and drops litter that is toxic to native amphibians and fish.

ecosystems, that concerned citizens maintain an Exotic Pest Plant Council to identify, monitor, and help state agencies control them.

The photos surrounding these pages provide examples of native and introduced flowers and weeds that are found in Florida. Animals and other living things fall into the same categories: some are native, some are introduced, and some are invasive. Among animals, one other important category exists: migratory species that predictably pass through, or reside in, Florida in certain seasons. Migratory birds come to mind. Some are spring and fall visitors; some are winter residents that fly north in summer to breed; and some breed here during summers and spend winters farther south. Migratory butterflies also cross Florida in big flocks. Other migratory species include marine mammals such as manatees and whales; sea turtles; and many species of fish. These populations, while here, are contributing members of local natural communities and interact with permanent residents.

Natural ecosystems, then, consist of populations of native species that have assembled and evolved on Florida's terrain and in its interior and nearshore waters. Examples are pine forests, hardwood hammocks, prairies, marshes, swamps, and aquatic communities in rivers, lakes, and coastal waters. Each ecosystem occupies a particular landscape feature (such as a sandhill, limestone plain, lime sink, cave, basin, streambed, or coastal bay), and each occurs repeatedly across the landscape. Thus, from a bird's-eye view, natural Florida is a mosaic of natural ecosystems.

Today, many lands and waters lie between the extremes of completely artificial and wholly natural ecosystems. They are altered from their his-

Florida native: Seaoats *(Uniola paniculata)*. This native grass is adapted to life on shifting, salt-sprayed sands. It is vitally important in stabilizing coastal dunes.

RPP

Florida migrant: Painted bunting (*Passerina ciris*). This bird's breeding (summer) range includes parts of northeast Florida and the Panhandle. Its winter grounds include parts of south Florida, and it resides the year around in a small strip of territory halfway down the peninsula.

torically natural state but resemble natural ecosystems in some ways. Examples are farms, orchards, ranches, pine plantations, parks, manmade lakes, and restored natural areas. From time to time, this book makes reference to altered ecosystems and explains how they resemble, and how they differ from, more natural ones. The comparison is based on the ways natural ecosystems perpetuate themselves, on their species composition, and on the valuable "services" they perform, such as purifying air and water, abating floods, and stabilizing soil.

DBM

Natural ecosystem: A hardwood hammock. The light-green trees are American beech, the dark ones are southern magnolia. Beech-magnolia forests flourish on many slopes in north Florida and support diverse animals.

FLORIDA'S ECOSYSTEMS CLASSIFIED

There are many ways of classifying ecosystems and many ways of counting them. No system is perfect; all have both merits and drawbacks, but for our purposes, the scheme used by FNAI works well enough. FNAI's *Guide to the Natural Communities of Florida* is listed in the Bibliography.

FNAI has identified sixty-nine ecosystems that occur repeatedly across the state. They fall naturally into six categories: interior uplands, wetlands, and waters; and coastal uplands, wetlands, and waters. The Appendix shows how FNAI's communities are distributed into this book's chapters.

Natural ecosystems possess many valuable qualities and provide a useful model for man-made or artificial ecosystems to emulate. To demonstrate the qualities of natural ecosystems, the next section presents a contrasting ecosystem: an artificial nature exhibit.

ARTIFICIAL ECOSYSTEMS

Florida's theme parks, zoos, and aquariums are examples of non-natural ecosystems. These establishments are important to Florida's economy. The tourists who flock to them enjoy superb entertainment and some of the highest-quality imitation nature in the world. Some of these places also help teach people to care about parts of the natural world. Many a visitor who never before thought about panthers, manatees, or eagles has become a dedicated conservationist after meeting these animals up close and personal in Florida's animal theme parks.

Beneath the surface, though, artificial displays of nature are but high-maintenance versions of the real thing. Many require heat, air conditioning, piped-in water, mowing, weeding, trash pickup, and artificial light. They cannot tolerate normal local weather conditions, much less weather extremes. If damaged, they can't recover. If the walls are breached, disaster follows—either the alligators eat the flamingos or the flamingos fly away. In short, such systems depend completely on people for their maintenance; we have to fight nature to keep them going. Stop the pumps or cut the power, and the water goes bad. Filth piles up. Plants and animals die.

Ironically, keeping these exhibits going is hard on the natural environment. They produce no resources, they consume them. To meet the demands of these operations, millions of gallons of water are pumped from aquifers, depleting local groundwater reserves. For power, huge quantities of coal, oil, and gas are burned, which pollute the region's air. Tons of human and animal waste must be transported, treated, and dumped somewhere. Truckloads of trash are hauled away to landfills. Natural ecosystems contrast dramatically with nature exhibits in all of these respects.

Natural ecosystem: A cypress swamp. Wetlands soak up rain, thereby restraining the oscillations between floods and droughts, and they support numerous native species.

NATURAL ECOSYSTEMS

Like nature exhibits, Florida's natural ecosystems attract tourist dollars into the state and help to educate the public, but environmentally, they cost

Natural ecosystem: Coastal scrub in St. Lucie County. Dry, sandy soil soaks up rain and returns it to the underground aquifer. Florida beaches and coastal scrubs provide much sought-after recreation for both residents and tourists.

much less, and contribute far more. They require less maintenance—in fact, none, if their surroundings and the terrain on which they occur have not been altered. No fossil fuels are needed to supply their energy; no pipes or pumps must deliver their water; no trucks must deliver their fertilizer or food. They clean themselves up; they generate no wastes for processing plants to dispose of because they recycle their resources. They run on renewables: sunlight, soil, air, water, and the work of living things. And for as long as they run, natural ecosystems produce all kinds of plants and animals, feed and house them, and enable them to reproduce, as well as rendering many services to humankind: abating floods, generating food, purifying air and water, and many more.

How do natural ecosystems manage to do so much with so little? If they owe their success to any single secret, it is that their resident populations are genetically adapted to grow and reproduce in local settings. They have evolved in these or similar environments over hundreds or thousands of generations and they are equipped to cope with both normal conditions and normal extremes. In short, they are the native species referred to earlier: they have "local know-how."

Florida's natural ecosystems fail to maintain themselves only if confronted with extremes they have not encountered historically—extremes to which their resident species have not evolved adaptations. If fire-dependent ecosystems are deprived of fire, their species populations decline. If the air, water, and soil become too badly polluted, natural communities deteriorate. If populations of key species are extinguished or if exotic species invade and replace the natives, important relationships vanish, leaving communities weakened and less able to renew themselves. If the terrain is drastically altered, water flows change, and the vegetation changes, becoming less diverse and less able to adjust to further change. If the climate warms

Natural ecosystem: A tidal marsh. The plants of tidal marshes capture the energy of sunlight, grow, and become food for vast quantities of fish, shellfish, and other marine life.

Natural ecosystem: A coral reef in the Florida Keys National Marine Sanctuary. Reef systems provide shelter, food, and mating and nursery grounds for a host of marine creatures including many that people use as food.

or cools too rapidly, plant growth and productivity diminish. Barring such extremes, however, the species populations of natural ecosystems can survive and reproduce indefinitely in local natural settings, under normally varying conditions.

Within these limits, thanks to the adaptations of its native species, a longleaf pine–wiregrass community doesn't have to be watered during the dry season or protected from lightning-lit fires; its trees and other native residents adjust to normal droughts and fires and continue growing and reproducing. The seagrass beds off Florida's coasts don't need protection from summer heat, winter cold, or tropical storms; they tolerate normal hot, cold, and turbulent weather and recover. When fires, floods, droughts, heat, or cold approach extremes, the living members of natural ecosystems may be stressed, but they rebound at the first opportunity, provided only that invasive exotics are kept out and that repeated injuries are prevented.

Natural ecosystems can even recover after natural "disasters" such as hurricanes and floods, which may appear to overwhelm them temporarily. Some ecosystems promptly reassemble in the same place where they grew before. For example, lakeside vegetation regrows after flood waters recede. Some ecosystems regroup elsewhere as suitable new places become available. For example, a hurricane may completely erase a beach dune with all its vegetation, but the same species of plants will regrow on sand eroded from that dune and deposited on another. In short, natural ecosystems tend to persist, provided that the physical stresses they encounter are within the bounds to which they are adapted.

Besides maintaining themselves largely without outside help from people, natural ecosystems conduct processes that support human life, and indeed all life, on the planet. Ecosystems regulate the climate, control concentra-

Natural ecosystem: A seagrass bed in St. Joseph Bay, Gulf County. The seagrass beds in Florida's near-shore waters are among the ocean's most productive ecosystems.

Cypress (a *Taxodium* species). Like all native species, cypress is adapted to, and interacts with, its environment. On this river bank, the tree's roots not only enable the trees to withstand powerful flood waters, but also stabilize the bank.

tions of atmospheric gases, circulate and purify water, produce food, and more. Chapter 23 describes these processes further.

ALTERED AND RESTORED NATURAL AREAS

Altered lands and waters are here to stay, and some are of poor environmental quality. However, some altered areas make the environment more pleasing, safer, or more profitable for people while still providing natural services to an extent. We need to appreciate the benefits these areas can deliver and learn to design them to maximize both their economic and their environmental values.

Some altered areas retain some of their original functions. An example is a second-growth forest of native pines managed for timber production using nature's materials and methods: native pines and ground-cover plants, periodic fires, selective harvesting, and cultivation of uneven-aged stands of trees. Such forests still filter water, purify air, conserve soil, stabilize land contours, and continue to provide habitat for native species. Some man-made lakes are also in this category. The normal oscillations between high and low water levels, which help to keep natural lakes healthy, no longer occur spontaneously, but can be simulated by lake managers. Moreover, if lake shorelines are planted with natural wetland vegetation, they can still help abate floods, recharge ground water, take up excess nutrients from stormwater runoff, and support native wildlife and fish.

Other alterations are more disruptive and produce areas that are less natural. Examples among forests include commercial plantations of single

A city garden. Although not a natural ecosystem, a park embodies many of the same values and can be a benign use of the land.

species such as slash pine or sand pine; forests that ultimately take over abandoned farm fields; forests that have had major tree populations logged out; and fire-dependent forests that have not been allowed to burn. Yet even these forests can produce masses of living things; allow rain to filter into the ground; recycle water to the air; help to cycle oxygen, nitrogen, mineral nutrients, and wastes; help to filter air and water; and provide habitat for game and other wildlife.

At the far end of the spectrum are degraded lands and waters. They exist in all settings. Familiar to all are paved parking lots, roads, and most road rights-of-way, which occupy millions of acres of land and of course support very few native species. Less well known is that much of Florida's coastline and especially its bay shores are so densely populated today that the natural communities of estuaries and lagoons have largely broken down. To regain natural function these need more than management; they require restoration. Although living communities may still grow in them, they are severely impoverished in numbers of native species and may hold none at all. List 1–1 provides some well-known examples of other degraded areas.

In summary, although few areas are wholly natural any longer, many can be maintained in as near-natural condition as possible. Where we must alter the environment, modifications can be designed to sustain, rather than disrupt, natural community composition and processes. On large tracts of land such as farm, ranch, and forest lands, in some basins and river systems, and in some lakes and ponds, environmental benefits can be reaped by supporting and mimicking natural processes. Even small plots of land can help, especially if many are added together. In cities and suburbs, people's yards and home gardens can help meet the habitat needs of some native plants, insects, birds, and other animals. Roadside strips can be used to grow native wildflowers. Artificial reefs can provide high-quality substitute (and much needed) habitat. This book is devoted to those ecosystems that are thought to most closely resemble Florida's original ones, but some of their qualities can be recruited for additional uses in today's contexts, and all lands can have some ecological values. There is, however, one facet of natural ecosystems already mentioned, which no other lands and waters can quite match. Natural ecosystems consist of, and support, native species.

FLORIDA'S NATIVE SPECIES

For review, Background Box 1–1, below, offers a definition of species in general. Background Box 1–2 follows with a description of the five kingdoms of living things within which species are the basic unit.

BACKGROUND   1–1

Notes on Species

A species is a distinct group of organisms that reproduces its own kind. A species may exist as several populations occupying separated habitats, but for as long as the individuals readily breed with one another, they are considered to be members of one species. If two populations of organisms

(continued on next page)

LIST 1–1
Degraded lands and waters (examples)

Lands
Fire-dependent forests from which fire has been excluded
Clear-cut forests with their soils chopped up
Drained, dredged, or filled wetlands
Areas where solid and toxic waste materials are dumped
Roadside ditches
Areas that have been invaded by exotic species such as kudzu, punktree (Melaleuca), or Chinese tallowtree
Most construction sites
Coastal areas where seawalls, groins, and jetties have been constructed
Parking lots, highways, and road rights-of-way

Waters
Channelized (straightened) rivers and canals
Dredged channels in river floodplains, wetlands, and bays
Piles of dredged material
Waterways that have been overwhelmed by exotic weeds such as common water-hyacinth or water-thyme (Hydrilla)
Seagrass beds scarred by propellers or smothered in silt
Seafloors scraped bare by dragging nets
Polluted zones at river mouths, along shores, in the open sea, and on the seafloor

> A **species** is a distinct group of organisms that reproduces its own kind and is reproductively isolated from other such groups.

are similar but display no tendency to interbreed when brought together, they are distinct species.[a]

A species breeds true because every individual receives the same basic genetic blueprint. For example, all baby raccoons inherit the information necessary to make fur, four legs, a ringed tail, black eye patches, and so forth—and to keep on making new raccoons. The inherited traits may vary a little in detail due to genetic diversity, but the basic structures and functions will be the same in all.

If a species breeds true, then how do new species arise? It is known that a small change in a population's genetic material can occasionally produce a new species within a generation or two.[b] Often, however, speciation is extremely slow. It begins with reproductive isolation. A single population may be split geographically into two subpopulations that remain separated for long times. Then, over about 20,000 generations, the separated subpopulations may develop so many genetic differences that they become unable to interbreed.[c] Florida's seven pine species, three of which are shown in Figure 1–1, probably evolved that way, long before Florida itself existed.

Speciation is the process by which new species arise, and viewed from the perspective of a human lifetime, it is often a slow process. In people, assuming 20 years per generation, it might take 400,000 years of separation to produce a new species. Comparably long times are required for slow-growing trees and for animals such as manatees, panthers, and sturgeon, whose generation times (to sexual maturity of the young) are on the order of several years to a decade or so. In plants and animals that produce one new generation a year, speciation might take

Slash pine (*Pinus elliottii*). This pine grows well in low, seasonally wet flatwoods. Its bark's broad, reddish plates resemble rectangular tiles.

Longleaf pine (*Pinus palustris*). This pine grows best in grasslands that frequently burn. It has the longest needles and largest cones of any Florida pine.

Loblolly pine (*Pinus taeda*). This pine occupies moist woodlands and often moves in, together with hardwood trees, where other pines have been cut.

FIGURE 1–1

Three Florida Pines

All pine species are descended from a single ancestral population that was present tens of millions of years before Florida came into being. Seven of them now grow naturally in Florida, each in a particular habitat in which it can dominate other trees. Thus the genus *Pinus* has come to occupy large areas of the landscape.

Note: Florida's other native pines, featured later, are Shortleaf pine (*Pinus echinata*), Spruce pine (*Pinus glabra*), Pond pine (*Pinus serotina*), and Sand pine (*Pinus clausa*).

20,000 years. In mice, which produce several new generations a year, it takes about 5,000 years. Extremely short times are needed for bacteria, some of which can produce several generations a day.

Speciation rate also depends on the strength of natural selection, which can either speed it up or slow it down. Here, the 5,000 years estimated for mice are assumed to approximate an average for all species, but immense variation is the rule.

Over the 3.8 billion years or so since life first arose on Earth, hundreds of millions of new species have arisen. Nearly as many have also gone extinct. Until recently, however, the rate of production of new species has, on average, slightly exceeded the extinction rate. Thanks to the enormously long spans of time since the origin of life, net species gains over losses now add up to a total, for the world, of many millions of species—perhaps even ten or 100 million, no one knows.[d] In Florida, new species of several animals and plants are presently evolving in separated river systems (as explained in Chapter 13) and on the barrier islands offshore (Chapter 15).

Notes: [a]Under abnormal circumstances, members of two different species may mate and produce hybrids, but then the hybrids are sterile.

[b]A sudden, small change in chromosomal DNA sufficient to change a multi-limbed strain into a six-legged strain of animals has been observed in the laboratory.

[c]Speciation may occur in still other ways. For example, an accident during fertilization may produce offspring with more chromosomes than the parents possessed—polyploid individuals, the progenitors of a new species. This may be a sudden occurrence, but the point here is that speciation probably usually takes long times.

[d]About 1.75 million species have been discovered and described. There are at least 8 million and probably more species still to be discovered.

Sources: Mode of speciation from Palumbi 2002; rate from Norton 1986, 121–122; mechanisms from McCook 2002.

Florida native: Eastern indigo snake (*Drymarchon couperi*). This large, harmless, beautiful, almost iridescent snake once ranged widely across the southeastern United States, but has become very rare. Because it needs a large territory unbroken by roads or development, it now occurs in only a few sites.

BACKGROUND

1–2

Categories of Living Things

In the scheme used here, living things are members of five different kingdoms: animals, plants, bacteria, protists (single-celled organisms), and fungi. Each kingdom is subdivided into smaller groups as follows:

Kingdom—Phylum (or **Division**, for plants)—**Class—Order—Family** (family names always end in *ae*)—**Genus** (the plural is **genera**)—**Species** (and the plural is also **species**).[a]

Every species has a place within this scheme of relationships. The most familiar example is that of our own species.

(continued on next page)

Protists are single-celled organisms with true nuclei.

Bacteria are simpler single-celled organisms.

Fungi (FUNJ-eye) are multicellular organisms: molds, mildews, mushrooms, and their relatives.

In this book, "plants and animals" refers to the members of all five kingdoms. "Plants" refers to all organisms that can conduct photosynthesis: multicellular plants, multicellular algae, single-celled algae, and photosynthetic bacteria.

The human species is classified as follows:

Kingdom: *Animalia* (animals)

Phylum: *Chordata* (chordates)

Subphylum: *Vertebrata* (vertebrates)

Class: *Mammalia* (mammals)

Order: *Primata* (primates)

Family: *Hominidae* (ho-MIN-id-ee, hominids)

Genus: *Homo* (human)[b]

Species: *Homo sapiens* (humans).[c]

Florida native: Turkey vulture (*Cathartes aura*). This scavenger of carrion plays an important cleanup role in natural ecosystems.

Earlier in our history, our closest relatives were the other members of the genus *Homo* and other hominids, but all are now extinct. Today our closest relatives are other primates. Next are other mammals, then the four other classes of vertebrates: birds, reptiles, amphibians, and fish. Besides the chordates, the animal kingdom has some two dozen other phyla, to which we are more distantly related.

The plant kingdom includes five divisions. One is the familiar vascular plants. That division and the other four include several hundred thousand species worldwide.[d]

The other three kingdoms consist of so-called "simple" organisms: bacteria, protists, and fungi. Through ancestors remote in time, we animals are related to all of these as well.

Notes: [a]Species may be subdivided into still smaller groups (varieties, subspecies, or races). These are thought to be populations on their way to becoming new species.

[b]By an internationally agreed-upon convention, genus names are capitalized and italicized.

[c]All species are given two-word names: the genus name with the species name appended. The two-word name is in italics, the genus name is capitalized, and the species name is never capitalized, even if derived from a proper noun (such as *americanus*).

[d]The other four divisions of the plant kingdom are the red algae (40,000 species); brown algae (1,100 species); green algae (7,000 species); and mosses, hornworts, and liverworts (23,500 species).

Florida native: Golden silk spider (*Nephila clavipes*). This spider strings up its giant webs in shaded woodlands all over the southeastern United States.

The Names of Species. One way to become familiar with a species is to know its scientific, as well as its common, name. Those featured in this book's photos are given both names in photo captions, but for ease of reading, most species mentioned in the text are given just their common names. Every common name is an "approved" one, though: authorities have specified what it should be. Thus the bear that is native to Florida is the Florida black bear; the two species of sturgeon that swim in Florida's rivers are the Atlantic sturgeon and the Gulf sturgeon. Using those names, all species are listed in the Index to Species, which precedes this book's General Index, and the scientific name for each one is also shown. The Reference Notes and Bibliography, after Chapter 23, list the authorities for the naming of species.[3]

Becoming acquainted with Florida's native species is a mind-expanding experience, somewhat like stepping through a small door to discover a multi-story mansion of a thousand rooms. Many a bird watcher is thrilled at first to see a dozen kinds of birds at the backyard feeder, but then goes on to learn that 30 times that number of bird species can be seen in Florida. At the St. Marks Wildlife Refuge alone, nearly 300 species of native birds have been observed. Moreover, birds are only one class of vertebrates (animals with backbones), vertebrates are only a small group among animals, and animals are only one of the earth's five kingdoms of living things. If one were to try to enumerate all of Florida's species, though, the vertebrates would be a good group with which to begin because they are familiar and closely related to us.

Florida native: Turkey-tail mushroom (*Trametes versicolor*). This fungus is widespread across the world, and like its fellow fungi, it digests dead wood.

A Count of Florida's Species. A list of all of Florida's land vertebrates—frogs, snakes, lizards, mice, birds, and all other terrestrial animals with backbones—would number some 700 species. In addition, at least a thousand fish species swim in Florida's surface and nearshore waters, about one fourth of all the fish species known in the western hemisphere north of the equator.[4]

The invertebrates—animals without backbones such as snails, dragonflies, and earthworms—greatly outnumber the vertebrates, however. Florida has about 30,000 species of land invertebrates, and thousands more in the rivers, lakes, and near-shore waters. Taken all together, the native plants, animals, and small and microscopic living things that occur within the state and in its near-shore waters probably number in the tens of thousands. Moreover, many species remain to be discovered in deep ocean waters and on seafloors offshore.[5]

In addition to its animals, Florida possesses more than 4,000 native species of trees, shrubs, and other flowering plants. It has more than 100 species of native orchids (compared to Hawaii's two species). Its ferns, numbering some 150 species, are the most diverse in the nation, and besides all these, there are many mosses, hornworts, liverworts, and algae. Three other whole *kingdoms* of living things are required to accommodate the protists, bacteria, and fungi.[6]

The photographs in this book are intended to give a sense of the diversity of Florida's native species. In proportion to the numbers of species in each kingdom, however, animals (especially birds) and plants (especially flowering plants) are over-represented. Florida is noted for the abundance of beautiful, conspicuous specimens of both.

Florida native: Horseshoe crab (*Limulus polyphemus*). This marine invertebrate visits beaches to mate and lay its eggs.

Florida's Endemic Species. Native species are of two types. Some are widely distributed; some are limited to small geographical areas. The live oak is an example of the first type. Live oak grows naturally in coastal and near-coastal areas all the way from Virginia, across Florida, and well into

Florida willow (*Salix floridana*)

FIGURE 1–2

Endemic and Near-Endemic Florida Species: Examples

The Florida willow is a near-endemic: it grows only in Florida and very nearby. The thickleaf waterwillow is endemic: it grows exclusively in Florida.

Thickleaf waterwillow (*Justicia crassifolia*)

Texas. The gopherwood (*Torreya*) tree is an example of the second type: it grows only in north Florida and very nearby in Georgia and is a "near-endemic" species (see Chapter 6). Figure 1–2 shows examples of a near-endemic and an endemic species.

Several hundred species and several natural communities are endemic to Florida. Among the endemic plants are about 300 plants and many algae. More than 400 endemic invertebrate animals are known; and among the vertebrates, more than 40 are endemic to Florida. No one knows how many endemic bacteria, protists, and fungi there are. Counts of Florida's endemic animals are in List 1–2, and the state's endemic natural communities are in List 1–3.

Florida's native species, and especially its endemic species and communities, are ours to protect or lose. An objective of the chapters to come is to make their value clear.

FLORIDA'S SPECIES TODAY

Figure 1–3 shows that Florida's species are among the most endangered in the nation. This is largely due to destruction of their habitats. When the last member of a species' last population dies out, the information necessary to produce new individuals of that species vanishes forever. The total information available for making living things also diminishes. Upon extinction, the information accumulated in a species over hundreds of millions of years is gone for all time.

Because new species have arisen more frequently than existing ones have gone extinct, the world has accumulated a net living reserve that today amounts to several tens of millions of species. Now, however, that trend has reversed. It still takes approximately 5,000 years, to produce a new species,

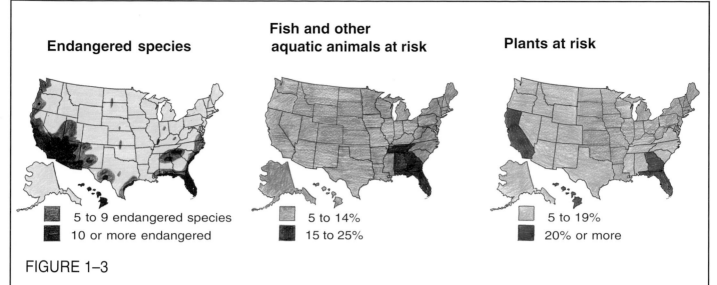

Endangered species

Fish and other aquatic animals at risk

Plants at risk

- ▨ 5 to 9 endangered species
- ■ 10 or more endangered

- ▨ 5 to 14%
- ■ 15 to 25%

- ▨ 5 to 19%
- ■ 20% or more

FIGURE 1–3

Florida Ranks High in Threats to Species

The maps show that Florida, Hawaii, California, and Georgia rank highest in the nation in numbers of species of plants, animals, and other living things becoming threatened and in danger of going extinct.

Sources: Adapted from Rodriquez and Roberts/Environmental Defense Fund 1997; Flack 1996; Doyle 1997.

on average, but species losses today, by current estimates, are amounting to dozens a day. Moreover, the extinction rate is *accelerating*.

The pace of Florida's species losses matches that of the rest of the world and also is accelerating. However, most of Florida's species have not gone, and need not go, extinct. If the people of Florida know and value local native species sufficiently, they will take what steps they can to protect local ecosystems.

* * *

This chapter has shown that Florida's ecosystems are very diverse. The next chapter shows why this is so—because the ecosystems are tied to a physical landscape which, itself, is extraordinarily diverse.

LIST 1–3
Natural communities endemic to Florida

(P) indicates communities in which the main plants are endemic. (C) indicates those in which the combination of plants is unique to Florida.

Dry prairie (C), widely scattered statewide
Oak scrub (P), in Highlands, Orange, Osceola, Polk
Pine rockland (C), in Miami-Dade and Monroe
Rosemary scrub (C), widely scattered across Florida
Sand pine scrub (P), in peninsular Florida
Seepage slope (P), in Hendry, Highlands, Osceola, Polk
Slash pine scrub (P), in Highlands County
Upland glade (C), in Gadsden and Jackson counties

Notes: Slash pine scrub may also be called Slash pine sandhill. Oak scrub may also be called Inopina oak scrub. Dry prairie may also be called Palmetto prairie.

Source: Adapted from Muller and coauthors 1989.

CHAPTER TWO

THE CONTEXT:
CLIMATE, LAND, SOIL, WATER

Adaptation is the key to the success of Florida's native plants and animals. They are adapted to the physical environment: the climate, topography, soils, and waters of Florida, and so they are able to survive the normal extremes in these factors over many millennia. The nonliving components of Florida's ecosystems are the physical context that has molded every species from the time of its arrival to the present.

CLIMATE AND WEATHER

Florida receives floods of life-sustaining sunlight and experiences mild weather that, for the most part, varies gently in temperature from season to season. Many living things thrive easily in such conditions. Stresses are presented, however, by alternating wet and dry seasons, punctuated by periodic tropical storms and hurricanes. Each species has developed adaptations to cope with the range of conditions it encounters, from hot to cold, sunny to cloudy, wet to dry, and stormy to calm.

Temperature and Humidity. Because the peninsula is so long, north to south, climatic conditions vary somewhat. Florida lies within the temperate zone, but the climate is subtropical in south Florida and tropical in the Keys. Winter cold fronts often penetrate north Florida, but seldom reach farther south, so north Florida winters have both more varied temperatures and more rain than winters in the rest of the state. This is one of many reasons why the species composition of ecosystems differs between north and south Florida.

The Gulf of Mexico also influences Florida's climate. Elsewhere in the world at Florida's latitude, deserts predominate, with wide variations in temperature and exceedingly dry air. Florida, though, has water on three sides. Warm water moves constantly along Florida's Gulf coast, around the peninsula's southern tip, and along the whole east coast. Proximity to

Florida native: Wormvine orchid (*Vanilla barbellata*). South Florida's warm and humid climate fosters the growth of tropical beauties like this rare orchid, which grows in mangrove swamps and coastal hammocks of Miami-Dade and Monroe Counties.

The **weather** is the state of the atmosphere at particular times and places.

The **climate** is the long-term state of the atmosphere defined by its averages and extremes over time.

OPPOSITE: A typical sunny day in central Florida. Bright sunlight pours down among scattered clouds and bathes sand pines in the Ocala National Forest, Big Scrub.

19

Frost at Alachua Sink, Alachua County, in central Florida. Frost has occurred nearly everywhere in Florida at one time or another in the past few decades.

water on three sides produces high humidity and abundant rain. The rain helps to maintain the interior's many lakes and wetlands, and these help to keep temperature variations to a minimum.

Although the climate is relatively constant over long periods and the weather is mild on average, day-to-day weather conditions in Florida vary tremendously, as weather does in most places. Temperatures can range in some areas from lows of below freezing on winter nights to highs of 90 degrees Fahrenheit or above on summer days. Snow, sleet, and hail, as well as temperatures above 100, although rare, are not unknown.

Rain. The quantity of rain that falls on Florida sometimes seems stupendous. Even a single summer shower, if it runs down a slope and collects in a low place, can quickly become several feet deep. A stalled tropical storm can drop up to 40 inches of rain in just a few days, and nine inches of rain can easily fall on a single location within a single afternoon.[1]

Heavy rainfalls are not the rule, however. Most rain showers bring down an inch or less. Only because they are so frequent do they add up to some four and a half feet a year. If all that rain soaked into the ground, it could easily sustain ground water levels against current and future demands for eternities to come. Actually, however, about three-fourths of the rain returns to the sky and about one-fourth runs off to the sea. Only about one inch each year soaks into the ground and stays there for a while. That one inch compensates for the one inch of ground water that runs out to sea each year by way of seepage and through springs.

Although the average rainfall in Florida amounts to about 53 inches in a typical year, areas of the state vary. The western Panhandle receives somewhat more rain than does the rest of Florida—60 inches, on average. The east coast receives less—50 inches. And wherever it is measured, the rainfall deviates greatly from the average in many seasons.[2]

PRICELESS FLORIDA

Sandy depression before and after a heavy rain. An inch or two of rain can fill a large depression with several inches of water for a time.

Any given area of Florida has, on the average, some 120 rain events a year. Most are showers, but some ten to twelve of these events are storms, including tropical storms and hurricanes. To stay in place, any plant must be well rooted. Especially on slopes, storms would cause tremendous erosion, were it not for the trees, shrubs, and ground-covering plants that hold the soil in place. A single shower may drop more than a billion tons of water in one little area. Each raindrop can dislodge material where it falls and carry that material downhill. On clay-rich, sloping land, raindrops quickly gather in surface indentations, form channels, and run off into streams and rivers. On sandy, level land, water sinks down, but only until it reaches an impermeable layer (a hardpan, which often underlies sand near the surface). Then the water accumulates and stands on the surface, sometimes for long spans of time.[3]

The most dramatic of Florida's storms cause rivers to surge out of their banks, sweeping their floodplains clean of accumulated litter. At the coast, these storms bring surges that rearrange masses of sand, annihilating beaches, depositing bars, shortening and elongating islands, and shifting submarine shoals from place to place. As inconvenient as these events may be for human beings, coastal and marine ecosystems, if not interfered with, continue to adjust as they have for millions of years.

Small and large rainfalls produce different benefits. Besides cleaning leaf surfaces and adding to groundwater reserves, small rains offer relief to Florida's dry-land plants during times of stress. Plants are adapted to withstand dry periods, but do not go altogether dormant, and prolonged droughts can set them back or even kill them.

Rainy and Dry Seasons. Alternating wet and dry seasons have characterized the region's weather patterns for thousands of years. Summers are wet all over the state, throughout June, July, and August (Floridians call August's rainy days the "dog days"). Winters are dry; in fact, October, November, and December are the driest months of the year. Other months' rainfalls differ between north and south Florida. North Florida has considerable rain in late winter (December to April), and then May turns dry. South Florida has little rain all winter except briefly before cold fronts.[4]

Florida native: Green arrow arum (*Peltandra virginica*). This water-loving Florida native grows best in swamps and hydric hammocks.

Plant and animal life cycles are in synchrony with rainy and dry seasons. Many plants drop their seeds just before seasons of heavy rainfalls. Animals such as salamanders and frogs breed during rainy seasons. They carry the timing within their genes, and their lives depend on the rains' accustomed schedule.

Lightning. Fires ignited by lightning strikes are an important force shaping many Florida ecosystems.

For wetlands, seasonal rains play a special role: they provide periodic times of inundation alternating with dry times. Wet times have laid standing water on low areas twice a year for so many millennia that many plants, once rooted and growing, are able to remain alive even when standing in water for weeks or months at a time. Some plants tolerate standing water for three months every year; some five; some are adapted to almost continuous inundation. The effects of periods of wetness (known as hydroperiods) on vegetation are described further in the chapters on wetlands, Chapters 8 through 11.

Some animals need equally specific wet-and-dry periods to breed and feed and raise their young. Without the accustomed hydroperiods, both plants and animals are stressed. In fact, without alternating wet and dry times, many plants and animals cannot reproduce or even survive. And extreme wet times each year—that is, annual floods—confer many benefits on wetlands and river bottomlands.

Upland plants are as adapted to mostly dry conditions as wetland plants are to periodic inundation. Under natural conditions, the ecosystems that top Florida's sand and clay hills are pine and scrub communities that not only tolerate fire, but depend on frequent burning for their renewal. Lower on slopes shy of wetlands, hardwood hammocks are dominated by fire-resistant plants.

Dry conditions combined with lightning make fires especially likely, and lightning is a frequent visitor to Florida. In fact, as it happens, lightning storms generate more strikes per square mile of land surface in Florida than in any other part of the country. In Tallahassee alone, 40 lightning strikes hit each square mile every year. Most lightning storms occur in summer, that is, in the months from May through August, and since May is dry, that is the month in which fires most easily ignite. Accordingly, Florida's fire-loving plants are especially well adapted to May fires. In natural, fire-adapted ecosystems, fires started by lightning burn off fire-sensitive plants, leaving fire-resistant plants in possession of the territory. May fires

Fire burning through pine flatwoods in Jonathan Dickinson State Park, Martin County. Fire is the agent of natural regeneration in these ecosystems.

stimulate these plants to flower, and by the time their seeds drop, the ground is bare and nutrients released by fire are available as fertilizer.[5]

Tropical Storms and Hurricanes. Windstorms are another force affecting Florida's ecosystems, and all species populations must have ways of surviving them. Tropical storms produce winds of more than 40 miles per hour and hurricanes are the most extreme of tropical storms, having winds greater than 70 miles per hour. A hurricane may produce rain and sometimes dangerously high water over an area hundreds of miles in diameter, while its wind gusts may reach speeds of 200 miles per hour and spawn tornadoes. Figure 2–1 shows the paths of all the tropical storms and hurricanes that struck Florida during the last century; one can hardly see the state's outline beneath the mass of storm tracks.

The winds of tropical storms and hurricanes pull weak limbs from trees, break brittle trees off midway up their trunks, and topple strong trees, exposing their roots. The result is a massive overhaul of forests. Bare earth lies exposed for new seedlings to take root. Sunlight floods formerly shaded ground, permitting seedlings to grow. Seeds that fall on the tops of broken-off stumps and start growing there are perched high above the next flood where they can grow without rotting. The tipped-up roots of fallen trees expose crevices in which many animals can make homes. The opportunities for new life to take hold in a forest following a storm are a major means of forest regeneration—not all at once, as forestry practice so often plans it, but piecemeal, summer after summer, now here, now there.

Besides rains and winds, a tropical storm brings a storm surge to the

> A **cyclone** is a spinning wind. A cyclone is called a **tropical depression** when its wind speeds are below about 40 miles per hour. A cyclone with winds between 40 and about 70 mph is a **tropical storm**, and when the winds exceed 70 mph, it is a **hurricane**.

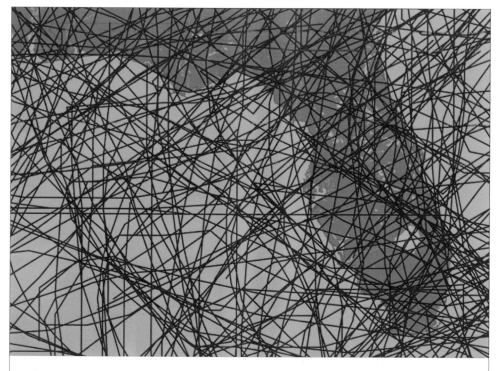

FIGURE 2–1

Tropical Storms and Hurricanes Striking Florida in 100 Years

Source: National Oceanographic and Atmospheric Administration.

SFWMD

Natural force: Hurricane winds. Several tropical storms and hurricanes strike Florida every summer, and native species populations must be equipped to withstand or recover from them.

After a storm. The toppling of a tree makes a varied surface of bare, soft earth available for new seeds to spring up on. The mass of upraised roots and earth, known as a tip-up mound, may persist for many decades and is an important feature of forest ecosystems.

coast. A hurricane expert describes storm surges as "big waves on top of water that is high above normal, a dome of water, maybe 50 miles wide, that sweeps across the coastline near the point where the storm makes landfall."[6] A storm surge may move a high wall of water several miles inland. The surge hits the beach with maximum energy, rushes over the first dune, and slows down as it moves inland.

Like rain and wind, storm surges wreak changes on the land. Where there are buildings, jetties, and seawalls, erosion is especially intense and rapid, but winds are constantly rearranging coastal sands anyway. Where vegetation has taken firm root and held on, dunes and islands may remain stable through major storms, but in hurricanes, huge dune systems, sandbars, barrier islands, and beach ridges may shift, break up, and re-form in wholly different places. Even then, however, native plants and animals re-establish their populations. In fact, some species are so well adapted to these types of changes that they require them to survive.

In summary, Florida's living things are adapted to temperatures, incoming sunlight, rains, droughts, fires, winds, and even storms as they have occurred through past millennia. Native ecosystems are also adapted to variations over time, such as wet and dry spells of several years.

SURFACE MATERIALS OF FLORIDA

Besides the climate and weather, the physical materials of which Florida is made profoundly influence the character of its ecosystems. That there are deeply cut ravines and high banks in one area, broad flat lakes in another, and caves and sinkholes and springs in still others, is due to the diversity of the sediments of the southeastern U.S. Coastal Plain.

The Southeastern U.S. Coastal Plain. This region, within which Florida lies, is shown in Figure 2–2. It has well-defined physical boundaries. The innermost boundary is the base of the Piedmont, which flanks the southern Appalachian mountains. The outermost boundary is the coastline.

Three physical characteristics distinguish Florida (and indeed the whole southeastern U.S. Coastal Plain) from the regions to the north. First, *marine* sediments (limestone and dolomite) lie in thick layers all over Florida, at or below the surface. These calcium-containing sediments were deposited during times when Florida lay under the sea and they profoundly influence the region's topography. Second, the whole Coastal Plain is deeply layered with *clastic* sediments (clay, silt, sand, and gravel). These sediments have been transported from the Appalachian mountains over the past 65 million years and generally lie on top of the marine sediments. Third, a layer of organic soil lies on or is mixed into the surface sediments. This layer may be nearly imperceptible on high, dry sand hills, but is inches to tens of feet thick beneath all kinds of wetlands and water bodies from small, shallow ponds and seepage slopes to large swamps and lakes.

Soils, which are the foundation material of most of Florida's living communities, are mixtures of these three basic types of sediments. Plants have to be adapted differently to grow on different kinds of soils. Well-drained

A **piedmont** is a rocky skirt that surrounds mountains. "The" Piedmont referred to here surrounds the southern Appalachian mountains.

The **continental shelf** is the part of the continent that lies submerged, offshore. At its far edge, the **continental slope** drops off to the ocean basins.

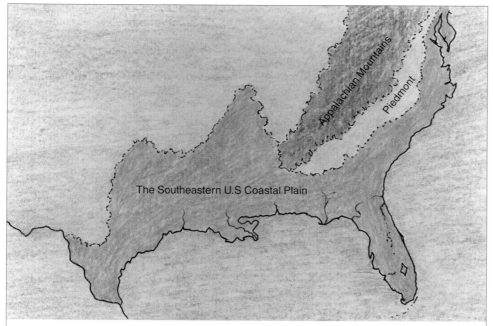

FIGURE 2–2

The Southeastern U.S. Coastal Plain

The southeastern U.S. Coastal Plain is the apron of loose sediments that spreads from the foot of the Piedmont to the shoreline.

Source: Brouillet and Whetstone 1993, Ch 1.

soils challenge plants to obtain and retain water. Water-saturated soils exclude air with its life-giving oxygen, so that plant roots have to be specially adapted to obtain oxygen, which they need for growth and metabolism. Moisture and drainage characteristics of soils are so important that ecosystems are often classed by them. For example, xeric, mesic, and hydric hammocks consist of trees growing on soils that are, respectively, dry and well drained, moist, or wet and waterlogged.

Depending on what they are composed of—limestone, clay, silt, sand, gravel, peat, muck, humus, or mixtures of these—Florida's soils may be fine or coarse, acid or alkaline, fast-draining or water-holding. Altogether, some 300 different types of soil occur in Florida, contributing much to the diversity of the state's ecosystems.[7]

Florida's Fluctuating Area. The Florida peninsula's edge, the present coastline, fluctuates relatively rapidly. After all, what we think of as the Florida peninsula today is simply the above-sea-level portion of Florida, and its area contracts and expands dramatically as the sea level rises and falls. In the distant past, Florida was altogether submerged beneath the sea, while at several more recent times, it has been exposed above sea level all the way to the edge of the continental shelf and even partway down the continental slope. Figure 2–3 shows that the land area of Florida varies considerably depending on sea level.

Today, only about half of the Florida platform is above water. The last time it was completely exposed was 19,000 years ago, when the seas were some 300 feet lower than they are now. When the first native Americans settled in Florida some 14,000 years ago, the peninsula was still much

The term **soil** applies to the particulate material lying on top of the land, which is capable of supporting plant life. Soil includes inorganic sediments; organic components such as peat, muck, and humus; living things including microscopic organisms and earthworms; and air and water.

The terms *xeric, mesic, and hydric* describe moisture-related characteristics of soils that are important determinants of what will grow in an area.

Xeric (ZEE-ric) soils are dry and contain ample oxygen to meet plant needs. Rain water drains rapidly from xeric soils.

Mesic (MEE-zic) soils are moist but do not become waterlogged; they hold oxygen and drain well.

Hydric (HY-dric) soils are wet and low in oxygen content.

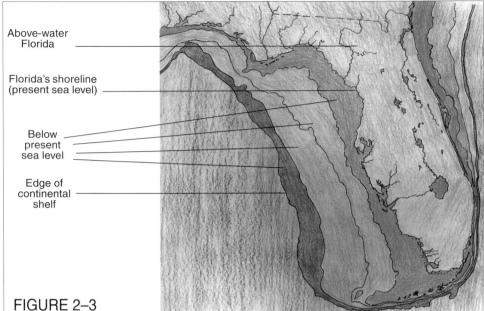

Above-water Florida

Florida's shoreline (present sea level)

Below present sea level

Edge of continental shelf

FIGURE 2–3

Florida's Varying Sizes and Shapes

The Florida platform extends all the way to the edge of the continental shelf, and at times the whole land mass has been dry. Chapter 22 tells the story of Florida's fluctuating sizes and changing ecosystems over millions of years in the past.

The successive contours shown outside Florida's present shoreline are at 60, 120, 300, and 600 feet below sea level.

The continental shelf is 35 to 100 miles wide off Panhandle Florida, 150 miles wide west of Tampa, less than 5 miles wide at Miami, and about 60 miles wide at Jacksonville.

Source: Adapted from Fernald and Purdum 1992, p. 14.

The **Florida platform** is a mass of sedimentary rocks deposited on the seafloor over millions of years. In the north, river-borne sediments have accumulated on the platform. In central and south Florida, marine sediments have accumulated.

broader than it is today—especially on the west side, where the continental shelf is nearly level—but the ocean was continuing to rise. Today, beneath the calm waters of the Gulf of Mexico, archaeologists can point to the evidence of early Indian villages among the remains of drowned forests, swamps, and marshes.

Today, the sea level is still rising. In the past hundred years, it has risen several inches around Florida. Largely as a result, waterfront towns and resorts all along the coast are frequently damaged by storm tides, and beach houses often fall into the sea.[8]

Regions of Florida. As testimony to Florida's earlier changes in area, many old shorelines remain both on top of mainland Florida and on the continental shelf offshore. None is continuous; all have been breached by storms and eroded by wind and rain since they were first created by the sea, but many earlier shorelines are still recognizable today.[9]

The old shorelines help to define three regions that are distinguished, roughly, by elevation and distance from the present coastline (see Figure 2–4). The *highlands*, *ridges*, and *upland plains* are highest and are farthest inland. The *coastal lowlands* are lower and are closer to the coast. The *coastal zone* is nearly at sea level.

The Floridan Aquifer System. For depths down to some 3,000 feet or more, Florida's subsurface limestones are porous and capable of holding water. For as long as Florida was below sea level, all water within the limestone was salty, but once the land was above sea level, rain fell on Florida for millions of years and fresh water worked its way down into the limestone, pushing the earlier salt water down and out to sea. Rain water is somewhat acidic as it falls from the sky. As it runs over the land picking up decomposed plant matter, the water becomes more acidic. The acid water eats into limestone, which is alkaline, percolates down into cracks and fissures in the limestone, dissolves spaces, and enlarges them. Enormous volumes of fresh water have moved through the Floridan aquifer system for millions of years and are still moving through it today.

> An **aquifer** is a water-holding layer of earth.
>
> The **Floridan aquifer system** is the great water-holding limestone system that underlies nearly all of Florida. (It is pronounced "FLO-ri-dan," not "Flo-RID-ee-an.")

TOPOGRAPHY

The land's topography, or sculpting, is profoundly influenced by the materials of which the land is made, and helps determine what sort of ecosys-

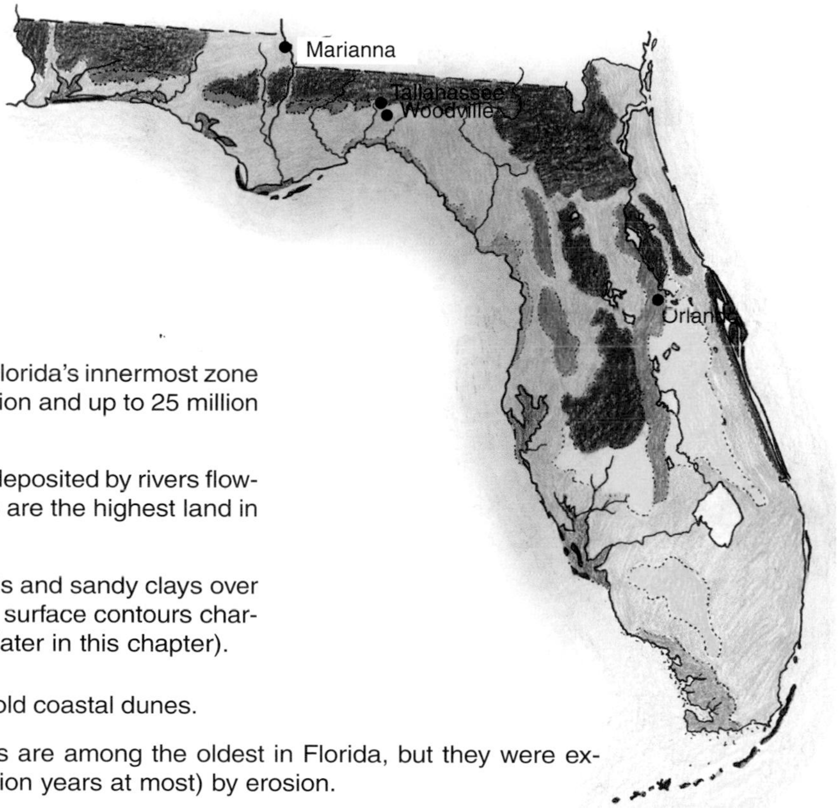

FIGURE 2–4

Florida's Three Zones

1 HIGHLANDS/RIDGES/UPLAND PLAINS. Florida's innermost zone has been above sea level for at least 6 to 7 million and up to 25 million or more years.

The **highlands** are mostly clay and sand deposited by rivers flowing down from the north. These highlands are the highest land in Florida.

The **upland plains** consist of clayey sands and sandy clays over deeply buried limestone. They display the surface contours characteristic of karst topography (described later in this chapter).

The **ridges** are high, sandy remnants of old coastal dunes.

2 LOWLANDS. The materials of the lowlands are among the oldest in Florida, but they were exposed only recently (during the past 1 to 2 million years at most) by erosion.

The **lowlands** are lower and more level on the whole, and the name of their predominant ecosystem, *flatwoods*, or *flats*, describes their topography. *Coastal lowlands* occupy the entire periphery of the state. *Interior lowlands* include the karst region around Marianna known as the Marianna Lowlands and the Woodville Karst Plain just south of Tallahassee. See "Karst Topography" in the text.

3 COASTAL ZONE. The third zone is the strip of land that lies along the coast and includes the current beaches, coastal marshes, and swamps that are presently being worked by the sea. Coastal ecosystems are constantly influenced by salt spray. Most of the coastal zone is too narrow to be shown at this scale, but two parts are apparent:

Major salt marshes and mangrove swamps.

Florida's major estuaries, where salt and fresh water are mixed.

Source: Adapted from Fernald and Purdum 1992, 15.

tem will form in a given area. Where there are clay-rich sediments, rain erodes gullies and rivers form. Where there are broad plains of sand, instead of running off, water soaks in. Where limestone lies in thick layers just below the surface of the ground, rain water dissolves the stone, causing sinks, springs, and caves (karst features) to develop. In various combinations, these factors apply to all regions of Florida. Clay-rich parts of the state are divided into many stream valleys with hardly any lakes; parts with deep sand are dotted with numerous lakes and ponds and have few streams; and limestone-dominated areas have many caves, sinks, and springs.

Clay and Sand Hills and Plains. The so-called ''clay hills'' of north Florida are deep layers of sandy clays and clayey sands deposited over millions of years by rivers carrying sediments down from the mountains to the north. Clay particles are so fine that it takes only 10 percent of clay in a soil to completely plug the pores and essentially prevent movement of water down into the soil. Limestone beneath clayey sediments does dissolve into ground water over long time periods, and there are karst features in clay-hills landscapes, but on the surface, water runs over the top of the land, carrying sediment downstream and carving gullies. Eventually, broad valleys and basins form, and these hold swamps, marshes, streams, rivers, lakes, and ponds.

Sand hills are ancient barrier islands and dunes—hills of almost pure sand that were built at coastlines now long since stranded inland by retreating ocean waters. Because the ocean has risen to cover parts of Florida and retreated again so many times, sand hills are scattered all over the landscape. Sand hills lie parallel to the Panhandle's coastline and all down the central peninsula, wherever shorelines have lain in the past, and they support ancient beach communities known as scrubs.

Sand, or sand mixed with varying amounts of other clastics, is also spread all across today's lowlands, which form nearly level plains, all around the periphery of Florida. These are ancient offshore seafloors, and their geologic history involves successive episodes of erosion, rearrangement, and deposition over thousands of years. These plains, the coastal lowlands, support communities known as flatwoods, or flats, as well as depression marshes, swamps, and ponds.

Clay. Exposed on a road cut through a high pine grassland, this clay has given the highlands of north Florida the name *Red Hills*. This is Pine Tree Boulevard, just north of the Florida-Georgia line in Thomas County.

Sand. Composed of particles of a larger size than clay, sand is the predominant material of Florida's beaches, sand hills, and coastal lowlands. This beach is in Deer Lake State Park in Walton County.

Limestone outcrops in the Waccasassa Bay State Preserve, Levy County. Accumulated on the seafloor mostly from the skeletons, shells, and bones of marine creatures, limestone gives evidence of Florida's submarine history.

PRICELESS FLORIDA

Karst Topography. Karst regions occur all over Florida. Outcrops, depressions, ponds, caves, caverns, tunnels, sinks, solution holes, and springs all testify to the presence of limestone at and beneath the ground surface. They hold living communities different from those in non-karst areas. Chapter 7 explains how karst features form and describes several ecosystems associated with them.

Coastal Features. Because Florida is bounded by water on three sides, it has diverse coastal realms. Coastal features vary, depending on the energy with which ocean waves attack them. This energy in turn varies with the slope of the shoreline. High-energy shores have steep slopes with energetic waves; low-energy shores have gentle slopes with little wave action. Where wave energy is high, ocean water washes away fine sediment particles (silt and clay), leaving predominantly sand and sand-sized pieces of shells. Where wave energy is low, a mixture of clay and silt (mud) remains. High-energy coastlines are lined with sandy beaches; low-energy coastlines hold tidal marshes and swamps.

Florida's present coastal features have a history numbered in only a few thousand years at most, and are in a constant state of flux. Especially along high-energy coastlines, the sands of beaches and barrier islands move around ceaselessly. People who live along the shore can see increments of change in the coastline within weeks or even overnight, as winds and waves rearrange coastal sands and erode or add to beaches. In contrast, along low-energy coastlines, where coastal marshes and swamps have developed, there is little or no transport of sediment, but a rising or falling sea level can drown or drain a coastal ecosystem in just a few hundred years.

Florida's offshore seafloors vary, too. Oceanward of high-energy shorelines, seafloors are of sand and are too dynamic to support stable communities, but where wave action is gentler, permanent life can take hold. Parts of the continental shelf, like karst areas on the land, have limestone outcrops and springs, while on other parts, living coral communities are still building reefs. Off low-energy shores, seafloors are of mud and support living communities such as seagrass beds and mollusk reefs.

Florida native: Red buckeye (*Aesculus pavia*). This native shrub thrives on calcium-rich soils, and therefore flourishes in karst areas.

FLORIDA WATERS

Finally, the water bathing Florida's ecosystems is highly variable. Coastal communities grow in water of every degree of saltiness, from fresh to brackish to saline. Interior water that wells up from the ground and is filtered through limestone and sand is alkaline and often as clear as window glass. Water that collects in swampy areas is acidic and stained a deep red-black by plant acids. Water flowing down streams from clay hills is clouded with loads of sediments. Water temperatures vary as well. Water from under ground is a cool 68 to 70 degrees all year long, whereas shallow surface water may be upwards of 100 degrees in summer and, in north Florida, may freeze in winter.

***

Anyone who knows these few general facts about Florida's geography and geological history can literally sense the past underfoot and can under-

Mud. Where wave energy is minimal, ocean waters leave small clastic particles, clay and silt, in place. Quiet shores support salt marshes like this one in the Lower Suwannee National Wildlife Refuge in Levy County.

Natural ecosystem: A coral reef. Nutrient-poor marine water bathes the reefs off the Florida Keys and provides the environment in which the corals can thrive.

stand why different natural communities flourish in different places. Whenever you walk on upland loam or coastal sand, you know that the earth beneath your feet came from the mountains. Near a sinkhole, you sense the ocean's deposits of limestone and can almost see the ancient seafloor alive with sand dollars and sea urchins, or decked with a beautiful coral reef. On the beach, you can watch yesterday's sand hills washing away— or tomorrow's, taking shape. The signs are everywhere: along highways, in state parks, on river bottoms, even in cities. The roadside bank of red clay, the flat plain of sand, the ponds, caves, cliffs, outcrops—all speak of the region's origins.

Among the best places to observe evidence from the past are the state parks. The natural features they exhibit are described in trail markers and guidebooks that can help Florida travelers appreciate what they are seeing.

This chapter has laid the groundwork, literally, for appreciating Florida's living communities. Additional details will be added in the chapters to come, which now turn to the natural ecosystems themselves.

PART TWO

INTERIOR UPLANDS

Interior Florida is all of Florida except for the strip right next to the sea. It holds upland, wetland, and aquatic ecosystems, and its waters are fresh. The term **upland** describes areas whose soils are dry and well-drained, or moist, but not saturated with water except after rains. Uplands are not all high lands: many upland regions occur in the coastal lowlands. The **interior uplands** are the terraces, plains, and divides between stream systems.

Pine Rocklands, Everglades National Park, Miami-Dade County.

CHAPTER THREE

HIGH PINE GRASSLANDS

Five hundred years ago, north and central Florida looked altogether different from today. A vast, continuous pine grassland covered nearly all the high ground and spread across the flatwoods almost to the coast. The pine trees were giants, widely spaced, and the ground was carpeted with a sea of grasses. Other plants were numerous and diverse, but pines and grasses dominated the scene. Pine grasslands like this had covered more than 20 million acres in Florida on and off for millions of years.[1]

Today, the vast and mighty pine forests that cloaked most of Florida for so many millennia are almost completely gone—logged out during the nineteenth century at a rate faster than the tropics are being logged today. Today, hardly anyone knows what they looked like. To today's Floridian, the word *forest* conjures up an abandoned farmers' field thick with newly growing trees, or a pulp farm, or a logged-over forest remnant. These are not natural, self-maintaining forest ecosystems, but the products of human interference with nature. The few remnants of the original ecosystem are mostly of second growth, but they are still prized for their extraordinary diversity and complexity. Figure 3–1 shows the change in extent of the original pine forests from preColumbian times to today.

ANCIENT PINELANDS

Natural high pine grasslands are so few and far between today that most people don't know what they look like—or, seeing them, don't recognize their significance. An authentic old-growth pine grassland is shown in the photo, opposite. It is strikingly different from the commercial pine plantations that have overtaken most of its territory today. The canopy is open. The biggest pines are more than 100 feet tall, straight, and spaced well apart. If allowed to grow to their natural age, upwards of 500 years, they will be three or four feet across at breast height. The original, pristine longleaf pine community had millions of monster trees.[2]

OPPOSITE: A high pine grassland in its natural state. Florida's high pinelands looked somewhat like this before the Europeans came. They were open, grassy lands where lightning was a frequent visitor. Lightning-set fires swept through them nearly every year, biting back small hardwood shrubs and trees, and clearing the way for grasses and wildflowers.

Florida's natural pine grasslands have an open canopy of pines and a floor of wiregrass or other grasses. Based on the tree cover, they are called **forests**, but from the ground cover, they are **pine grasslands** or **pine savannas**.

Other terms referring to Florida's high pine grasslands described in this chapter are **upland pine** (on clay hills) and **sandhill** (on sand hills). Florida's low pine grasslands, flatwoods and prairies, are described in Chapter 4. The pinelands of extreme south Florida are **pine rocklands**, and are treated in Chapter 7.

PKS

Florida native: Lady lupine (*Lupinus villosus*). This plant is widespread in the dry soils of longleaf pine sandhills.

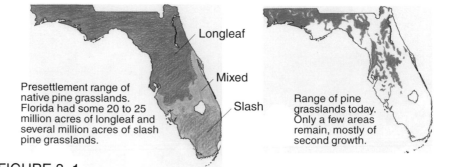

Presettlement range of native pine grasslands. Florida had some 20 to 25 million acres of longleaf and several million acres of slash pine grasslands.

Longleaf

Mixed

Slash

Range of pine grasslands today. Only a few areas remain, mostly of second growth.

FIGURE 3–1

Native Pine Grasslands Yesterday and Today

Originally, native pines and grasses occupied most of the southeastern U.S. Coastal Plain, spreading in an almost unbroken expanse all the way from eastern Texas to Virginia, an area 85 to 90 million acres in extent. Shown here is the presettlement range of longleaf and slash pine grasslands in Florida, compared with the range today.

Other pine species occupied small areas. Sand pine topped coastal dunes and a few high sand hills in interior Florida. Slash pine grew along the coastal borders of north Florida's longleaf pine savannas and in low areas transitional to wetlands. Pond pine grew in wet areas in pine flatwoods.

Hardwood trees dominated other areas, where forest fires (which favored pines) occurred infrequently: hardwoods grew in closed-canopy forests along river valley bottoms, in deep ravines, on steep slopes, and in low-lying swampy depressions. Strands and circles of cypress also outlined some streams and ponds, but pine grasslands held all the rest. Vaster than the Everglades, they were old, well organized, diverse, and Florida's most important natural ecosystem.

Sources: Drawings adapted from Platt 1999, 24 (adapted with the permission of Cambridge University Press) and Myers 1990, 175. Information from Anon. 1991 (Longleaf pine communities vanishing); Derr 1989, 110–114; Means 1994 (Longleaf pine forests).

The **overstory**, or **canopy**, of a forest consists of its tallest trees.

The **understory**, or **midstory**, consists of trees and palms that grow below the canopy.

The **ground cover** consists of grasses, other herbs, and shrubs that grow beneath the trees, the lowest layer of greenery.

A tip-up mound in a pine grassland. Such mounds are also significant in hardwood forests, as explained in Chapter 6.

Other longleaf pines are of all ages and sizes, evidence that the forest is continually reproducing. There are fat, bright-green baby pines; all sizes of saplings; and maturing trees of 20, 60, 100, and even 120 feet in height. Even the youngest trees have no branches near the ground. The oldest trees are branchless to a considerable height, making it easy to see for great distances across the grassy forest floor.

Dead standing trees—"snags"—are present, too, and some trees are blown down by wind with their roots upraised as tremendous, tangled tip-up mounds. Some dead trees lie full-length on the ground in the last stages of decay. Huge pine cones, 8 to 12 inches long, are scattered on the ground, dropped by the pines or by fox squirrels. Pine straw litter hangs from the branches and lies strewn on the grasses.

One has to wonder how such a place could ever come to be. Why is this forest so open? Sunlight is plentiful, yet no other trees are growing in the spaces between the pines. Is that because the grasses are too thick? And where did these thick bunch grasses come from anyway? You don't see grasses like these in today's younger forests; rather, you see a thick tangle of hardwoods mixed with pines, shrubs, and vines. But in the few old-growth pine grasslands of today, open space is the rule. Why? Read on and discover how fire shapes this ecosystem.

FIRE IN PINE GRASSLANDS

It is a dry, breezy May afternoon. Embers are smoldering inside an old pine tree that was struck by lightning in a storm a day or so ago. Now the ground is dry, and when an ember falls to the ground, flames spring up in the pine straw at the foot of the tree and begin to spread rapidly through the wiregrass. A sudden gust of wind makes the fire flare up in a crackling, five-foot-high wall for a few minutes, but then the breeze changes direction and the fire's leading edge creeps along with flames only a foot high, leaving behind black ashes that quickly cool.

The line of fire sweeps steadily across the ground surface. A red-tailed hawk eases from tree to tree working the fire line, swooping down to catch grasshoppers and mice, which are moving about to avoid the flames. Finally, in the waning evening light, the fire moves over a rise out of sight.

In the morning, the ground is covered with black ash. Only a few wisps of white smoke still rise from dead snags and large fallen branches on the ground. Several days later, the old tree that was smoldering earlier has burned down to a hard, resinous snag, the "fatwood" or "lighterwood" cortex of the tree. Impregnated with resins that protect the wood from rot, this snag will burn and reburn for decades.

The scaly pine bark is black to several feet above the ground. Some of the pine needles are scorched, but many trees still have healthy green needles and all are still alive. The wiregrass clumps are circles of black stubble, the other herbs seem to have been vaporized completely, and the running-oak stems are brittle and break off easily, but beneath the soil,

Florida native: Spurred butterfly pea (*Centrosema virginianum*). Butterfly pea grows naturally in many Florida ecosystems, including pine grasslands.

Fire in ground cover in the Apalachicola National Forest, Liberty County. Not much litter has accumulated on the ground since the last fire, and there is little midstory growth, so the fire burns at a low temperature and moves across the surface without significantly damaging the pines. In presettlement times, surface fires worked their way around ponds and the heads of streams and could progress freely for hundreds of miles across Florida, Georgia, and southern Alabama.

Land managers recognize different kinds of fires on the land. Described here is a **surface fire**, or **ground fire**, one that burns through litter and undergrowth. Longleaf pine communities typically experience ground fires.

A **crown fire** is one that burns through the tops of trees. Such fires occur in sand pine scrub as described in Chapter 5.

The term *ground fire* is sometimes used to mean a **peat fire**, one which burns deep into accumulated peat in the ground. Peat fires may occur in peaty wetlands during severe droughts.

Slow ground fire in virgin ground cover. At night, it is easy to see that the fire is burning mildly but steadily.

Longleaf pine stump still burning after recent fire. When it finally burns out, this stump will leave a deep, branched hole in the ground that will serve as habitat for many native animals.

All nonwoody flowering plants, both grasses and broad-leafed types, are **herbs**.

Grasses have narrow blades and distinctive flowers that identify them as members of the family Poaceae.

Forbs are all of the herbs other than grasses.

Some forbs that have narrow blades and look like grasses (sedges, lilies, palms, and others) are called **grasslikes**.

the roots of all these plants are already resprouting. Only trees that are not adapted to survive summer fires, namely hardwoods such as southern magnolia and American beech, are completely killed. Those trees are absent from this ecosystem. If they start to invade, they are eliminated by the next fire that will come a few years hence.

A close look reveals that the community has burned unevenly. The high, dry ground is totally scorched. Lower, damper spots are less intensely burned, and unburned patches are common. Different plants will dominate these areas, adding to the ecosystem's diversity.

In September, a return visit reveals an astonishing scene. The grassy floor is alight with vibrant, new life. The wiregrass carpet has put up millions of slender, haylike, golden flowering stalks, waist-high, and soft to the touch. Myriad wildflowers, earlier unseen, are blooming: pink, lavender, purple, red, blue, and yellow. Insects are hopping, buzzing, and crawling everywhere, making a feast of juicy young plants, and cross-pollinating the flowers as they move about. Small birds, lizards, and frogs are pursuing the insects. As predicted, the plants are unevenly distributed: there are patches of legumes here, goldenrod there, berry bushes elsewhere. Fire creates a mosaic, and because in different years fires come at different times, even the patches differ somewhat from one year to the next.

The reason the forest floor is so open is now clear: fires keep the woody plants down. Prior to European settlement, fires swept the surface nearly every summer. That may seem improbably frequent, but a single summer day may see a hundred fires start up naturally in Florida thunderstorms. For millions of summers in the past, lightning-set fires swept unimpeded across the pinelands of the Coastal Plain.[3]

It was only after people came on the scene that this steady state began to change. When the early Indians began to practice agriculture, they

High pine grassland with ground-cover flowers in bloom. This forest burned in June and different flowers bloomed every month thereafter. These are October blooms in the Wade Longleaf Forest in Thomas County, Georgia.

Unnatural system: Pine grassland deprived of fire. This forest was kept from burning for twelve years, enabling water oak, laurel oak, and sweetgum to encroach on the pines. Ironically, the forest is now dangerously fire-prone, because flammable pine litter has piled up. Native animals are not well adapted to living in an unburned pine forest, and many cannot survive in this environment.

cleared some fields to plant crops, hindering fire's spread, but they compensated by setting fires intentionally so that they could see their enemies, control mosquitoes, create spaces for recreation, or hunt game. Later, the Europeans cleared more land for fields, pastures, roads, and towns. They, too, burned patches of pine grasslands to maintain them, but fires no longer swept freely across most of the land.

Today, roads, agriculture, and developments block or suppress the spread of fires. Freed from control by fires, shrubs, scrub oaks, and other hardwoods quickly grow into a dense subcanopy that shades out the ground cover and young pines. Finally, the pine grassland gives way to a tall, dark, broad-leafed hardwood forest that will not burn again, for hardwood litter resists burning. Today's hardwood forests are far more extensive than they were in presettlement times.[4]

Fires in pine grasslands not only suppress hardwoods, they also help the community's plants to reproduce. Fire stimulates grasses and some forbs to flower and release their seeds. Fire also exposes patches of bare soil, which permit new seeds to take root, and the sunlight that floods the ground after a fire fosters the growth of new seedlings. Without fire, bare sun-drenched soil is hard to come by, because the ground cover and pine litter are so dense. Bare ground is available only on an occasional tip-up mound, or where a gopher tortoise or pocket gopher has dug a burrow. Fires also turn accumulated litter to a residue of ash, which contains the phosphorus, nitrogen, and other nutrients that newly growing plants need. Rains wash the ash fertilizer into the soil just as new seedlings are starting to grow. And fires preferentially burn trees that are already weakened by infection and have started to rot. Elimination of these trees helps to control disease agents.[5]

Because fires sweep the terrain frequently, they also keep dry litter to a

Natural ecosystem: A longleaf pine–turkey oak community. Turkey oak (*Quercus laevis*) is kept low to the ground by frequent fires.

minimum. This prevents major, hot, pine tree–killing wildfires. The animals benefit, too. Young plants that arise after fires make tender and especially nutritious food for newborn grazing animals. Open spaces permit fruitful hunting for hawks and ease the gathering of seeds by quail and other seed eaters.

The season of burn is important. It might seem that fires at any time of year would do equally well, but fires in winter are usually not effective at killing hardwood trees. In winter, the plants' nutrient reserves are stored down in the roots, where fires cannot destroy them. If young hardwoods are exposed to a winter fire, even if it burns them to the ground, they can resprout from their roots the next spring.

The fires that maintain the ecosystem best are those that occur in the season when lightning storms are most common in Florida—that is, early in the growing season, from April to July. Summer fires stress hardwoods more effectively, both because they are hotter and because they burn the plants' stems and leaves while the nutrients are in them. Also, most grasses and some forbs only flower and produce seeds in response to early, growing-season fire. If forest managers burn the forest in winter, although they may clear out the ground cover just as well, they will induce fewer plants to flower and release their seeds.

In short, historically, the natural fire cycle was governed by consistent summer lightning storms. Today these have to be simulated by intentionally set fires known as controlled, or prescribed, burns.

All of the ecosystem's plants exhibit adaptations that make them tolerant of fire. The longleaf pine itself is a paragon of fire adaptedness. It has been evolving for millions of years in a fire-dominated realm, and today it is by far the most robust and fire resistant of Florida's pines. Not only that; it actively promotes the igniting and spreading of fires, thus perpetuating the conditions that favor its dominance of the landscape.

LONGLEAF PINE, FIRE STARTER

Figure 3–2 shows that longleaf is superbly adapted to a fire-frequent environment. One secret of its success: it can withstand fire as a seedling. Slash pine and other pines can, when mature, withstand fire better than most hardwoods but, at least in north Florida, they cannot out-compete longleaf.

While maintaining superb defenses against fire, the longleaf pine also tolerates soil infertility better than hardwood trees do, and this favors its supremacy on the infertile, sandy soils of the Coastal Plain. Hardwood trees grow better on fertile soil. Another contrast: the needles, twigs, and bark dropped by longleaf pines are highly flammable, whereas hardwood leaves resist fire. Even after they have fallen on the ground and turned brown, hardwood leaves, having no flammable resin, tend to hold moisture and refuse to burn.

Longleaf also holds onto the land tenaciously enough to withstand hurricane-force winds. Its taproot plunges deep into the ground to reach a steady water source and at first is as stout as the pine tree's trunk. The

FIGURE 3–2

Longleaf Pine Adaptations to Fire

SEEDLING

The longleaf pine seedling hugs the ground while the stem and bark grow thick. Only a bushy spray of needles is exposed above the ground for the first two to ten years. The needles are thick, more than 12 inches long, and moist. Silver scales on the growth bud help reflect the heat of fire.

While it hugs the ground, the tree is said to be in its "grass stage." During the grass stage, it puts down a deep taproot and stores energy and nutrients in it. The taproot grows well below the fire zone and never burns.

FIRE SURVIVORS

If a fire comes through while the tree is in the grass stage, the long, moist needles give off steam. This keeps the growth bud at a temperature no higher than the boiling point of water. The needle tips may be scorched, but the growth bud is protected. Only in extremely hot, dry, windy conditions can fire kill young longleaf pines.

SAPLING

Now that it has a stout taproot with a good store of water, energy, and nutrients, the sapling wastes no energy making side branches, but shoots up fast, rising above the fire zone within two years. The first year, it grows 1 foot, the next year, 2 feet more. After three more years of growth, it is 5 to 6 feet tall. This is the tree's so-called "broom stage."

The young tree has a sturdy trunk and thick, corky bark. It grows lateral branches only after the growth bud at its top is above the zone in which a mild surface fire can damage it.

The mature tree's bark is scaly, and when it burns, the scales flake off, and carry heat away. The trunk can resist brief, intense heat, even that of an acetylene torch.

YOUNG TREE

MATURE TREE (not shown): Mature at 40, longleaf pine then can go on to live for 300 or 500 years or longer. It dies, usually, due to lightning strike followed by insect attack.

Sources: Croker and Boyer 1975; Wahlenberg 1946; Muir 1916; Myers 1990; Clewell 1981/1996, 104–107.

Slash Pine

Slash pine growing in south Florida has a grass stage and exhibits other similarities to longleaf pine. It tolerates fire well.

In north Florida, however, slash pine has no two- to ten-year-long grass stage. At 1 year, it is already 6 inches tall, with a spindly stem, paper-thin bark, and needles all along the stem. Even a mild grass fire will kill this seedling.

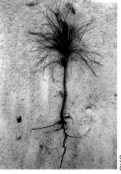

SEEDLING

As a sapling, the northern slash pine has a slender trunk with thin bark and needle-laden branches close to the ground. It is readily killed by fires, even mild surface fires.

MATURE TREE (not shown): Slash pine matures at 40, and dies at 100 or 150 years, typically; earlier in bay swamps and probably much earlier in sandhills.

SAPLING

Longleaf pine taproot. Shown partially exposed here, the longleaf taproot sometimes penetrates as far below the ground as the tree is tall.

Root-associated fungi are known as **mycorrhizae** (my-co-RIZE-ee).

Trees that yield huge crops of seeds at intervals of several years are said to be **masting**. Hardwoods such as beech and oak trees exhibit this pattern of seed production and the term **mast** refers to their nuts and acorns. Longleaf pine, too, exhibits masting and the mast is its cones.

taproot puts out lateral roots, and from these, secondary taproots. Fine feeder roots from all of these branch prolifically beneath the ground surface. This stupendous anchoring system is more effective, even in sand, than that of any other tree. At maturity, the root system's wide-ranging penetration of the underground space enables the tree to stand firm against storms and collect water and nutrients from a circle some 60 feet or more in diameter.

Associated with the root system are several species of underground fungi that extend the tree's nutrient-gathering system severalfold more. Webs of fungal hairs intertwine with, and even penetrate, the roots and deliver minerals to them. Some root-associated fungi contain bacteria which capture nitrogen from the air and bind it into usable compounds. Other fungi produce hormones that stimulate tree growth. You can easily detect such fungi in an old, undisturbed ecosystem. Gently lift the mat of litter around the base of a tree. See how it is knit together by a web of nearly invisible threads. This is a fungal network that links the litter to the soil.

Thanks to its land-grabbing root system, longleaf competes fiercely with other trees, even its own seedlings, for water and nutrients. If growing too close together, young longleaf pines cannot survive for long. In an undisturbed old-growth tract, each mature tree releases hundreds of thousands of seeds during its lifetime, but very few (perhaps only one or none) grow to maturity and reproduce. Ironically, where patches of bare soil become available, young longleaf pines may begin to grow in clusters, but competition eventually thins them out.

The longleaf's open spacing produces further advantages. Because the tops are not in contact with each other, they cannot carry crown fires. Even in a single tree, the branches are so irregularly spaced that fire cannot jump from branch to branch. And the trees' open spacing also admits abundant sun, which favors the growth of ground-cover plants. Given such favorable conditions, the wiregrass that naturally accompanies these trees on undisturbed sites overlaps in a continuous sea of fine fuel that guarantees the spread of fire across the ground (see next section).

To retain dominance, the trees have to produce immense numbers of seeds. Numerous animals eat longleaf seeds, typically devouring 99 percent of them or more every season. About every seven years, however, a stand of longleaf turns out vastly more seeds than the average population of seed eaters can consume in a year. The intervening years of less abundant seed production then keep seed-eater populations in check.

What more could the trees possibly do to perpetuate conditions that would favor their own survival? They could actively promote the igniting and spreading of fires—and they do. Once mature, longleaf pines become host to an in-dwelling fungus that softens their heart wood, a condition called heart rot. When a tree's core with heart rot catches fire, it can smolder within the tree for days, even through heavy rains. Then, after the rains have passed and the humidity is low, sparks from the smoldering fire can jump to the ground, where the needle litter is. Longleaf pine needles also burn better than those of any other pine native to the southeastern United States. The burning pine straw ignites dry wiregrass leaves, and the fire spreads. Thus longleaf both invites fire and helps it to spread.

The more old trees there are and the longer the time that has passed since the last fire, the more likely fires become. Long intervals between fires, which might permit hardwood trees to invade, are thus made highly unlikely.

Finally, the tree's own life cycle favors its renewal. Seed drop is in fall, after the summer lightning season. No other pines release seeds in this season, so longleaf seeds have free access to whatever bare soil is available. The seeds then germinate in winter, supplied with fresh ash fertilizer, and have a fire-free season in which to start growing.

Thanks to all these adaptations, longleaf pines, once established, tend to perpetuate themselves naturally. Foresters may choose not to plant longleaf because other pines produce more wood at first, but recent studies show that after the first two decades, longleaf catches up and thereafter produces more, and higher-quality, wood than other pines. In short, longleaf is the tree best adapted to grow in Florida's pine grasslands.

Heart rot in pine.

Heart rot is the softening of a tree's heart wood, caused by an indwelling fungus. Wood affected by heart rot burns slowly and, because it is protected from rain within the tree, sometimes keeps a fire going in wet weather.

WIREGRASS, FIRE SPREADER

The natural ground cover in a pine grassland consists primarily of huge fountains of bunch grasses—in north Florida mostly wiregrass; in south Florida other grasses. The grass plants are ankle-high and nearly a yard across. They feel springy underfoot, but one can easily turn an ankle on the hidden center of a clump under the long, hairy leaves.

Nestled among the grasses are dozens of other low-growing plants such as running oak, saw palmetto, bracken fern, and numerous flowering forbs. These form the base of a food chain that consists of a vast variety of small insects, insect eaters, and their predators. But the grass is obviously dominant and is the most important plant. It is an indicator species signifying that the whole, integrated community is likely to be present. Without wiregrass, many of the other species of ground-cover plants will be missing, too, and a stand of pine will be just a tree farm, poor in species. The natural ground cover produces the seeds, flowers, fruits, and leaves on which the community's tiny animal members feast, these animals become food for other animals, and in turn for the whole community. High plant diversity supports still greater animal diversity.

An **indicator species** is one whose presence in an area signifies that a whole suite of other species is present along with it. It is a sign of an intact ecosystem.

The sea of grass that carpets a pine grassland withstands and promotes fire just as the trees do. Like the pines, the grasses of pine grasslands deal well with infertile soil and competition; they have become adapted to these conditions over millions of years. The roots are safely buried under the soil and escape the heat of ground fires. The plants are remarkably evenly spaced across much of the terrain, forming a continuous mat, almost a grid, of overlapping plants. Their blades, as they reach full length, die and turn brown, but stay in place. They readily catch fire from burning pine needles and neighboring clumps of grass. Once ignited, the dry grasses alone carry fire for long distances among widely spaced pines.

The grasses of pine grasslands, like longleaf pine, require fire to thrive. They flower and drop their seeds, just as the pines do, in fall after a fire in the lightning season. Grasses love sunlight. Wiregrass produces more than

Florida native: Wiregrass (*Aristida stricta* var. *beyrichiana*). Wiregrass is one of several native, flammable grasses that spread fire in Florida's pine grasslands.

Florida native: Red-banded hairstreak (*Calycopis cecrops*). This butterfly has eye-like spots at the ends of its wings. The spots tempt predators to strike inaccurately, and this often enables the insect to escape.

20 times more seed under an open canopy than in the shade. Without summer fire, wiregrass makes few seeds and fails to grow, even if the nearby earth is exposed by plowing. After a summer fire, however, the grass seeds colonize the exposed soil and guarantee the dominance of this tenacious and amazing grass over all other species.

That wiregrass survives at all in environments where people have prevented summer fires is a tribute to the tenacity of its roots. In a natural, old-growth pine grassland, individual wiregrass plants may live and keep growing for several hundred years. In an unburned forest, as woody shrubs and hardwoods invade, wiregrass plants may persist for a long time, but they gradually shrink and die in the shade.

As longleaf pines tend to fend off other trees, wiregrass plants compete aggressively with other grasses and forbs wherever fire is frequent and sunlight abundant. The roots of a healthy old clump of wiregrass form a dense, shallow underground mat up to 18 inches in diameter that captures nearly all available water and nutrients. Above the ground, the blades are arranged somewhat like a funnel, so that they channel rain water down to the center of their own root system, leaving their competitors dry. After growing upward, the long grass blades fall over and lay several inches of thick wiry growth in a three- to four-foot circle on top of other plants, hogging the sun and pitilessly shading out everything underneath. Other plants' seeds often can't find a way to the ground, or if they do, can't germinate in the shade.

OTHER GROUND-COVER PLANTS

Despite the dominance of bunch grasses, other plants do grow successfully among them—thanks to fires at the right time of year. Growing-season fires burn back the grasses, eliminate the dense pine litter, and allow seeds to reach bare soil in fall, before the grasses can recover. Many of the grassland's ground-cover plants require summer fires, too, to stimulate them to flower (see List 3–1). After fires, perennial plants also regrow from their roots and rise high enough to get above the grass. Bracken fern springs up, blueberry and dwarf huckleberry bushes bloom, greenbrier vines and gallberry shrubs resprout.

Fires bring out legumes, too—herbs such as milkpea and lespedeza, whose root-associated bacteria can capture free nitrogen from the soil and incorporate it into their own tissues. They return nitrogen to the food web just after fires have freed much of the nitrogen from earlier plants and plant litter. After fires, legumes increase tenfold in small patches and become an important food for wildlife. Later they are shaded out, and when they die and decompose, they leave nitrogen in the soil where other plant roots can absorb it.

Between one fire and the next, the running oaks, which have been burnt back to the ground, grow new shoots a foot or two tall. Knee-high running oak stems in a pine community may outnumber the pines, and their extensive root systems may be hundreds of years old, but the stems always stay low to the ground. One often sees hundreds of running oak stems in a

LIST 3–1
Ground-cover plants of the longleaf community: a sampling

Arrowfeather threeawn*
Big threeawn*
Bush goldenrod
Croton
Dogfennel and several
  relatives*
Dwarf huckleberry*
Flaxleaf aster*
Florida calamint
Florida dropseed*
Florida greeneyes
Gayfeather*
Goat's rue and several
  relatives*
Honeycombhead
Lespedeza, several species*
Milkpea, several species*
Mohr's threeawn*
Partridge pea

(continued on next page)

dense green patch. These are all the same plant, which are connected by their root system. Their ability to come back again and again from their roots is a successful adaptation to a fire-frequent environment. Turkey oaks also are clones of stems growing from an underground network of roots. Normally, the stems are kept low to the ground by fires, but in the absence of fire, the stems can become small, scraggly trees that can live for 50 to 60 years or so. They can quickly regrow after fires by drawing on the pool of energy stored in their extensive root system. If frequently burned, both running oak and turkey oak shoots yield acorns after only two years.[6]

The scrub oaks that grow scattered among longleaf pines in sandhill environments also respond adaptively to fire. If left unburned, they become small trees and don't live long, but if frequently burned, they remain healthy.

WEBS OF RELATIONSHIPS

Pines, grasses, and ground-cover plants form the basis of a well-integrated, complex ecosystem, whose intricate interactions have evolved over countless generations. Every member is connected in myriad ways with many others. Consider a single old-growth pine: a dozen or so species of fungi are connected to its roots. Other species of fungi may grow on the litter around its base, others on the bark, and still others in the heartwood. The tree may support a hundred different plants: algae, mosses, ferns, and vines.

A single tree also houses hundreds of species of insects and other small invertebrates, which eat the needles, bore in the bark and under it, lay eggs along the needles, and feed on the pollen: ants, beetles, caterpillars, flies, and more. In one high pine community, researchers fogged just two longleaf pines with insecticide and collected 50 species of arthropods.[7] A few of the insects associated with longleaf pine are named and described in Background Box 3–1, below.

BACKGROUND   **3–1**

Insects that Live on and in Longleaf Pine

On just a single pine tree, these insects eat the needles:

- The caterpillar of the pine devil moth

- An inchworm, the caterpillar of another moth

- The pine webworm, which hides inside a nest made of caterpillar droppings, clips off needles, and pulls them in to eat

- Two species of leaf beetles

- Pine scale insects, which suck sap out of the needles

- Woolly pine scales, which live under a layer of fleecy white wax and eat needles

- Tortoise scales, which suck sap from growing shoots

(continued on next page)

Florida native: Pine bark borer (*Tyloserina nodosa*). The beetle breeds in the bark of dead and dying pines and deposits its eggs there. Once hatched, the larvae feed in the bark and destroy the larvae of other bark beetles.

Florida native: Black carpenter ant (*Camponotus pennsylvanicus*). These ants have excavated tunnels beneath the bark of the tree and are tending a herd of aphids there.

Florida native: An arboreal ant (*Crematogaster ashmeadi*). This ant spends its entire life in longleaf pine trees and is a major food item for the red-cockaded woodpecker.

These insects eat the needles (continued):

■ Sawflies, several species in two genera, which lay their eggs inside pine needles, form pupae under the bark or in the soil beneath the tree, and emerge as flying adults in a cycle that turns over four or five times a year in Florida.

These eat other parts of the tree:

■ Pine tip moth caterpillars, which eat growing tips

■ Another moth caterpillar, which eats male catkins and young female cones

■ The deodar weevil, which attacks twigs and stems, living in small holes in shoots or chambers in the sapwood

■ Reproduction weevils, which feed on pine seedlings and produce larvae that tunnel through the inner bark leaving loose reddish-orange fluff behind

■ Mole crickets and May-June beetle grubs, which eat the roots

■ Wood-boring beetles of several genera, and other beetles that prey on them

■ Carpenter and harvester ants, which forage on the tree and rear herds of aphids there. The aphids suck up fluids from the tree and the ants then "milk" the aphids

■ An arboreal ant, which conducts its entire life cycle in galleries under the bark left behind by wood-boring beetles. These ants never touch the ground. The queens colonize new trees by flying to them.

Sources: Minno and Minno 1993; Hahn and Tschinkel 1997.

Thanks to the many insects, insect eaters can thrive. Foremost among them are other insects and spiders, but many larger creatures eat insects too, including sapsuckers, other woodpeckers, nuthatches, bluebirds, kestrels, wood ducks, flying squirrels, frogs, salamanders, lizards, snakes, skunks, moles, shrews, foxes, opossums, and bears. In fact, that is why all those hungry insects don't eat the tree up altogether: their many predators eat them first. On a pine plantation where the top priority is timber production, managers may deem it necessary to kill insects with pesticides, but in a natural old-growth pine community, numerous biological controls are acting. The birds alone are so important in controlling insect populations that they have been called "the keepers of forest health."[8] A single woodpecker eats several times its weight of insects in a day, and all trees are monitored.

Still other animals prey on the insect eaters. Among them are raccoons, hawks, shrikes, owls, indigo snakes, rat snakes, pine snakes, gray foxes, weasels, bobcats, and (in the past) panthers and wolves.

After woodpeckers drill holes in search of their prey, other birds and squirrels follow to eat from, or live in, the holes; then snakes climb the tree to go after the nesting young. Other snakes, lizards, salamanders, toads, and frogs hide and hunt other prey among the roots.

The tree serves the ecosystem in still more ways. The pine straw prevents erosion, preserves soil moisture, protects small plants, spreads fires, and slowly releases nutrients. Tip-up mounds expose bare soil, otherwise rare except after fires, on which seedlings can take root. Downed snags and rotted stumps provide moist shade, food, support for fungi, and habitats for mice, voles, toads, and other ground animals. And even after the tree itself has burned or rotted away, it benefits other community members. Its ash is the ideal fertilizer for the next generation of trees. The deep, branched holes that remain after the roots have burned make superb tunnel systems in the ground. At least ten species of animals use these stump holes, and the eastern diamondback rattlesnake is known to prefer them 60 to 1 over gopher tortoise burrows (which themselves are first-class accommodations).[9]

Besides providing habitats, the pine, its attached plants, and the plants of the ground cover produce food for many plant eaters. The fare includes flowers, leaves, stems, seeds, nuts, berries, acorns, bark, tubers, and rhizomes. These nourish insects, quail, squirrels, rabbits, deer, gopher tortoises, pocket gophers, cotton rats, mice, and numerous others. Even beaver and rabbits eat parts of the pines, and deer eat lichens that grow on the bark.

Clearly the resident populations of Florida's ancient pine grasslands exhibit many adaptations not only to the land's character and to fire but to each other. They have evolved side by side for so many generations that innate interdependencies have arisen — evidence of coevolution. All members benefit from the stability these relationships confer on them. The next three sections describe the webs surrounding a squirrel, a woodpecker, and a tortoise.

THE FOX SQUIRREL

The sight of this large, beautiful squirrel running across the ground with its long, dark, flowing tail rippling behind it like a small puff of smoke is almost diagnostic for old-growth high pine. When you see a fox squirrel, look around. You will usually see many other features of the classic environment.

The fox squirrel, the longleaf pine, and some of the underground fungi associated with the pine are knit together in a three-way relationship that suggests they have been coevolving for a long time. The squirrel derives food from both fungi and pine cones and helps to propagate both by spreading their spores and seeds. The fungi and the pine also interact, with the fungi taking sugar from the pine's roots, and the pine deriving soil minerals from the fungi.

Many aspects of this relationship illustrate why the natural, old-growth pine forest provides the best environment for its resident species. For example, green longleaf pine cones are a key support for the squirrel at the end of the hungry winter season, when other foods are in short supply. Green cones are produced reliably only by old-growth pines. It also hap-

Florida natives: Lichens on tree bark.

A **lichen** (LIKE-un) is an organism in which an alga and a fungus cooperate so closely that neither can survive without the other. The alga, which can capture the sun's energy, makes sugars that the fungus can use. The fungus dissolves mineral nutrients from the substrate into fluid that the alga can take up, and provides some protection from drying.

Coevolution is the process by which two or more species develop genetically determined adaptations to each other. A familiar example is that of flowers that have evolved scents and color patterns that attract certain insects and insects whose anatomy fits the flowers and ensures pollination.

Florida native: Fox squirrel (*Sciurus niger*).

Longleaf pine cones.

pens that green cones are so tough that the fox squirrel is the only squirrel that can tear them apart to get at the seeds while they are still viable. In the process, the squirrel scatters a few seeds, and they have a chance to germinate. Months later, the cones will finally open and fall to the ground, but by then beetles have already eaten most of the seeds and those that remain are no longer viable. The squirrel's early attack on the cones is a service to the pines.

The sound of a squirrel feasting on a green longleaf pine cone is arresting. The animal ravenously tears it apart like a demolition expert on deadline, and shredded parts fly everywhere. After the feast, you can tell it was a fox, not a gray, squirrel that was at work because the cone has been torn apart and chewed to the core. The squirrel's size and strength help it to protect its food, too. A longleaf pine cone is heavy, but if a predator is near, a fox squirrel can carry the cone up a tree to the end of a branch, and eat it there. The squirrel can also run fast and far over ground where the trees are spaced well apart.[10]

The squirrel also eats fungi—both mushrooms and the fungi that produce their fruiting bodies underground. In the process, the squirrel ingests millions of fungal spores. Later, as it roams far and wide, the squirrel excretes the spores, undigested, in its droppings—most often where it spends most of its time—near the base of a longleaf pine. When a pine seed sprouts and becomes a tree seedling, the fungal hairs are already present in the soil to intertwine with it and form the close association that will benefit both throughout their lives. The three-way relationship described here holds true for eight or more whole genera of underground fungi.[11]

In late spring, summer and fall, high pine provides a cornucopia of nourishing foods within easy reach: pine seeds, low-growing acorns, and the nuts, buds, and bulbs of other native plants. The squirrel's population size varies with the availability of these foods, which are most abundant in regularly burned pine grasslands. Different foods nourish the squirrel in each season, as shown by an eight-year study whose results are summarized in List 3–2. Take away a plant species and a food will go missing—possibly an important food. This accounts for the rarity of fox squirrels except in old-growth pine.[12]

The squirrel also requires a nearby closed-canopy forest where it can breed and raise its young, and where the branches are close enough to permit safe travel from tree to tree. The squirrel builds a platform nest of sticks in the oaks, pines, or cypresses of hardwood hammocks, but returns to the nearby pine grassland to forage for food where it can move about on open ground without fear of unseen predators. The natural landscape provides an ideal arrangement: hardwood hammocks and cypress stands embedded in a pineland matrix.

Each fox squirrel has a home range of up to 50 acres, though some 13 to 60 squirrels can share a square mile of good habitat. (There are 640 acres in a square mile.)

Squirrels native to Florida today include the flying, gray, and fox squirrels, and the fox squirrels are the strongest and fastest. The fox squirrel is represented by some nine different subspecies, each occupying a different cluster of pine habitats.

THE RED-COCKADED WOODPECKER

The red-cockaded woodpecker's worth to its natural ecosystem far exceeds its own presence. It is an indicator species, in that its presence signifies a native community. It is adapted by tens of thousands of years of evolution to court, breed, nest, forage, and raise and defend its young in surroundings that offer numerous insects and an open, fire-pruned understory in which to hunt them. It is also a keystone species, in that its activities enable many other animals to thrive.

The bird's requirement for a nesting site is exacting: it is the only woodpecker in Florida that drills its cavity in a living, native pine. No other woodpecker can excavate a nest cavity in a tough, living tree, and even the red-cockaded finds the task hard to do. The pine must have heart rot to make it soft enough to drill, and that means it has to be an old pine. As a result, the woodpecker is found only on sites with trees older than about seventy years.[13]

The bird meets the challenges of its difficult life by forming clans. Each breeding pair produces its eggs in the spring, and non-breeding family members stay nearby throughout the summer to help find food for the hatchlings. The pair drives away the female clan members in the fall, but may keep the males around for several years to help feed, house, and defend successive families. Clan members cooperate in drilling several cavities at a time. It may take from half a year to several years to complete any one cavity—one bird's roosting site. Not every bird rates a cavity; some roost in scars or crotches already available in trees.

Insects and other small invertebrates are the woodpeckers' food of choice: they eat a varied diet of spiders, ants, roaches, centipedes, and insect eggs, helping to control populations of these creatures. Lightning-struck pines make for especially bountiful dining, because they are infested first by beetles and later by ants that move into the beetles' abandoned burrows. (The ants are the arboreal species shown and described on page 44.) At times, up to 80 percent of the birds' diet consists of these ants, which they easily find under the bark of dead pine branches. If need be, the birds can also eat blueberries, cherries, wax myrtle berries, and other fruits.[14]

The red-cockaded woodpecker requires a large foraging area. Ideally, a single clan will be surrounded by at least 125 undisturbed acres of mature, live trees nine inches or more in diameter. If the site is of lower quality, a clan may need several hundred or even a thousand acres. The presence of several thriving clans of red-cockaded woodpeckers in a given area suggests that the habitat must be somewhat similar to that in which the species and its co-inhabitants evolved.

Red-cockaded woodpeckers once were common. Now they are rare, because their habitat, the pine grassland, has so greatly dwindled in area. Isolated colonies tend to die out because females, once they have left their original clans, fail to find mates nearby. The largest known population of red-cockaded woodpeckers living anywhere in the world consists of several hundred clans in Florida's Apalachicola National Forest.[15]

If a red-cockaded woodpecker's cavity comes available, many animals

A **keystone species** is one that many other species depend on, so that its loss spells the loss of many others.

Florida native: Red-cockaded woodpecker (*Picoides borealis*). Belying the description that its name implies, the red-cockaded woodpecker is a black and white bird, with an almost undetectable spot of red above each eye, in males only.

The woodpecker's nest cavity. The bird defends its home by pecking small wells around the entrance, so that sticky resin will drip down the trunk all around it. The resin deters invasion, particularly by tree-climbing rat snakes that live in the same territory. For as long as a bird uses a cavity, it keeps freshening these injuries on the tree.

Florida native: Gray rat snake (*Elaphe obsoleta spiloides*). Rat snakes readily climb trees to feed on baby squirrels, baby birds, and bird eggs. This one is climbing a longleaf pine.

compete to move into it. Among would-be tenants are snakes, several types of bees, the broad-headed skink, and eleven kinds of birds, including the red-bellied woodpecker, the pileated woodpecker, the red-headed wood-pecker, the great crested flycatcher, the American kestrel, and even the wood duck, which may nest hundreds of yards from water. Mammals use the cavity, too, among them the flying squirrel.

To this point, the strands in the pine grassland's web of relationships have all been attached directly to the pines. But there are other webs, and one involves a ground animal, the gopher tortoise.

THE GOPHER TORTOISE

The gopher tortoise is an ungainly, slow-moving, bland-looking creature, but like the red-cockaded woodpecker, it is a key member of the pine-grassland ecosystem. For one thing, now that bison and wild horses no longer graze the ground cover, the gopher tortoise is the community's most significant native grazing animal. For another, the tortoise's burrow serves as a sort of community safe house. A northern bobwhite being chased through the woods by a cooper's hawk can fly directly into a tortoise bur-row and hide until the hawk has gone away. A rabbit may dash into a burrow to escape a fox. A Bachman's sparrow may flit down a burrow to hide from a hawk when the ground cover has recently been shorn by fire.

Gopher tortoises typically live at a maximum population density of only three animals per acre, each maintaining about eight to ten burrows. The burrows are pivotal to other animals' survival on that acre. Only one ma-ture gopher tortoise lives in a burrow, but members of some 300 to 400 other species may reside there, too, finding needed shelter from sun, fire, and extremes of heat and cold. For at least three or four species of verte-brates and for many arthropods, the presence of the gopher tortoise may today spell the difference between survival and extinction. Three dozen species live there the year around, including the gopher frog, the Florida mouse, the gopher cricket, and the threatened, nonvenomous, and beau-tiful indigo snake (which locals may call the gopher snake). Other per-manent residents include blind and colorless beetles and flies that are adapted to life in the dark and cannot survive in the outside world. Visitors may include many kinds of salamanders, rattlesnakes (which don't bother the tortoises), pine snakes, skunks, opossums, armadillos, burrowing owls, lizards, frogs, toads, and many invertebrates. North America's largest tick lives on the tortoise and can use no other host. It takes a long span of years for such an exclusive relationship to evolve. The tick may not be a charm-ing creature, but it does testify to the antiquity of *Gopherus polyphemus* as a species.

For plants, too, the tortoise's burrow digging offers mounds of bare soil in the sun—and in the absence of fire, these, together with tip-up mounds, are the main sites on which new pine seedlings can start their lives. For all these reasons, the presence of a stable and thriving tortoise population in its natural habitat signifies a healthy natural community and its absence indicates ecological decline. Where gopher tortoises today make their homes on grassy road shoulders or in the backyards of houses in subdivisions,

Florida native: Common gar-ter snake (*Thamnophis sir-talis*). This snake uses high pine grasslands and many other ecosystems as its habi-tat.

REA

The tortoise burrow. The modest, tortoise-sized entrance offers no clue that the burrow descends well below the surface of the ground. Burrow lengths vary from three to 52 feet.

Tortoises may be either right-handed or left-handed. Those that have a stronger left claw will dig a burrow that curves to the right, and vice versa.

Source: Ashton and Ashton 2004.

Florida native: Gopher tortoise (*Gopherus polyphemus*). The gopher tortoise's ancestors have been present in North America for at least 55 million years and have produced 22 different species of land tortoise. Four species remain today, and the gopher tortoise is the only one remaining east of the Mississippi River.

Being vulnerable to predation by bears, wolves, coyotes, and big cats, the tortoise digs burrows in which to hide. It is well adapted to this endeavor. The burrow's entrance is always on high, dry ground, but the tortoise digs down to just above the water table and so achieves a comfortable humidity and temperature. Except during heavy rains, the burrow is dry all the way to the bottom. Like people, tortoises can inherit homes from their forebears from several generations back. Some live in burrows known to be several decades old.

they may survive for a time, but most of the animals that occur with them in natural sites will be missing.

Summer fires are as vital to the gopher tortoise as to the other members of the high pine ecosystem. The tortoise mates during April and May, and several weeks later the female lays about six eggs in a sunny place, usually in the sandy mound in front of her burrow. Nine years out of ten, the clutch is destroyed by predators such as armadillos, raccoons, foxes, skunks, snakes, and fire ants, but now and then the eggs survive for the whole 80 to 90 days until they hatch. They hatch in August when, under natural conditions, summer fires have already burned across the landscape and new plant shoots are just coming up. The tortoise hatchlings are only about two inches long, and they struggle through ground cover and debris looking for their first food. They need to find young, tender, and abundant grasses and forbs right at ground level where they can reach out with their short necks and nibble the leaves. Dense, thigh-high ground cover is an impenetrable jungle to a hatchling tortoise. A pleasing sight after a summer burn is the tiny burrow made by a baby gopher tortoise.

Tortoises move to find grassy areas. They tend to disappear from areas that remain unburned, where shade and litter have built up. Their numbers increase in areas recently burned.

Once past the vulnerable hatchling stage, gopher tortoises are large enough to deter predation and too tough-shelled to eat. Thereafter they

DBM

Florida native: Gopher frog (*Rana capito*). This frog is named for the gopher tortoise, whose burrow it frequently uses as its home.

BM

Florida native: Florida mouse (*Podomys floridanus*). A gopher tortoise burrow makes a secure home for this little rodent.

Florida native: Six-lined race-runner (*Cnemidophorus sexlineatus*). This speedy reptile thrives in xeric habitats.

have few enemies other than people and dogs, and are thought to live for 50 years or more. They are slow to reproduce: they become sexually mature at about 18 to 25 years and only then can pass on their special traits to the next generation of tortoises. Their slow reproduction rate, like their dependence on a habitat that is shrinking, makes them vulnerable to extinction, and today they are protected throughout their range.

From the gopher tortoise's point of view, three wishes must be granted to secure happiness: well-drained, sandy or loamy soil in which to dig a burrow; abundant low-growing tender plant life for food; and open, sunny areas in which to bury nests of new eggs. Fire-maintained high pine communities fill the bill.

OTHER RESIDENT ANIMALS AND VISITORS

Many other animals native to Florida's high pine grasslands could have been chosen for emphasis in this chapter. Beneath the litter layer, on and in the soil, are many other habitats and strands of the food web. Earthworms, grubs, millipedes, moles, ants, and others keep turning the soil, returning to the surface nutrients that might otherwise leach down in rain, and freeing those nutrients for reuse. In addition, besides the gopher tortoise, several dozen species of amphibians and reptiles are found within the range of longleaf pine, and several are endemic. Examples are listed in List 3–3.

Some birds spend their whole lives in Florida's pine grasslands. The red-cockaded woodpecker has already been mentioned. Northern bobwhites nest on the ground; wiregrass is ideal cover and lends protection as they scurry along the ground with their chicks behind them in single file. The American kestrel and the brown-headed nuthatch, too, require old-growth longleaf or a close facsimile for their habitat: they are absolutely dependent on its open spaces in which to hunt their prey. List 3—4 itemizes birds observed pairing and raising broods in a north Florida longleaf pine community, as well as birds that spend winters there. It is just a sampling of the birds that use north Florida longleaf forests. There are many more, and other, more tropical birds use south Florida's pine grasslands.

Winter birds are far more numerous than resident birds. They fly south to use the pinelands as a winter haven, then return north for the summer to breed. Winter-bird numbers are higher in longleaf pine than in any other kind of Florida forest, and more bird species are found there in winter than in any other season. More than any other factor, the amount of forage determines bird population sizes, and the foraging is good, especially in natural, old-growth pine grasslands. Communities 70 years old and older are the richest in bird life: they have both the most species and the most numerous birds. Stop in an old longleaf pine community in winter, get out of the car, and stand still. At first, the treetops seem devoid of life. But wait until the ripples you've created die down, and suddenly birds are sprinkled all over the trees, flitting from branch to branch, chipping and chirping, picking at the bark and pine cones, finding a feast of seeds and insects that earth-bound humans would never know was there. Oth-

LIST 3–3
Amphibians and reptiles native to Florida's high pine grasslands—a sampling

Amphibians
Barking treefrog
Gopher frog
Narrowmouth toad
Oak toad
Ornate chorus frog
Pine woods treefrog
Spadefoot toad
Striped newt
Tiger salamander

Reptiles
Coachwhip
Corn snake
Eastern diamondback
  rattlesnake
Eastern fence lizard
Eastern glass lizard
Eastern indigo snake
Eastern pine snake
Gopher tortoise
Gray rat snake
Ground skink
Mole skink (endemic)
Short-tailed snake
Six-lined racerunner
Southeastern crowned
  snake
Southeastern five-lined skink
Southern black racer
Southern hognose snake

ers rustle about in the ground cover, eating tidbits from the grasses and shrubs, and pulling worms and grubs from among the roots.

Other migratory birds use Florida's pine grasslands as stopping places between their summer and winter homes. They fly through in spring, on their way to breeding grounds farther north, and then return in fall on their way to wintering grounds in South America. Like other birds, these migrants seek out old-growth sites when they are passing through—those sites best meet their needs for food and shelter. Florida's few remaining natural high pine communities, especially the old ones, are vital to many of the birds of eastern North America.[16]

Bird expert Roger Tory Peterson remarked that the variety and abundance of bird life in an environment is an index of its health. By that index, Florida's high pine grasslands are healthy indeed. Moreover, because natural high pine communities support migratory birds, their integrity supports the bird life of other ecosystems elsewhere.

VALUES OF HIGH PINE COMMUNITIES

Originally, Florida's ancient pine grasslands were valued for their timber. The loss of nearly all of them was due almost entirely to logging. The once mighty trees became masts for tall sailing vessels and the timbers for lodges, mansions, and dwellings all over the continent and in Europe. The ecosystem's lumber brought substantial profits to the South, but today, other ecosystem values are being recognized as perhaps even greater than that of the timber.

Viewed simply as masses of living plants, pine grasslands, like all terrestrial ecosystems, help regulate local and global environmental conditions. They capture carbon dioxide and store carbon in their tissues, helping to prevent global warming. They take up water from below the ground and release it into the air, contributing to regional rainfall. They absorb heat, moderating the climate. They keep soils porous and able to absorb water and prevent runoff. They immobilize or detoxify pollutants that would otherwise taint surface waters. They serve as a conduit for massive flows of materials (carbon, oxygen, nitrogen, phosphorus, and others) through natural cycles from earth and water to air and back again. Thanks to these services, some economists now agree with ecologists that "forests are worth more standing than logged."[17]

In addition to these values, which are common to all forests, the natural pine grassland has unique values of its own. It is a truly ancient ecosystem. Pine savannas have existed on the Coastal Plain for more than 20 million years, the species that grow in them are superbly well adapted to the landscape, and many are endemic.

Fossil records show that many species have been present in this ecosystem, essentially unchanged, for at least two million years, and their close interdependencies suggest that they have been evolving together for a longer time than that. When species are removed from this interdependent system, or when the management pattern is changed, all species are affected and many populations die out.[18]

LIST 3–4
Birds that use north Florida pine grasslands

American crow
American robin
Bachman's sparrow
Barred owl
Blue grosbeak
Blue jay
Blue-headed vireo
Brown thrasher
Brown-headed nuthatch
Carolina chickadee
Carolina wren
Chipping sparrow
Common grackle
Common yellowthroat
Downy woodpecker
Eastern bluebird
Eastern kingbird
Eastern meadowlark
Eastern towhee
Eastern wood-pewee
Fish crow
Great crested flycatcher
Great horned owl
Hairy woodpecker
House wren
Loggerhead shrike
Mourning dove
Northern bobwhite
Northern flicker
Palm warbler
Pileated woodpecker
Pine warbler
Red-bellied woodpecker
Red-breasted nuthatch
Red-cockaded woodpecker
Red-headed woodpecker
Red-winged blackbird
Ruby-crowned kinglet
Summer tanager
Swamp sparrow
Tufted titmouse
White-breasted nuthatch
Wood duck
Yellow-bellied sapsucker

Note: Some of these birds breed in the forest in summer, some spend winters there, and some do both.

Sources: Engstrom 1982, Repenning and Labisky 1985.

Florida native: Northern bob-white (*Colinus virginianus*). The bobwhite's ideal habitat includes both open, bare ground on which to scratch up seeds, and thick tufts of wiregrass or the like in which to nest and hide. Longleaf pine communities managed naturally with summer fires best meet its needs.

Source: Bailey 1998. Need of summer burns from Tall Timbers staff 1996.

High pine savannas help to introduce fire into other communities nearby. Fires start up most frequently in high pine, and then sometimes spread into neighboring communities, which require fires at longer intervals to maintain them. Sand pine scrub sometimes ignites from high pine. Pond pines that grow in wet depressions in pineland grasslands depend on the heat, and possibly smoke, from fires in the surrounding pine community to open their cones. Herb bogs are wet and often nearly treeless, so they are not prone to lightning-struck fires, but fires spreading into them from nearby high pine communities permit them to regenerate. At long intervals, fires from high pine burn into red-cedar groves and sand pine stands and promote their reproduction. Even the gopherwood tree, found in deep ravines in another kind of forest altogether, may benefit from its proximity to nearby pine communities. Periodic exposure to smoke from forest fires is thought to help control a fungus that attacks the gopherwood and is now driving it toward extinction due to lack of fire.

Thus, if pine grasslands decline, other communities decline in concert. If the grasslands are restored, neighboring communities may recover as well. The whole, interlocked system of high pine, sand pine scrub, lowland pine, bogs, and adjacent wetlands is a single, fire-dependent system.

Beyond all of that, the pine grassland, in its natural state, is very species-diverse, with some 50 to 250 species of plants per acre. One researcher has identified more than 50 herb species within just a few square meters. And the mix of ground-cover plants is not the same in north Florida as in south Florida, Virginia, or Texas. Florida's pinelands also are a refuge for dozens of plant species that are endangered or threatened throughout their ranges.[19]

Finally, some people find spiritual inspiration when they visit places such as this. They hold that old-growth communities have a "transcendent aesthetic and religious value in the inner landscapes of natives and newcomers alike" and that "their majesty inspires comparison with . . . great cathedrals."[20] Perhaps we have some innate way of sensing intricacy, stability, and harmony. Perhaps something deep in us knows and needs these qualities.

MAINTAINING HIGH PINE

It is a relatively new and welcome idea that forest ecosystems can be managed for values such as biodiversity instead of, or together with, timber extraction. In the case of high pine communities, management can center on maintaining a few species such as the red-cockaded woodpecker. Because the red-cockaded woodpecker's requirements are so exacting, they practically define the mature, healthy Florida pine grassland. If caretakers manage the ecosystem to provide optimal habitat for the birds, this benefits the whole community. Red-cockaded woodpecker habitat contains old trees, well spaced, that are suitable for cavities, and younger trees that will become good cavity trees later. It is frequently burned in summer and has few or no hardwood trees in the understory. It has many different plants bearing flowers and fruits at different times of year; it has healthy insect and other invertebrate populations——in short, it provides an environment

Florida native, endemic: Mock pennyroyal (*Hedeoma graveolens*). Mock pennyroyal grows on sandhills, in flatwoods, and around pond margins.

in which all of the community's inhabitants can thrive and perpetuate themselves.[21]

Timbering is not incompatible with such management, provided that the ground cover is maintained relatively undisturbed and that different patches are selectively logged every year, leaving mature and variously aged trees for natural regeneration and later harvest. To make sure most pine seedlings survive, managers can note areas where seedlings have survived in late winter and protect these from fire for at least a year, but otherwise, burn frequently in summer. Management of several large and significant longleaf pine communities in the Apalachicola National Forest, at Torreya State Park, and in Eglin Air Force Base proceeds according to these guidelines. What cause to celebrate, that this magnificent ancient ecosystem, after dwindling to a fraction of its former range, may still be there for our children and grandchildren to enjoy and learn from.

* * *

Fire exerts its influence on ecosystems all across the highlands, upland plains, and coastal lowlands, and even into wetlands such as bogs and marshes. The next chapter is devoted to the high pine grasslands' major neighboring communities—the flatwoods and prairies that surround them, which extend nearly all the way to the coast.

DBM

Florida native, endemic: Florida beargrass (*Nolina atopocarpa*). This delicate plant, found only in Florida's pine grasslands, has become very rare.

CHAPTER FOUR

FLATWOODS AND PRAIRIES

Come down off the rolling sand and clay hills, now, and view the expanses of flat lands that skirt them. These, too, are part of Florida's pine grass-lands—in fact, they are the larger part. They occupy the whole, broad band of sand and marl plains that once lay under water off Florida's shores—that is, the coastal lowlands (shown in Figure 2–4 on page 27). In the past, they were the most extensive grasslands in the southeastern United States and covered about 50 percent of Florida. Today, although the underlying level terrain is still there, it is largely developed. The natural systems have all but disappeared.[1]

Flatlands communities go by many names: flatwoods, savannas, prairies, grasslands, plains—but because they all are level in topography, it is suggested that as a group they be described as *flats*. Some parts of the flats are predominantly dry, some moist, some wet, and at the wet end of the moisture gradient, flats support wetlands, as described in the chapters of Part Three.

Despite their various names, there is no mistaking flats visually, as is clear in Figure 4–1. They have widely scattered trees, or no trees. They may have some shrubs, notably saw palmetto and gallberry, but under natural conditions, most shrubs are inconspicuous or absent in the flats. They normally have a diverse ground cover of grasses and forbs, including legumes, but they may have only patchy ground-cover vegetation. What defines them all is their flat expanses across which you can see long distances.

From their level appearance, it is easy to picture how these flat lands formed. During the past two million years, after the seas had receded from Florida's higher lands, ocean waters moved onto and off the lowlands many times and spread sand all across them. All of the flats have these features in common: fine sandy soils that are poor in nutrients, low flat topography, and low rates of runoff and percolation to lower soil layers. All are slightly to strongly acidic. (An exception is marl prairies, which are treated with marshes in Chapter 10.)

Florida native: Rosebud orchid (*Pogonia divaricata*). This orchid grows naturally in Florida flatwoods and bogs.

The **flats** (or **flatwoods** and **prairies**) are the predominant natural communities of the coastal lowlands. They are level in topography with a continuous carpet of grasses (often wiregrass or its close relatives) and herbs.

The flats called *flatwoods* have scattered trees; those called *prairies* are treeless or nearly so.

OPPOSITE: Longleaf flatwoods in April after winter burn. Pine flatwoods were popularly called "pine barrens" in earlier times, but in truth they are far from barren. It is now known that there may be more than 100 species of plants, and many more species of insects and other small animals, in an acre.

FIGURE 4–1

Various Flats Communities

A = A central Florida pine grassland abloom with wildflowers
B = A cabbage palm–wiregrass prairie near Kissimmee
C = A shrubby grassland in Kissimmee Prairie
D = A longleaf pine–saw palmetto flatwoods in the Panhandle

In north Florida, the trees on natural flats are mostly longleaf pines except at low-elevation borders, where slash may hold a strip. In south Florida the pines are more often slash pines; midway down the peninsula they are mixed. Pond pine often grows in depressions. South Florida flats have cabbage palms and some hardwoods as well as pines.

Flatwoods appear monotonous at first sight. Who would guess that they support vast numbers of fascinating plants and animals? One reason is that they are connected to many other kinds of communities. Taken all together, these communities provide many different combinations of habitats that meet animals' needs for food, shelter, nesting, and nursery grounds.

THE FLATLANDS MATRIX

Flatwoods and prairie communities form a matrix within which other systems lie. High pine grasslands stand on rises within them. Lower areas hold hardwood hammocks, marshes, swamps, ponds, and streams (all of these are treated in later chapters). Many animals conduct their lives using several different communities within such a mosaic. Figure 4–2 shows a set of communities embedded in a Panhandle flat.

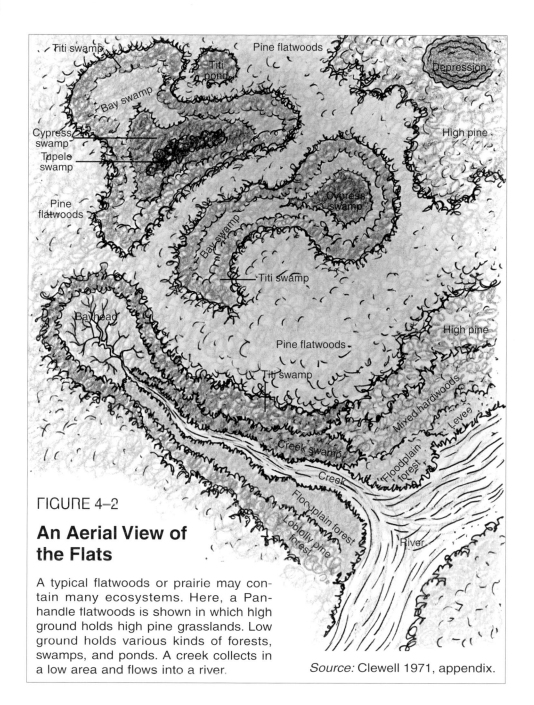

FIGURE 4–2

An Aerial View of the Flats

A typical flatwoods or prairie may contain many ecosystems. Here, a Panhandle flatwoods is shown in which high ground holds high pine grasslands. Low ground holds various kinds of forests, swamps, and ponds. A creek collects in a low area and flows into a river.

Source: Clewell 1971, appendix.

FIRE AND RAIN ON THE FLATS

Fire, water, and lack of water all influence what grows on the flats. Under natural conditions, fires frequently clear away shrubs, favoring an open, grassy aspect. Rains are seasonal, and the land receives them differently than in high pine grasslands. In high pine, rain runs off or sinks into the soil relatively quickly, but in flats, soils drain poorly and rains leave behind standing water. During rainy seasons, flats remain soggy for weeks at a stretch, which for many plants would make survival difficult. Another stress, drought, alternates with rainy seasons and sometimes lasts for several years. To live and reproduce in flatwoods and prairies, plants have to be adapted to standing in water, to seasonal droughts, to decades-long droughts—and even to fires. Remarkably, many native plants possess all of these adaptations. Plant diversity in flatwoods and prairies is high.

Florida native: Pawpaw (an *Asimina* species). Many pawpaw species grow in pine flatwoods throughout Florida.

Three Florida natives: At left, narrowleaf blue-eyed grass (*Sisyrinchium angustifolium*). Center, tuberous grasspink (*Calopogon tuberosus*). At right, white birds-in-a-nest (*Macbridea alba*), an endemic species.

Before people altered the landscape, fires probably swept across the broad expanses of the flats only slightly less often than in high pine communities. Fires typically started in summer, when lightning struck pines on ridge tops. Then the flames swept down into the flats. Sometimes, the fires burned out at the edges of bogs and swamps, but in times of drought, fires might burn all the way to the borders of water bodies. A summer fire in Florida's flatwoods is shown on page 22.

FLATWOODS AND PRAIRIE VEGETATION

Trees are sparse, spindly, and stunted in the flats; sometimes there is only one slender tree per acre. Conditions are unfavorable for trees. The sandy soil is underlain by a water-retardant hardpan that tends to keep the soil waterlogged. That, and the soil's infertility, usually stunt the trees to well below 100 feet in height. Speculation has it that because the trees are more slender than in the sand and clay hills, they are more vulnerable to fire and require longer times to recover from scorch. The completely treeless prairies of central and southeastern Florida were once thought to be logged-over flatwoods, but they may in fact have been treeless all along, due to frequent fires.[2]

A **hardpan** is a layer of water-impermeable substances below the surface of the soil. It consists of accumulated organic materials and aluminum and iron compounds that have leached down from the sediments above it.

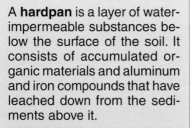

Endemic: Leavenworth's tick-seed (*Coreopsis leavenworthii*). The several plants in the genus *Coreopsis* bloom all year in moist to wet flatwoods and marl prairies as well as in disturbed areas. Collectively, they are Florida's "state wildflower."

Native: Chapman's rhododendron (*Rhododendron minus* var. *chapmanii*). This rare plant grows in Panhandle flats.

Native: Adam's needle (*Yucca filamentosa*). This showy flower blooms in spring in the drier parts of the flatwoods.

If shrubs get a foothold in flatwoods, frequent fires keep them low to the ground. Where shrubs are numerous or large, it seems likely that fire has been excluded for unnaturally long periods; shrubby flatwoods are probably the result of fire exclusion for ten to 25 years or more. Once present, though, shrubs can survive repeated fires: their root crowns and taproots store water and nutrients underground, and they can draw on these stores to regrow after their tops are burned.[3] List 4–1 names the main shrubs of flats.

The grasses and forbs of flatwoods and prairies are mostly fire-loving plants, and as in high pine grasslands, many flourish best only if burned during summer every few years. They respond to fire by rapidly regrowing from their roots and by flowering and releasing seeds. Many, being highly flammable themselves, also help spread fires.

Hardly anyone, casually viewing a natural flatwoods or prairie, would guess that there were upwards of 100 species of ground-cover plants on an acre. Only a few are apparent at any one time; they become conspicuous only when they bloom—and many don't bloom unless the flat has been recently burned. After a burn, though, the floral variety is astounding. Many of the ground-cover plants of the flats are widespread across the southeastern United States, but a few are endemic and rare.[4] Endemic species are named in List 4–2.

Florida native: Hairawn muhly (*Muhlenbergia capillaris*). This fire-loving grass, with its reddish seedheads, grows in flatwoods and coastal swales all over the eastern and southern United States and in other parts of the world. Here, it is growing in Caladesi Island State Park in Pinellas County.

Flatwoods and Prairie Animals

What kinds of animals might make use of a flatwoods or prairie? It seems an unlikely place for animals of any kind. Birds can't roost where there are few or no trees. Large animals such as deer, bear, bobcats, and foxes don't spend much time there, because they need cover. But simple looks deceive. The great variety of plants make food for many kinds of animals including a multitude of tiny insects and other invertebrates and small vertebrates that eat them. On a tiny scale, the animal life is rich and varied, especially when legumes and other forbs are flowering in response to burns.

Invertebrates. The buzz of insects in a flatwoods or prairie bespeaks a horde of tiny things that live in, and eat, all parts of the plants, and serve the plants by cross-pollinating them. Attention to detail in the grasses may reveal many grasshoppers, beetles, and other small insects. Spiders are numerous: many species live on the ground and many others in foliage. Inspection of saw palmettos may disclose other beetles, walking sticks, and paper wasps. Scrutiny of the soil might well reward the searcher with any of four species of ants, while at least two species of termites are busy disposing of fallen litter. A centipede and a millipede are also active in Florida flatwoods. Unseen, hundreds of other invertebrates are also present.[5]

To appreciate how many animals there may be in a flatwoods or prairie, consider just the earthworms: a single acre of land may contain some 50,000 earthworms or more. One earthworm may not appear very impressive, but 50,000 can do an impressive amount of work. Earthworms ingest the soil grains, digest bacteria and other microbes from them, and then excrete the grains as small, crumbly deposits called castings. Each earth-

LIST 4–2
Endemic herbs in flatwoods and prairies (examples)

Cutthroatgrass*
Fallflowering ixia
Mock pennyroyal
Rugel's false pawpaw
Scareweed
Yellow milkwort

Note: *Cutthroatgrass requires seepage water and "cutthroat seeps" are wetlands. See Chapter 9.

Source: Adapted from FICUS 2002.

Florida native: An earthworm of the flatwoods (*Diplocardia mississippiensis*).

Florida natives: Walking stick (a member of the order Phasmatodea, left) and Eastern lubber (*Romalea microptera,* right). These insects and hundreds of others occur naturally in multitudes in Florida flats. The lubber grasshopper is a young specimen. Adults are black with yellow markings and red on their wings.

worm excretes about a teaspoon or so of earth on the soil's surface each day, so that an acre of worms turns a prodigious ten to twenty tons of soil a year.

Processed by earthworms, flatwoods and prairie soils stay loose and inviting for plants to grow in. During droughts, earthworms, seeking moist soil, dig deeper. Then plant roots find the moisture they need by growing downward within the pores that the worms have left behind. Live worms serve as the dinner of choice for many slightly larger animals. Dead ones serve as food for decomposer organisms and then as nitrogen-rich plant fertilizer. Earthworms do more work, and of more different kinds, than any plow can do, but they burn no fossil fuel and leave behind no pollution.[6]

The earthworms of Florida's flats are, by worm standards, stupendous in size. They are also numerous. A collector can gather a thousand in a day just by vibrating a stake in the ground (called "grunting" for worms). The vibrations drive the worms to the surface.

On learning to "grunt" for worms, one writer reported, "I bore down, surprising myself as I suddenly produced just the resonant worm-tingling grunt I was after . . . and pale legions of annelids magically began to emerge from their tunnels. I had become a pied piper of worms. The finest bait in the entire South—the finest, some say, in the whole of the planet—was surfacing all around me."[7]

Hundreds of other invertebrates dwell in the flats, each species in almost unimaginably large numbers and performing equally important tasks. A lot more goes on in the flatlands than people see.

Amphibians and Reptiles. Consider what an amphibian (say, a salamander) is up against when living on a Florida flatwoods or prairie that may be altogether dry for long times, and then soggy wet for equally long times. Salamanders easily meet the challenges of the alternating flatwoods seasons: they spend their early lives in the water and their later lives on land, takng refuge underground. Many species of frogs also spend their lives in flatwoods. Like salamanders, they breed in ponds and puddles and their tadpoles mature there; then, as adults, they live on higher ground or in nearby trees. These small animals make attractive prey for garter, eastern ribbon, and hognose snakes, as well as for visiting birds. Two species of turtles are also common—the mud turtle and box turtle.

Florida native: Flatwoods salamander (*Ambystoma cingulatum*, left) and its larva (right). This salamander lays its eggs in depressions at the bases of clumps of wiregrass in October just before the winter rains fill them with water. The larvae then hatch and grow in the water. They metamorphose to adults in April or May, as the dry season begins.

Florida native: Ornate chorus frog (*Pseudacris ornata*).

Salamanders are safest from predation in ponds that dry up every year (see Chapter 10). The dwarf and mole salamanders also breed in puddles and ponds in the flats.

Birds. Some birds nest in the flats, while others nest in adjacent swamps and hammocks and then fly into the flats to forage. Migratory birds use the flats as refueling places. Open, largely treeless expanses invites insect-eating birds to skim low over the grasses and pick out morsels among them. By day, the eastern meadowlark and red-winged blackbird hunt this way, and by night, the common nighthawk does the same. These smaller birds become prey for the American kestrel, the white-tailed kite, and the short-tailed hawk, which are often seen soaring over the flats. More rarely, over Florida's central dry prairies, the peregrine falcon, the grasshopper sparrow, and the bald eagle may be seen. List 4—3 presents a partial list of birds often seen in Florida's pine flatwoods.

Three species of large birds nest in flats (especially in prairies): a crane, an owl, and a vulturelike hawk. In spring, the Florida sandhill crane executes its thrilling, ages-old mating dance on central Florida's prairies; then it nests in embedded marshes. The burrowing owl makes its home in the abandoned burrows of gopher tortoises and other digging animals on central Florida's flats. The crested caracara, a scavenger, nests and forages in cabbage palm flatwoods.

Wild turkeys illustrate especially dramatically how well-adapted native species can be to their natural environments. Wild turkeys have inhabited the flats for millions of years. Observer Joe Hutto spent a season living and roaming with a flock of wild turkeys and reports that their instinctive behaviors are molded to give them every advantage under natural conditions. They notice everything that relates to their survival and take appropriate action, and by the same token, they ignore things that are unimportant (to turkeys). They run from large predators and fend off small ones. They recognize venomous snakes, warn each other, and steer clear of them, but waste no energy being alarmed by harmless snakes. They alert each other to danger, stick together in a flock, call in straying members, and find nourishing foods in every season.

Turkeys find the many foods they need all year in the flatwoods. In spring, young turkeys eat spiders and green grasshoppers; in summer, they turn to ripe blackberries and, later, blueberries. By fall, when fully grown, they eat fox grapes, lizards, and frogs together with earthworms and doz-

LIST 4–3
Birds seen in Florida's pine flatwoods: a sampling

Bachman's sparrow
Bald eagle
Black vulture
Brown-headed nuthatch
Carolina wren
Chuck-will's widow
Common yellowthroat
Eastern towhee
Henslow's sparrow
Northern bobwhite
Northern cardinal
Pine warbler
Red-bellied woodpecker
Red-cockaded woodpecker
Red-shouldered hawk
Red-tailed hawk
Turkey vulture
Wild turkey
Yellow-throated warbler

Source: Adapted from Stout and Marion 1993.

Florida native: Burrowing owl (*Athene cunicularia*). Close by its burrow, this engaging small creature bobs, cackles, preens its mate, and periodically leaps up to hover 20 feet above the ground. It has a rich vocabulary of some 13 different calls, one of which mimics the rattlesnake's rattle and scares potential predators away. Florida populations are non-migratory and most pairs mate for life.

Source: Haug and coauthors 1993.

Florida native: Crested caracara (*Caracara cheriway*). Unlike vultures, this scavenger spends most of its time on the ground.

Florida native: Florida sandhill crane (*Grus canadensis pratensis*). One of Florida's largest native birds, the crane uses a mosaic of habitats in the Okefenokee Swamp, in basin wetlands such as Paynes Prairie near Gainesville, and in parts of the Kissimmee Prairie north of Lake Okeechobee. These ecosystems offer both shallow ponds, in which the crane builds large conspicuous nests, and expanses of dry prairie where it forages.

Each pair mates for life and produces one to two young per year. The courtship ritual is a famous, dramatic mating display.

Source: Stys 1997.

ens of kinds of insects. (List 4—4 shows what their fall menu might consist of.) Hutto sees in wild turkeys "a remarkable preexisting genetic understanding of the various aspects of their environment . . . it seems they are born ancient."[8]

The wild turkey uses every part of the flatwoods mosaic—not just the hardwood hammocks or the flats, but all parts, including open prairies or flatwoods, hammocks, marshes, swamps, ponds, and streams. The turkey nests in thickets on dry ground burned a year earlier. As the young are mature, they begin to roost in the trees. Turkeys stroll in bands through flatwoods and hardwood hammocks in search of food, seek cover and rest in thickets, and quench their thirst in streams and ponds.[9]

Mammals. Several species of small mammals are numerous in flats, among them the cotton rat, the cotton mouse, the harvest mouse, the least shrew, and the short-tailed shrew. The rodents thrive on the living salad provided by the diverse ground-cover plants. They use all parts: roots, stems, leaves, flowers, and seeds. The shrews eat the multitude of insects that swarm among the plants. In turn, these small animals are a favorite food of larger mammals, snakes, and birds of prey.

That small mammals can withstand the constant pressure of such heavy predation is a tribute to the rapidity with which they multiply. Take the hispid cotton rat, for example: it has been called "the ultimate baby machine." In theory, a single newborn female cotton rat can produce more than 150,000 descendants within a year of her own birth. The cotton rat's gestation period is a brief four weeks; the newborns number up to a dozen

Wild Turkey (*Meleagris gallopavo*). Turkeys use all parts of the flats—pine flatwoods, hardwood hammocks, thickets, ponds, and streams. In earlier times, they roamed freely all over Florida, but today only the more cautious ones have remained and produce equally cautious young.

(averaging seven) at each birth; they are weaned and independent within a week; and the mother can immediately become pregnant again. Such rapid reproduction rates enable rats and mice to serve as the community's main converters of plant to animal protein. All of the local carnivores can stuff themselves with rats and thrive. Clearly, although a flat may not look like much, it is a highly productive ecosystem.[10]

Mammals that visit Florida's flatwoods and prairies from nearby include the raccoon, opossum, cottontail rabbit, gray fox, bobcat, white-tailed deer, Florida black bear, and even Florida panther. The fox squirrel is occasionally seen.

Prior to 11,000 years ago, a species of bison (*Bison antiquus*) also roamed central Florida's flats and played an important role as the ecosystem's major grazing animal. It maintained shortgrass prairie plants:

Florida native: Hispid cotton rat (*Sigmodon hispidus*). For its rapid reproduction rate, this little animal might well be called "the mother of all mothers."

Florida native: Cotton mouse (*Peromyscus gossypinus*). This is another prolific and important prey animal for predators of flatwoods and prairies.

Florida native: American bison (*Bison bison*), restored to Paynes Prairie, Alachua County.

cutting grasses and forbs stimulates their growth. Bison also served as a dispersal agent for plants because their thick fur readily picked up barbed and sticky seeds. Their hooves conditioned the sod to ease rain infiltration, and their wallows caught water in rainy times and served as small reservoirs in dry times. Their extinction in Florida, together with the demise of many other large animals, may have been caused by overpredation by early human hunters (see Chapter 22).

After the original bison died out, another species, the American bison (*Bison bison*) migrated in from the Great Plains. These bison penetrated as far south as Hillsborough County, but they, too, were overhunted, this time by European and African settlers. The last of these bison was shot in 1875. As a conservation measure, small herds of plains bison have been reintroduced to several locales in Florida.

VALUES AND MAINTENANCE OF FLATS COMMUNITIES

Not many of Florida's original flatlands ecosystems remain in natural condition. A few notable flats communities that remain are listed in List 4–5.

Because natural flatwoods and prairies are so much reduced in area, those that remain are especially important ecologically. The ecosystems that naturally occur on the lowlands have been present and evolving for millions of years. They are part of a landscape mosaic, linked to other communities at both higher and lower elevations adjacent to them, and the whole array hosts enormous diversity of both plants and animals.

Fire links these neighboring communities together. Fires at intervals of from one to five years are indispensable in maintaining flatwoods and prairie plant communities, and once ignited in one community, fire can spread into the others. Animals, too, link Florida's flatwoods and prairies to neighboring communities. The fox squirrel breeds and nests in hardwood forests but forages in pine grasslands. Many frogs and salamanders require ponds in which to reproduce and nearby higher, dryer land for their adult lives. The Florida sandhill crane nests in marshes and forages in prairies. The wild turkey uses all parts of flatlands mosaics. For these animals and many others, the integrity of the landscape is important. The few natural flatlands communities still found in Florida today deserve preservation.

Florida native: Tiger swallowtail (*Pterouras glaucus*, yellow form).

CHAPTER FIVE

INTERIOR SCRUB

From a rural road somewhere in the interior of Florida, a traveler turns onto a nameless, sandy byway to explore what looks like a natural area. The road ascends gently to a hilltop where one might expect to find a pine grassland, but instead a beachlike scene appears. Even though the ocean has been nowhere near this place for thousands of years, there seems to be the tang of a sea breeze on the air and the sound of seagulls calling. Just as on a coastal dune, the sand is powdery and largely bare, and indeed, dune vegetation is growing on it: weathered small oak trees, saw palmettos, rosemary shrubs, and sparsely scattered patches of dry, pale greenish-gray lichens. Sea oats are the only missing element.

How did this pile of powdery sand get here? Where did this array of plants come from? This natural community must have originated at the coast. Figure 5–1 shows how a coastal feature may become an inland sandhill.

The sand tells more of the story. If it were coarse and unsorted, then a likely guess would be that this was a beach ridge, sand bar, or spit piled up by water. But because this sand is sorted, fine grains only, it must have been carried here by wind. The plants confirm this guess. These plant species took hold on coastal dunes, and today they remain in the interior of the Panhandle on patches of dune sand that have persisted, undisturbed, for 10,000, 100,000, or even a million years or more.

The name *scrub* is given to this ecosystem to describe the scrubby plants that characterize it. Despite this humble name and their often unassuming appearance, scrubs are rare and special ecosystems. Their remarkable plants and animals and webs of relationships are only now becoming known, even as the land is overtaken by pine plantations, citrus groves, and development.

Florida native: Rosemary crab spider (a *Misumenops* species).

Scrubs are xeric communities growing on well-drained, infertile sand formations of marine origin in both coastal and interior Florida. Evergreen or nearly evergreen scrub oaks or rosemary shrubs or both predominate in some scrubs, while in others, sand or slash pines are present, sparsely or abundantly.

Coastal scrubs grow along today's coastal strand and are influenced by salt spray.

Interior scrubs grew at the coast in the past, then were left in the interior as a falling sea level exposed the land around them. They differ from coastal scrubs in that salt spray no longer reaches them.

Small, scrubby oak trees are casually called **scrub oaks**. There is also a species of oak, known as **scrub oak**, which is endemic to Florida and grows in the Florida scrub.

OPPOSITE: Scrub scene. The floor is of powder-white, mostly unvegetated sand. The shrubs are false rosemary (foreground) and saw palmetto (background); the trees are slash pines. Alligator Point, Franklin County.

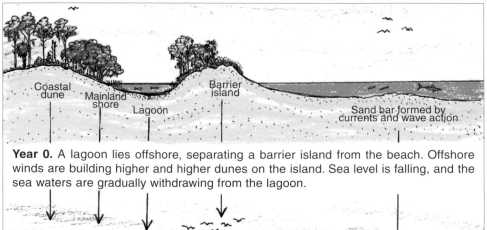

Year 0. A lagoon lies offshore, separating a barrier island from the beach. Offshore winds are building higher and higher dunes on the island. Sea level is falling, and the sea waters are gradually withdrawing from the lagoon.

1,000 years later. Sea level has fallen. The former beach and barrier island are now two long, parallel sandhills separated by a shallow valley. The old barrier island's outer shore is now the main beach, and a new barrier island is forming offshore.

FIGURE 5–1

How a Barrier Island Becomes an Interior Sandhill

FIGURE 5–2

Scrub Environments in Florida Today

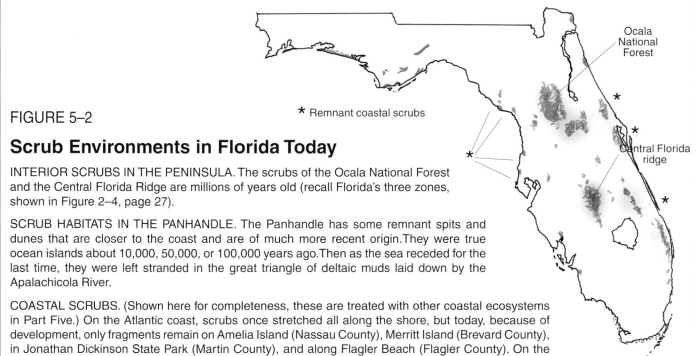

* Remnant coastal scrubs

INTERIOR SCRUBS IN THE PENINSULA. The scrubs of the Ocala National Forest and the Central Florida Ridge are millions of years old (recall Florida's three zones, shown in Figure 2–4, page 27).

SCRUB HABITATS IN THE PANHANDLE. The Panhandle has some remnant spits and dunes that are closer to the coast and are of much more recent origin. They were true ocean islands about 10,000, 50,000, or 100,000 years ago. Then as the sea receded for the last time, they were left stranded in the great triangle of deltaic muds laid down by the Apalachicola River.

COASTAL SCRUBS. (Shown here for completeness, these are treated with other coastal ecosystems in Part Five.) On the Atlantic coast, scrubs once stretched all along the shore, but today, because of development, only fragments remain on Amelia Island (Nassau County), Merritt Island (Brevard County), in Jonathan Dickinson State Park (Martin County), and along Flagler Beach (Flagler County). On the west coast, bits of scrub remain near Cedar Key (Levy County). Along the Panhandle, some relatively undisturbed scrubs persist both near the coast and on spits and barrier islands in Franklin, Gulf, Santa Rosa, and Escambia Counties. These include Alligator Point, St. Vincent Island, Santa Rosa Beach, and the Gulf Islands National Seashore. (The latter extends into Alabama.)

Sources: Tebo 1988; Puri and Vernon 1964, fig 5; Means 1977.

DISTRIBUTION OF FLORIDA SCRUBS

Scrubs remain today only in the small bits of Florida shown in Figure 5–2. The oldest of these bits are ancient sand ridges and dunes that run north-south along the central Florida ridge. Some 25 million years ago or more, there was no Florida; there was just a chain of islands in the sea off a shoreline that lay up in Georgia. Then the ocean retreated, leaving the islands connected, and Florida was born.

Today the scrub patches are still islands, but they stand in the Florida peninsula, surrounded by land, not water. The two largest blocks of ancient scrub that remain in relatively natural condition are in the Ocala National Forest in central Florida, and on the southern end of the central Florida ridge in Polk and Highlands counties.

In the Panhandle, too, there remain a few sand piles that were islands and dunes in the past, but they are not on the highlands like the central Florida scrubs. They formed much more recently, on the order of tens of thousands of years ago, when the waters of the Gulf of Mexico covered today's coastal lowlands. They now stand near the Apalachicola River as slender sand ridges parallel to the coast. Most of these formations have lost their original scrub vegetation, but traces of their island origins remain.

TYPES OF SCRUBS

Since they were first separated from each other, Florida's scrub patches have evolved independently of each other. As a result, today, different scrubs have somewhat different plant associations and appearances.

Several types of scrubs are shown in Figure 5–3. Rosemary scrub (Figure 5–3, A) is open to the sky, often a gardenlike scene that features beautifully spaced, round rosemary bushes with only a few or no sand pines and scrub oak trees. Oak scrub (Figure 5–3, B) has more oaks and few or no pines. On its way to maturity, an oak scrub may become an impenetrable thicket of Chapman's oak, myrtle oak, and sand live oak, together with a few other plants such as saw palmetto. When the canopy closes

D. Sand pine scrub with partially closed canopy. This is the typical aspect of many a sand pine forest. Wind has tilted the pines, but their interconnected roots have kept them standing.

A. Rosemary scrub. Shade is patchy. The white, sterile sand drains fast, and only drought-tolerant plants and burrowing animals can survive.

FIGURE 5–3

Florida Scrubs

B. Oak scrub. Stressed by drought, the oaks assume twisted shapes. Since the canopy has closed, most of the ground-cover plants and shrubs have been shaded out, leaving an inviting, open interior.

C. Sand pine scrub with closed canopy. The pines have formed a solid forest with deep shade, in which few other plants can grow. This is a mature sand pine forest. The biggest trees are about 65 feet tall and 15 to 20 inches in diameter at breast height.

completely, many of the ground-cover plants are shaded out, the interior opens up, and the scrub becomes an artful and inviting arrangement of small, twisted oaks over ground covered with gray-green deer lichen (often incorrectly called deer moss).

Sand pine scrubs vary. In some sand pine scrubs, the pines have bushy branches all the way to the ground. In others, the pines are tall and robust, forming a closed canopy with a dark, open interior beneath it. A sand pine forest of this type grows in the Ocala National Forest and is shown in Figure 5–3, C. Others have a more open canopy (Figure 5–3, D), and in still other sand pine scrubs, the pines are stunted and gnarled. Sand pines seldom reach more than 50 to 70 years of age, but if not killed by fire, they may occasionally reach 100 years. The biggest trees are generally only 65 feet tall and 15 to 20 inches in diameter at breast height. The champion tree, which is at Wekiwa Springs, is 85 feet tall, with a diameter of two feet and a crown spread of 40 feet.

Peninsular Florida is thought to have held predominantly rosemary scrubs at one time. Then, over thousands of years, oaks came to dominate these scrubs. Finally, particularly in the central peninsula, some became sand pine scrubs, but along the Gulf coast, rosemary and oak scrubs still persist on the driest dune ridges.[1]

FIRE IN SAND PINE SCRUB

Like high pine grasslands, sand pine scrubs are fire-maintained systems, but the nature of the fires and the responses of the plants contrast dramatically with those described in the last two chapters. Imagine this scene for a moment. A high, dry, open forest of sand pine simmers in the blistering sun. Among the pines are widely spaced rosemary bushes, each a sphere of dark green up to six feet across (see photo at start of chapter). Dry, fluffy lichens are scattered across the sand.

Storm clouds gather and a hot wind tosses the dry pine branches. Suddenly, lightning strikes, and within seconds a firestorm is racing through the shrubs and treetops, throwing hot sparks. Within half a day all the trees and the above-ground parts of the shrubs are dead.

Until today, no fire had touched this scrub for thirty years or more. Now everything seems to have been destroyed in a single afternoon. Scorched pine skeletons, dead shrubs, charred pine cones, and black ash piles are all that remain on the ground. Except for a beetle crawling on the hot sand, nothing remains alive—or so it seems.

An hour, a day, or a season later comes the rain. It drains quickly into the sand, leaving it as dry as before, but hints of new life begin to appear. Around each dead rosemary shrub, new seedlings pop up. Over each old lichen bed is a haze of green. Tiny pine seedlings sprout from living seeds that were released from the blackened cones. Thirty years from now, this place will look almost exactly as it did before. Then if lightning ignites another fire, hot winds will fan it through the treetops, and the cycle will begin anew.

WBSP

Crown fire in sand pine in the Waccasassa Bay State Preserve in Levy County. Fire readily climbs sand pines and torches their tops. Typically, scrub fires burn hot and fast, devouring everything in their path. Although the conflagration looks devastating, renewal of the scrub absolutely depends on such fires.

Scrubs depend on fire differently than longleaf and slash pine grasslands. In a pine grassland, thanks to its continuous ground cover, fires occur every few years, burn at a relatively cool temperature, and do not climb the trees. In a scrub, fire occurs only once every ten to 50 or more years, burns hot, and does climb the trees, which have a highly flammable varnishlike wax that coats their leaves and needles. Burning litter on the ground creates heat, the fire jumps to the tops of the trees, which explode into flame, and sparks fly, torching one tree after another. and all the trees die, all at once. The shrubs burn explosively, and in a closed-canopy sand pine forest, the fire races unbelievably fast through the treetops. One fire in the Ocala National Forest burned 35,000 acres of sand pine scrub in four hours.[2]

The crown fires that occur in sand pine scrubs, although they appear catastrophic, are exactly what is needed for renewal, especially in central Florida. Central Florida's sand pines do not release their seeds unless heated by fire. By the time they have reached their mature height, they are loaded with seed-filled cones (see Figure 5—4). Then when fire sweeps through, the cones open and scatter seeds all over the ground.

If there are oaks among sand pines, they, too, will appear to be killed by fire, but they actually remain alive. Scrub oak stems burn to the ground, but much of their mass is underground in the roots, alive, and packed with stored energy and nutrients. The oaks quickly resprout and regrow to chest height within three years.

Rosemaries die in fires, but their seeds remain viable in the soil and give rise to a new generation of healthy rosemaries right away. Saw palmettos recover within a season (see Figure 5—5).

It must be clear by now why, in a sand pine scrub, the pines so often are all of the same age. They all started up simultaneously from seeds thrown during a fire. The oaks and rosemaries preceded them by a short interval, providing the ideal setting for sand pine seedlings to start growing. Partial shade helps to protect them from the sun's killer heat at first. By the time the young pines have overtaken the shrubs, they are able to grow well in

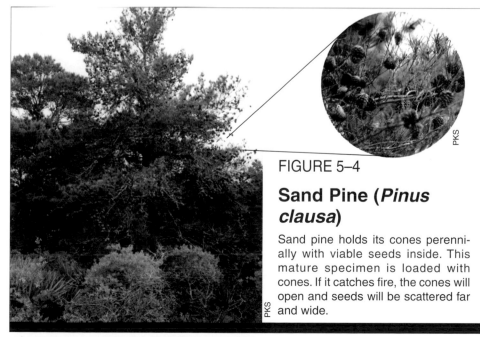

FIGURE 5—4

Sand Pine (*Pinus clausa*)

Sand pine holds its cones perennially with viable seeds inside. This mature specimen is loaded with cones. If it catches fire, the cones will open and seeds will be scattered far and wide.

After a fire, some saw palmettos will look dead,

whereas in some, there will be subtle signs of life . . .

and some will be obviously alive . . .

. . . but all will recover completely within a season.

FIGURE 5–5

Florida Native: Saw Palmetto (*Serenoa repens*)

intense sunlight and can tolerate widely fluctuating temperatures, exactly the conditions that they face. The replacement forest then becomes a stand of even-aged trees.

Scrub animals, too, all manifest adaptations to fire. Many animals simply dig escape tunnels down into the sand, where the fire's heat does not penetrate. Other ground animals retreat into their own or gopher tortoise burrows. Birds fly up, and if their nests are lost, they soon renest.

Fire scenarios play out in other ways than those described here. If a scrub does not burn within about 50 to 55 years after the last fire, the sand pines typically begin to decline due to heart rot and then gradually give way over several hundred years to live oaks. In contrast, if a scrub burns too frequently, its sand pines may not have a chance to mature, and longleaf pine may invade.

Scrubs in Panhandle Florida may regenerate other than by fire. North Florida sand pines commonly have open cones and can release their seeds without the stimulus of heat. If space opens up, colonization by seeds can proceed, although it is slow. Hurricane winds at times clear land for the regrowth of sand pines. Still, fire is probably a major agent of renewal under natural conditions.[3]

* * *

Two Florida natives: scrub lichens with no common names. Above: *Cladina evansii.* Below: *Cladonia perforata,* an endemic.

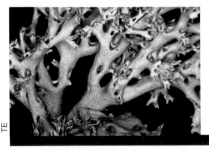

Of interior scrubs that remain today, those on the central Florida ridge are fairly well described and are under ongoing study. Those in the Panhandle have hardly been studied at all. Most of the information presented here is drawn from central Florida scrub communities in which sand pines, rosemary bushes, palmettos, and lichens are the most noticeable plants.

PLANTS IN CENTRAL FLORIDA SCRUBS

On the central Florida ridge, numerous islands of scrub lie in a matrix of sandhill communities and (today) orange groves and residential developments. Each scrub island is at most only a few miles wide, and each is a world unto itself.

On these ancient islands, hundreds of plant and animal species have become adapted in unique ways to the dry, porous, nutrient-starved soil they live on. About half of all scrub species on the central ridge are endemic, and they are among the rarest species known in Florida. In fact, the central Florida ridge today has more rare and endangered species than any other habitat in North America.[4]

The plants in these scrubs are well adapted, not only to periodic hot fires, but to the challenge of their extremely dry environment. They also are stalwart defenders of their territory, even against their own offspring. The many ways in which they cope with their stern environment are lessons in evolutionary adaptation.

Water-Conserving Adaptations. The sand on which scrub plants maintain their foothold drains extremely fast. Water droplets sink down so rapidly that the upper layers are always dry, even in rainy seasons. Most of the woody shrubs have roots of two kinds. One or a few tap roots reach deep down in the sand; while shallow root mats spread around the plants over areas that are broader than the plants' tops. The root mats grab all available rain practically as it falls. This root competition for water, and possibly growth-inhibiting chemicals released by the plants, limit the number of other plants that can grow and account for the expanses of bare sand that often surround those that are present. Few herbs can compete successfully with these water-grabbers; herbs collect less water and lose it more readily to evaporation, so they tend to die in droughts.

The shrubs and the few herbs that can hold their own among them are mostly evergreens with small leaves or needles whose waxy coats protect them from losing water by evaporation. They stay small and low to the ground. An observer notes that "In [oak] scrub, people [are] giants, their heads poking out above the treetops. Many plants grow only a few inches high, producing bouquets of flowers fit for a doll's house."[5] List 5—1 names some of the plants that are often found in scrubs.

Sand pines exhibit an additional, and remarkable, water-conserving adaptation. In stands where they grow close together, the roots of adjacent trees interweave and even graft to one another, with the result that the trees become, in a sense, a single organism. As such, they can share water, minerals, and the sugars generated by photosynthesis. They can also withstand storm winds, which are especially powerful along the coast where the sand pines first grew. Notice how often a sand pine scrub seems to consist of trees that are all leaning one way. You would think that many trees would fall in high winds, but they remain standing, thanks to their interconnected roots.[6]

Chemical Defenses. Many scrub plants produce chemicals that act as powerful defenses against competition and insect pests. Several shrubs produce chemicals that, when leached from their foliage into the soil by rain, inhibit the reproduction and growth of pine seedlings and grasses. This may account for the clean boundary that separates pine grasslands from scrub. The chemical released by the Florida rosemary may also keep

LIST 5—1
Plants found in scrubs throughout Florida (examples)

Lichens
Deer lichen
(Others)

Herbs
Beaksedge
Feay's prairieclover
Milkpea
October flower
Paper nailwort
Sand spike-moss
Tread softly

Shrubs and Trees
Adam's needle
Apalachicola false rosemary
Bush goldenrod
Carolina holly
Chapman's oak
Coastalplain staggerbush
Deerberry
False rosemary
Florida rosemary
Greenbrier
Hog plum
Myrtle oak
Red bay
Sand live oak
Sand pine
Saw palmetto
Scarlet calamint
Scrub hickory
Scrub palmetto
Scrub wild olive
Shiny blueberry
Silk bay
Sparkleberry
Yaupon

Sources: Adapted from Muller 1989; Myers 1990; Florida Biodiversity Task Force 1993, 2–3; Clewell 1981/1996, 184; *Guide* 1990, 6.

Florida native: Florida rosemary (*Ceratiola ericoides).* This is the species of rosemary that grows in most Florida scrubs.

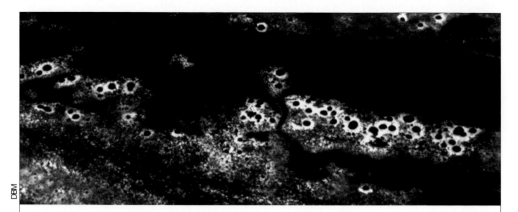

FIGURE 5–6

Haloes around Rosemary

This aerial photo of dunes along the Panhandle shows a circle of white sand around each rosemary bush. It is thought that each rosemary releases a growth inhibitor that prevents the growth of other rosemary plants nearby.

its own seeds from growing until a fire kills off the parent plant, thus minimizing root competition (see Figure 5–6). Two genera of rosemaries also produce insect repellents that help them to withstand plant eaters.[7]

Another insect-repellent plant is a sweet-smelling wild herb, the scrub balm (Figure 5–7). It was first noticed because its leaves showed little evidence of insect damage. An oil is contained in capsules on the leaves and when released, it serves as a powerful ant repellent. Laboratory studies have revealed that the plant produces twelve volatile aromatic compounds, all of potential usefulness to humans. One is noxious to cockroaches. No other plant is known to make these compounds, just this one species of mint, which grows on only 300 acres, the last remnant of its habitat on earth, which is now in a preserve. Intervening land uses have isolated this population from other habitats where it might be able to grow.[8]

Among its other interesting features, the scrub balm has twenty different species of fungi in its tissues, some of which produce antibiotic compounds that are toxic to other fungi. Fungi do not qualify for the Endangered Species list, but if they did, all twenty of these would be listed with their host plant. As the discoverer of this plant says, "In these days of biotechnology, we can access more than just the chemicals of nature. We can access the genes . . . for the . . . chemicals [and put them to use]."[9]

Fungi are everywhere in the soils of the scrub. Some of them decompose dead matter into its component water and carbon dioxide, gathering energy in the process. They capture the minerals liberated in the process and feed these to the scrub plants' roots through tiny filaments. Other fungi, which live inside the plants' roots, capture atmospheric nitrogen, and incorporate it into compounds that the trees can use. Thanks to their root fungi, the trees grow well even though the soil is dry and sterile.

Because of its plants' special adaptations, scrub is considered one of the most botanically important ecosystems in the United States. All of the ground plants are special. More than thirty are listed or under review for listing by state and federal endangered species agencies.

Scrub balm close-up

FIGURE 5–7

Florida Native, Endemic: Scrub Balm (*Dicerandra frutescens*)

ANIMALS OF CENTRAL FLORIDA SCRUBS

Scrub animals are as challenged by the scrub's harsh environment as are the plants, and like the plants, they are superbly well adapted to cope with heat, fire, lack of water, and their enemies. Many an investigator who began with a casual interest in scrub animals has ended up devoting his or her professional life to studying them. List 5–2 names a few animals that reside in the scrub. Others visit, and are enumerated later.

Insects, Spiders, and Their Associates. Subterranean life makes sense where there is little structure above the ground, and most small scrub animals live in the sand, where they can find the temperature they require by moving up, down, or sideways. There are 59 species of ants in Florida scrubs, 35 species of velvet ant (actually a kind of wasp), 70 species of bees, 80 species of scarab beetles, 45 species of bark and ambrosia beetles—the list goes on and on.[10]

Each subterranean invertebrate has distinct food preferences. Caterpillar and beetle larvae graze on tender, fine rootlets in the top sand layers. Termites maneuver just beneath the sand to find partly buried leaves and dead twigs. Tiny, pale yellow predatory ants hunt freely among the sand grains and scavenge animal carcasses. Minute larvae of bee flies and robber flies search for grubs and grasshopper eggs. Many burrow-making animals hunt and carry their prey back to burrows in the sand, where they can guard their kills. Hunting wasps hunt by day, wolf spiders by night. The profusion of actively hunting insects is unusual, but abundant prey is available for them.

The Florida harvester ant gathers and eats the seeds of many understory plants. It has developed a kind of basket on its body, made from a ring of long, curving hairs that extend from the jaw parts and the underside of the head. Using this basket, the ant hauls clumps of sand from three feet below the ground to make its nest (and in the process, returns nutrients to the surface, a vital service to the scrub that is described later more fully). A species of scarab beetle is so specialized that, like moles, it never comes to the surface. It has strong, heavy claws like those of a mole, it is blind, and unlike other beetles, it has no wings.[11]

The pygmy mole cricket is a tiny insect, less than a quarter-inch long, that lives in a system of branching passageways just below the surface of the sand. It chooses sites only in full sun, never beneath rosemary bushes or other shrubs. Researchers were mystified at first, wondering how the creature could find enough to eat there. Most other species of mole crickets live on the edge of wet places and graze on algae that grow in the soil or on wet leaves. If one habitat dries out, they fly or swim to another. But this cricket's habitat is dry nearly all the time, it is flightless, and it has no place to swim. Its habitat is open, unshaded, apparently sterile, bare sand.

The mystery was solved when the researchers discovered that a thin layer of microscopic algae grows in scrub sands just beneath the surface. Each time it rains, the algae catch water and grow. Between times, the few translucent sand grains above them protect them from scorching, hold in enough moisture, and admit enough light for them, but offer no hospitality to plants with greater moisture requirements. The cricket makes its living by grazing on these algae wherever they grow.[12]

An oak-leaf-crunching invertebrate, the Florida scrub millipede, may be necessary to the mole cricket's survival. Found only in Highlands County and not in nearby Polk County scrubs, it lives above the sand surface in oak leaf litter, which it chops into tiny fragments and eats. It has no close relatives anywhere else in the world, but in this community it is an important link in the chain of life. By keeping the sand clear of leaves, it permits sunlight to penetrate the top sand layers where the algae grow that become food for the pygmy mole cricket and other tiny creatures.[13]

Millipedes participate importantly in soil food webs. After pulverizing their chosen dead salad greens, they mix them with bacteria in their guts, then deposit fecal pellets. Other bacteria attack the pellets, fungi invade them, and small arthropods eat them. Bacteria in the arthropods' guts attack that material, then the arthropods deposit smaller fecal pellets. Still other organisms eat these, mix them with soil, and start another round of the same events.

Many scrub insects are hidden, but some are in plain sight and still virtually invisible. The rosemary has a specialized moth, whose caterpillar can take two entirely different forms, achieving perfect camouflage in every season (see Figure 5–8). Other moths, beetles, sap-sucking insects, and a grasshopper also appear to be totally dependent on the rosemary.

Study of endemic scrub insects and spiders has enabled researchers to piece together the probable history of their island habitats. Some are found only in the Ocala National Forest, some in scrubs on the north end of the Lake Wales ridge, some on the south end, and some species are restricted to single, tiny patches and are not found even in other nearby scrubs.

The wolf spiders provide a good example. Most young spiders depart from the parental web by "ballooning"—that is, they take to the air hanging from tiny parachutes made of spider silk, and they achieve wide dis-

Florida native, endemic: Florida scrub millipede (*Floridobolus penneri*). The beads of moisture, which contain iodine, deter predators.

Florida native, endemic: Lake Placid funnel wolf spider (*Sosippus placidus*). The spider's burrow entrance is a funnel-shaped web that traps small prey animals crawling over it. When an insect or other small creature falls in, the mother subdues it, and then she and the young feast on it together. Young wolf spiders of this species live with their mother until they are about half grown. One young spider is shown in the foreground by its mother's leg.

Source: Anon. 1990 (Wild Florida, Vol. 1).

Florida native, endemic: Rosemary grasshopper (*Schistocerca ceratiola*). This grasshopper depends solely on the rosemary for its food.

Twig form

Needle form

FIGURE 5–8

Emerald Moth Caterpillar on Florida Rosemary

The emerald moth (*Nemouria outina*) depends completely on the rosemary for its food. To hide from its predators it assumes different forms. In winter it is gray and looks like a dead rosemary twig (LEFT). In summer it is green and looks like a live rosemary needle (RIGHT). It stays completely motionless by day and eats at night when its predators (birds, mice, and others) can't see it moving.

persal that way. Probably, when their homes were on islands, though, young spiders that ballooned fell into the sea and drowned. Those that stayed near their birthplaces survived, and passed on their stay-at-home behavior to their offspring. The scrub islands where these spiders are found are now connected by land, but the animals have remained confined, with their offspring, to separate invididual patches of scrub ever since long ago.

A series of beetles and a series of grasshoppers exhibit the same pattern. They are closely related but distinct from one scrub to the next and they are designed not to disperse: the grasshoppers and the beetles are flightless, and the beetles are burrowers. Still other small invertebrates are no doubt endemic to scrubs and remain to be discovered.

The scrub balm was mentioned earlier as an example of a plant with interesting chemistry. It also has a special relationship with a particular bee fly that does 95 percent of its pollinating. The plant attracts the bee fly with a "nectar guide," a dark pink center in its flower. It has a lower petal that serves as a perch, and when the bee fly lands on it, its weight bends the flower so that the nectar cup closes. The bee fly has to thrust its head in, hard, to get nectar. That triggers an explosion of pollen which adheres to the bee fly's belly. There are many flowers, both male and female, on one scrub balm plant, but self-pollination does not occur, because the male flowers are open only in the morning and the female parts, the pistils, are accessible only in the afternoon. Any insect that has picked up pollen from one plant in the morning will be far away, depositing it on another plant,

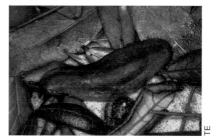

Florida native with no common name: a snail slug (*Veronicella floridana*). By mimicking an oak leaf, this animal escapes notice often enough to maintain populations in the scrub.

in the afternoon. In short, as the discoverer of this relationship sums it up, the scrub balm "accomplishes cross-pollination by teaching a common generalist bee fly to act as a specialist, in return for exclusive nectaring rights."[14] Other pollinators have equally mutually rewarding relationships with scrub plants.

Burrowing Reptiles. The vertebrate residents of the scrub are nearly as secretive as the insects, spiders, and others just described. Most are reptiles, which cope with desertlike conditions that would quickly kill most other animals. Many can "swim" through sand. An assortment of these curious animals is shown in Figure 5–9.

A visitor to the scrub might never see these animals, but would see mounds of yellowish sand, from a few inches to three feet in diameter, scattered over the smooth, powdery ground. These mounds are the piles of soil pushed out of burrows by animals who are hiding from the hot sun by day.

The sand skink is camouflaged light brown and has short, vestigial legs that fold back against its cylindrical body, giving it streamlined, submarinelike contours. It has smooth, slick scales and a wedge-shaped snout that is ideal for pushing its way through the sand. It even has lower eyelids that are transparent, like goggles, so it can keep its eyes covered and still see the termites, ants, and insect larvae in the half-light in which it hunts. The slim, long, elusive, and rare short-tailed snake also swims through the sand in search of its favorite prey, tiny peninsula crowned snakes, which it kills by constriction. The scrub lizard does not swim in sand but stays in the dense shade underneath rosemary bushes, other shrubs, and sand pine. If threatened, it can run extraordinarily fast in the powder-soft sand.

Endemic: Short-tailed snake (*Stilosoma extenuatum*). This constrictor is the rarest snake species in eastern North America. It spends most of its time underground.

Endemic: Florida scrub lizard (*Sceloporus woodi*). This animal occurs in scrubs on the central ridge and also along both coasts in south Florida.

Eastern slender glass lizard (*Ophisaurus attenuatus*). This legless lizard occurs in many eastern U.S. xeric habitats.

Endemic: Bluetail mole skink (*Eumeces egregius lividus*). This rare animal is found on one site only, on the Lake Wales Ridge.

Endemic: Sand skink (*Neoseps reynoldsi*). This animal, endemic to the central Florida scrub, is the only representative of its genus that remains alive in the world today.

Endemic: Florida worm lizard (*Rhineura floridana*). Not a worm but a reptile related to lizards and snakes, this burrowing animal is extinct everywhere in the world except dry habitats in central Florida.

FIGURE 5–9

Native Reptiles of the Florida Scrub

PRICELESS FLORIDA

One other reptile lives in scrub that should be mentioned. The gopher tortoise can easily dig its burrows there, although it finds suitable forage only in high pine habitat. Gopher tortoise burrows serve as "safe homes" in scrub just as they do in pine grasslands.

The Florida Scrub-Jay. The scrub-jay originated in the desert Southwest and expanded into Florida when sea levels were lower and coastal strands lined the shore all the way from west of the Mississippi River to the Florida peninsula. Then the ocean rose and covered the beaches, and the Florida jays evolved to become distinct from their western relatives. The Florida scrub-jay is now found almost exclusively in Florida and is considered to be a Florida endemic—the state's only endemic bird.

The scrub-jay finds its most suitable habitat in scrubs along the central ridge where scrub oaks and rosemaries abound under an open canopy of a few sand pines. Unlike its relative, the blue jay, the scrub-jay prefers and is adapted for life on the ground under the cover of the oaks where it scratches and digs with its strong feet for acorns, berries, seeds, insects, spiders, lizards, and small snakes.

The scrub-jay has developed some special adaptive behaviors as its habitat islands have dwindled in size during the last sea-level rise. A scrub-jay pair mates for life and maintains a small family group whose members cooperate in defending the pair's territory and raising the young. Each breeding pair requires about a 20- to 40-acre range in which to forage. Suitable habitat is scarce, and not all jays can pair up and claim territories. Rather, some maturing members of each brood postpone breeding for up to six years, until a place comes available for nesting. Until then, these relatives help a pair to feed each brood, keep the nest clean, and take turns serving as sentinels in the treetops to warn of approaching predators. A sharp whistle signals "Dive: hawk approaching," and a low cackle combined with a directed gaze means "Look out for that predator on the ground." The jays' apparent altruism toward their close kin has led some biologists to study their genetics in search of "generosity genes" that might run in social species.

Scrub-jays provide for their own future food supply. They are omnivorous, but in the fall, when insect food is scarce, each jay eats several thousand acorns and caches several thousand more in the sand.

The scrub-jay thrives in open scrub habitat that has burned within the preceding 10 to 20 years. It abandons patches that become overgrown with pine trees. When you see a resident scrub-jay family that is flourishing, then you will probably see other evidence that you are in prime scrub habitat, just as when you see an active red-cockaded woodpecker clan, you will often recognize your surroundings as a largely natural longleaf pine-wiregrass community.

Scrub Mammals. Like most of the scrub's other animals, the mammals are mostly burrowers. One, the Florida mouse, uses the gopher tortoise burrow as a main chamber and creates its own dwelling in tunnels that branch from it. Over years of evolutionary history, the Florida mouse has become a light tan color that closely matches the sand.

Florida native: Florida scrub-jay (*Aphelocoma coerulescens*). This is the state's only endemic bird species.

Florida native: Pocket gopher (*Geomys pinetus*). Several mounds spaced in a row are most likely to be the work of a pocket gopher, who always has sand all over his face.

Evidence of a pocket gopher's work. These mounds are all connected by a long, branching tunnel.

The Florida mouse is a member of the genus *Podomys*, the only genus of mammal endemic to Florida. It apparently evolved in Florida; no fossil records are known from outside the state. It requires a high, dry, well-drained habitat maintained by fire, and because it subsists largely on acorns, it flourishes best on sites such as sand pine scrub, which have abundant scrub oaks. The mouse easily scurries up scrub oaks to gather acorns.

The mouse has many predators and no special defenses against water loss. It uses the gopher tortoise burrow, when available, both for shelter and for protection from the sun. Because it digs its nest burrows off the main chamber, it has been called the gopher mouse. If pursued, it freezes and vibrates its tail. If a predator grabs the tail, the brittle sheath slips off, leaving the predator mouseless and the mouse free.[15]

Another scrub mammal is the pocket gopher, a fist-sized rodent with strong digging paws and pockets in its cheeks for storing food. It digs a long, horizontal tunnel system about 30 inches beneath the surface and periodically pushes its diggings out onto the ground surface. Several mounds in a row, equally spaced, are likely to be the work of a pocket gopher.

While the scrub serves as habitat for many permanent residents, it also provides a stopping and feeding place for migratory birds and many land animals. List 5—3 itemizes some of the many animals that use the scrub in passing.

PANHANDLE SCRUBS

Eight old coastal formations are still apparent around the Apalachicola lowlands. Their soils are of finely graded and sorted sand, reflecting wind-created dunes. Some now support the same longleaf pine—wiregrass community that grows on the sandy hills just north of them, but some still bear their original cover: sand pine scrub, slash pine scrub, and xeric hardwood forests, primarily oak.

The plant communities of these ancient beach formations have never been systematically inventoried, but may rival those of central Florida in rarity and uniqueness. They may contain the only living remnants of coastal plants that grew there when the land lay next to the ocean. The endemic Apalachicola false rosemary now grows on just one strip of sand from Liberty to Santa Rosa County that once was a barrier island. Due to clear-cut logging and pine planting over most of the Panhandle, the rosemary's only remaining habitat is in a few patches that have escaped extinction: the uppermost, dry edges of steephead ravines (described in Chapter 6).

The Panhandle sandhills support other rare plants, including some of the healthiest populations of Chapman's rhododendron remaining in the world. On some sandhills, a maritime forest grows, with live oak and sand live oak mixed with southern magnolia and saw palmetto. Animals are seldom seen, but white-tailed deer come to eat the acorns, and occasional birds, such as the eastern towhee and the great crested flycatcher, visit to forage for nuts, seeds, fruits, and insects. Harvester ants are numerous;

they gather and eat the seeds of many understory plants. Many reptiles and amphibians are present, but seldom seen.

A race of kingsnakes that is almost white exists in the Panhandle scrubs. Kingsnakes in neighboring areas are black with characteristic widely-spaced cream-colored bands, a pattern that makes them hard to see against the background in which they hide. But the members of this small population of "white" kingsnakes are protectively colored for hiding against a sandy background, rather than in dark woods (see Figure 5–10). The "white" kingsnakes presumably evolved in the distant past on a barrier island or spit that today is an inland sandhill. Additional evidence supporting the island origin of this sandhill is that at least 15 other endemic species occur in the same area.[16]

Most people in the region don't recognize the Panhandle's sandhills as ancient beach formations. Many have been bulldozed and plowed for logging and other uses. Other ancient barrier islands may remain in the Panhandle with endemic species that no one has yet discovered.

THE SCRUB ECOSYSTEM

Like Florida's natural pine grasslands, scrub is an ancient community, one in which every member both benefits from, and serves, the others. For example, while the burrowing lifestyle meets scrub animals' own survival needs, it also serves another function—that of recycling soil nutrients. Scrub occupies the tops of hills. Without active recovery, nutrients would drain down through the soil and be lost to the system. Swamps and marshes can catch nutrients that flow into them from uphill, but no living system lies uphill from the scrub. The scrub receives a few minerals from rain, but its main source of nutrients is retention via recycling.

Root-associated fungi have already been mentioned as important mineral recyclers, but what of the nutrients that escape the fungal mats and leach down out of reach in the sand? Large burrowing animals can't retrieve them; they lack the elbow room for digging where trees are crowded. But small burrowers are numerous in scrub. The many mounds of yellowish sand on the white powdery floor of the scrub represent burrowing animals and the yellow color signifies that they are mineral rich. A forester counted some 2,000 animal-burrow mounds in an acre, one day,

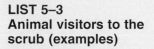

**LIST 5–3
Animal visitors to the scrub (examples)**

Amphibians
Barking treefrog
Gopher frog
Mole salamander
Oak toad
Slimy salamander
Southern toad
Spadefoot toad
Striped newt

Reptiles
Coachwhip
Eastern coral snake
Eastern diamondback
    rattlesnake
Eastern indigo snake
Six-lined racerunner
Southeastern five-lined skink
Southern black racer

Birds
Common ground-dove
Common nighthawk
Eastern towhee (winter)
Loggerhead shrike
Northern bobwhite
Northern cardinal
Palm warbler (winter)
Turkey vulture

Mammals
Bobcat
Florida black bear
Gray fox
Raccoon
Spotted skunk
Striped skunk
Virginia opossum
White-tailed deer

Sources: Muller 1989; Myers 1990; Clewell 1981/1996, 184; *Guide* 1990, 6.

LEFT: Eastern kingsnake (*Lampropeltis getula*, "black" form). This animal is black with creamy markings, a good camouflage in a shady environment.

RIGHT: Apalachicola kingsnake (*Lampropeltis getula*, "white" form). Kingsnakes in the Apalachicola River region are cream-colored, a good camouflage where the background is sandy. The color difference suggests that this group diverged genetically from the main population when isolated on an island in the ocean.

FIGURE 5–10

Eastern Kingsnake, Two Forms

Track of the sand skink. Tracks like these are a notable feature of the dry, sandy environment of the scrub, where animals "swim" just beneath the surface of the sand.

and reported that each mound, on average, weighed about two pounds. This much earth, brought to the surface in one year, amounts to about two metric tons per acre every year.[17]

Another animal activity important in scrub, as in all ecosystems, is what might be called cleanup. The sand-clearing activity of the scrub millipede was mentioned earlier; similar cleanup services are performed by other invertebrates. Plants drop litter on the ground; then beetles, grubs, mites, ants, and others add their wastes, mix them in, adjust the chemistry, loosen the texture, aerate the soil, and enable rainwater to sink into the ground.

Other animals perform other services. Insects disperse the seeds of, and cross-pollinate, many plants. Browsers such as white-tailed deer nibble twigs and branches, and the plants respond much as they do to a gardener's pruning: they branch and grow thicker, providing more shade and wildlife cover. Seldom do browsers overeat the food supply in any one area. In one study, one deer was found for every 101 acres of forest, even though the available food could have supported one in every 38 acres. This margin allows for other herbivores to coexist with the deer. Predation does not wipe out whole herds, but thins them and eliminates diseased and aging animals.

In short, the scrub system is truly a web, an ecosystem. There is overlap, too. Several species perform most vital functions.

VALUE OF THE SCRUB

Like all ecosystems, scrubs perform environmental services. Because their sandy soils let rain percolate quickly down to the water table, interior scrubs are vitally important in recharging the aquifer. Scrubs are also potential refuges for coastal species, because, if or when sea level rises, their cousins on the present coastal strand may be wiped out. Later, when the sea retreats again, interior scrubs can become sources from which plant and animal populations can expand to repopulate the coast.

Scrubs are also important to the understanding of evolution, because so many of their plants and animals are the last remaining remnants of ancient populations now all but extinct. The plants contain important information, because they are supremely successful in the scrub's harsh surroundings, where no other plants can perpetuate themselves for long. They have all evolved over millions of years to deal with wet summers, dry winters, and poor soil, much as farmers have to do in most of the world's agricultural areas. John W. Fitzpatrick, a scrub researcher, says the scrub is a "gold mine of genetic information. . . . These plants may be holding secrets to how we can end up genetically engineering our agricultural systems to improve them, especially in dealing with a warmer planet."[18]

Perhaps above all, though, the scrub is simply valuable in its own right. Its peculiar plants and animals are endlessly fascinating and promise to yield many more discoveries in future research.

This chapter concludes the series on north and central Florida's fire-dependent pine communities except for south Florida's pine rockland, which is treated in Chapter 7. Other fire-dependent ecosystems exist in Florida. Some wetlands, notably marshes, require fire to remain healthy as described in the chapters of Part Three. In contrast, hardwood forest systems are notable for their *exclusion* of fire, and it is to these that the next chapter turns.

A native moth (*Shinia gloriosa,* no common name) on an endemic plant, Feay's palafox (*Palafoxia feayi*). Active only at night, the moth "disappears" at dawn by settling on this plant. For years, because it was so well camouflaged, no one knew where this moth went during the day.

CHAPTER SIX

TEMPERATE HARDWOOD HAMMOCKS

Hardwood trees and shrubs have been growing on the southeastern U.S. Coastal Plain for more than 50 million years. During hot, wet times, the trees that flourished most abundantly were those of tropical rainforests. During times of cooler climates, temperate hardwoods tended to take over. Today, both types of trees comprise Florida's hardwood hammocks. Temperate hammocks occur all over north Florida, and tropical hammocks in extreme south Florida. Forests of mixed temperate and tropical composition are arrayed between these two extremes (see Figure 6-1, next page). This chapter deals with temperate hardwood hammocks and Chapter 7 discusses tropical hammocks.

In presettlement times, hardwoods naturally occupied only a few protected areas at the bottoms of steep slopes near rivers, lakes, and sinks, but they now grow both higher and lower on slopes. Wherever fire exclusion has prevailed, hardwoods have crept uphill to cover once pine-clad uplands, and wherever wet areas have been drained, hardwoods have spread downhill into former wetlands. Originally, hardwood forests and hammocks may have occupied only some 1 to 2 percent of Florida's land area, but today they hold ten times that much.

The driest hardwood forests are nearly as dry as high pine; the wettest are wetlands. The Florida Natural Areas Inventory (FNAI) classifies them as xeric, mesic, and hydric. Xeric hardwood forests are of oak, often live oak and laurel oak. Hydric forests often consist mostly of cabbage palm, swamp bay, and red cedar. Between these extremes are mesic forests containing many mixtures of trees (slope forests, upland forests, and mixed forests), and at the wet extreme are swamps known as floodplain and bottomland forests (these are treated in Chapter 11). Other hardwood assemblages include rockland hammocks growing around lime sinks in north and central Florida and in limestone-rich areas of tropical Florida (these are in Chapter 7).

Many variables besides soil moisture influence each forest's composition and appearance. Climate is a major factor: north Florida's forests are temperate; south Florida's are more tropical. Other factors include the pres-

Softwoods include pines and other conifers. Most are true evergreens: they retain their leaves (needles) all year, shed only a few from day to day, and add new growth at the branch tips in spring. **Hardwoods** are broad-leafed, flowering trees and most are deciduous: they shed their leaves in fall and grow new ones in spring.

Temperate hardwoods are those that grow best in the temperate zone (in which Florida lies). **Tropical hardwoods** grow best in extreme south Florida and beyond.

A **forest** is a large community of trees and shrubs. The name **hammock** is commonly given in Florida to non-pine forests. (Florida hammocks do contain a few non-hardwoods, such as cabbage palm and red cedar, however.)

Florida native: Northern mockingbird (*Mimus polyglottos*). Hardwood forests afford nesting sites for parent birds and insect fare for hungry hatchlings.

OPPOSITE: A hardwood hammock in the Apalachicola Bluffs and Ravines Preserve in Liberty County. Forests like this have hundreds of species of vascular plants and a diverse assortment of animals.

FIGURE 6-1

Representative Florida Hardwood Hammocks

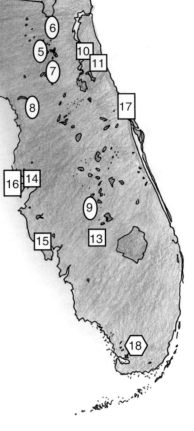

(A) Forests containing mostly temperate species, including many deciduous trees, occur across the Panhandle:

1. Eglin Air Force Base (Santa Rosa, Okaloosa)

2. Marianna Lowlands, Florida Caverns State Park (Jackson)

3. Apalachicola Bluffs and Ravines Preserve (Liberty and Gadsden)

4. Hammocks in the Red Hills (Leon and nearby Georgia counties)

(B) These forests contain mixtures of species in (A) and (C):

5. San Felasco Hammock (Alachua)

6. Goldhead Branch State Park (Clay)

7. Mud Lake, Ocala National Forest (Marion and Lake)

8. Pineola Grotto (Citrus)

9. Tiger Creek Preserve (Polk)

C Forests dominated by broad-leafed evergreens occur throughout the central peninsula:

10. Welaka Reserve (Putnam)

11. Anastasia Island State Park (Flagler)

12. St. Marks National Wildlife Refuge (Wakulla)

13. Highlands Hammock State Park (Highlands)

14. Alafia River hammocks (Hillsborough)

15. Little Salt Spring, Myakka River State Park (Charlotte and Sarasota)

D Mixtures of (C) and (E) predominate here:

16. Highlander Hammock (Pinellas)

17. Turtle Mound (Volusia)

(E) Tropical hardwood hammocks are scattered across the Everglades and Florida Keys:

18. Everglades National Park (Miami-Dade and Monroe)

Note: The Panhandle forests (A) may be called southern hardwood forests. The peninsula forests (C) may be called temperate broad-leafed evergreen forests. The tropical hardwood hammocks (E) may be called tropical rainforests.

Source: Adapted from Platt and Schwartz 1990, 200, fig 7.1.

ence of seepage or frequency of flooding (if any), steepness of slope, compass direction toward which the slope faces, soil texture and drainage, soil chemistry (especially whether limestone-influenced or not), and the forest's own past history and age. Each of these factors varies widely and many different combinations are found together, so one forest can be quite distinct from most others.

The expansion of hardwood forests into upland sites in the last several centuries has afforded opportunities for species to come together in places and mixtures not seen only a few centuries ago. "New" forests have sprung up wherever natural pinelands have been cleared, then used for a while for farming, silviculture, or human settlement and then abandoned. Today, many of Florida's abandoned farm fields, pastures, and neglected backyards have grown up into forests of pine, oak, and hickory that never existed originally on most upland sites. Pine-oak-hickory woods are the result of clearing followed by fire exclusion and invasion of temperate hardwoods, shortleaf pine, and loblolly pine into areas once dominated by high pine grasslands. Forests of this kind now occupy large areas of Florida.

However, the emphasis in this chapter is on living communities of ancient origin or of enduring types, the products of relatively long periods of ecologically stable conditions and evolutionary adaptation of their plants and animals. The hammocks identified in Figure 6-1 fit that description.

AN EXAMPLE: A BEECH-MAGNOLIA FOREST

Imagine standing in a tall, shady hardwood forest partway down a slope somewhere in north Florida—a forest that has grown here, constantly renewing itself, for thousands of years. It is summer. The trees have all leafed out, and the shade is deep, except in a sunny patch where a giant southern magnolia has fallen over. The ground is covered with leaves and there is little undergrowth. Listen: you hear several bird songs: the distinctive chicky-tucky-tuck of a summer tanager, the loud, clear call of a yellow-throated warbler, the chip-trill-chip of a white-eyed vireo, and the hoarse caw-caw of a crow. Squirrels are scurrying up and down the trees, and chasing each other across the canopy.

You recognize two of the biggest trees from their bark and from the leaves they have dropped around their bases. Southern magnolias have gray bark, smooth to the touch, and large, dead, oblong leaves piled a couple of inches thick beneath them. American beeches also have smooth, gray bark, but only small, crinkly, rusty-tan leaves on the ground. Looking up, you see no sunlight through the myriad light-green leaves of an American beech; looking down, you see few or no plants growing beneath it. The shade it casts is so dense that nothing can find light here.

The forest is named a beech-magnolia forest for these trees, but there are many other tree species including several kinds of oaks, two hickories, American holly, spruce pine, and others—altogether, 25 to 35 species of broad-leafed flowering trees (hardwoods) mixed with a handful of conifers. Some have attained the canopy with the beeches and magnolias, some are midstory trees, and a few low-growing shrubs and herbs have found spots to grow on the forest floor. List 6-1 names some of the canopy and midstory trees that grow in forests of this type.

This forest is strikingly different from the pine grasslands and scrubs described earlier. What makes it possible for so many trees to grow so close to one another? How do they all obtain water, sunlight, and places to reproduce? What animals live in, on, and around them, and why?

Gap Succession. Perhaps the greatest distinction between this forest and a pine forest is that this forest resists burning. Its thickly vegetated edges and closed canopy hold moisture in and shield the interior from wind.

LIST 6-1
Tree layers in hardwood forests: a sampling

Overstory trees
American beech
Basswood
Black walnut
Carolina ash
Florida maple
Hickories (several species)
Laurel oak
Live oak
Loblolly pine
Shortleaf pine
Southern magnolia
Southern red oak
Spruce pine
Sugarberry
Swamp chestnut oak
Sweetgum
Tuliptree
Water oak
White oak

Midstory trees and shrubs
American holly
American hornbeam
American plum
Cabbage palm
Common serviceberry
Common sweetleaf
Eastern hophornbeam
Eastern redbud
Flowering dogwood
Green ash
Hawthorn (several species)
Red buckeye
Red maple
Red mulberry
Silverbell
Southern arrowwood
Sparkleberry
Wax myrtle
Wild olive
Winged elm
Witchhazel

Source: Adapted from Wolfe and coauthors 1988, 131.

wind. The trees drop moist litter that rots slowly, keeps the ground cool and damp, and compacts the fuel beds beneath it, excluding air. Within this deep shade, saplings that need abundant sun such as longleaf pine cannot grow, so there are no fire starters or carriers. Faced with these obstacles, even if an occasional fire should, during a drought, creep into a beech-magnolia hammock from a nearby fire community, it would not burn hot or far across the ground and could not leap to the layers above. (This is true of all hardwood hammocks, with only a minor exception. Mild surface fires clear the undergrowth in mature oak woods from time to time.)

If fires do not kill trees and open the canopy for new growth, then how do the trees in a beech-magnolia forest renew themselves? The answer is by gap succession—that is, by the periodic death of mature trees. Trees are snapped off or blown down by windstorms or die from injury and disease, and this opens gaps in the canopy and on the ground. They are succeeded at first by a mass of young trees of many species; these undergo intense competition for sunlight and resources; many die; and finally, the gap is refilled by one or two mature trees of the same species as those nearby. Renewal by gap succession takes place continuously. Diverse tree seeds are present at all times in the soil, having been dropped by the trees and by birds and other animals, especially squirrels. Whenever a tree loses major branches or dies and falls over, young trees start up immediately in a flood of sun. Small, fast-growing saplings are the first to shoot up, so they get a head start, but slower-growing, shade-tolerant evergreens soon follow. Finally, even slower-growing and more shade-tolerant trees overtake the others and close the canopy above all the others. A natural hardwood hammock displays all the stages of gap succession: the trees are always diverse and in varying stages of growth. This ongoing process permits many different tree species to find places in the forest.[1]

Plant Adaptations. Because hardwood forests are multi-layered forests in which trees compete for sun at every level, each tree species has somewhat different ways of obtaining sun. The trees that grow fast as saplings command most of the sunlight in a gap at first while they attain their mature height. Slower-growing trees manage to grow in the sparse shade beneath them; they develop more massive, stronger trunks and can ultimately attain more height. Some of the adaptations of magnolias and beeches, which are slow-growing trees, are mentioned in Figure 6-2.

Trees also display great variety in the ways they capture sunlight. Because the canopy trees reflect most of the green wavelengths of light, lower-growing plants live in a bluish shade and are adapted to catch light in that range of wavelengths. Moreover, because some canopy trees are deciduous, midstory trees and shrubs can compete for sun below them by holding their leaves later in fall. These tardily deciduous trees do much of their growing after most of the trees above them are bare. Other midstory trees, such as hollies, are evergreens. They gather a lot of their energy in winter when all the deciduous trees are dormant. Also, at Florida's latitude, especially in winter when the sun is at an angle to the earth, sunlight sneaks into the interior of forests from their edges and enables a low layer of herbs, shrubs, and small trees to leaf out and flower in early spring. In effect, there are many different habitats for plants in a forest, and there are plants to take advantage of each one.

On the floor of a beech-magnolia forest, the herbaceous ground cover

American beech (*Fagus grandifolia*).

Southern magnolia (*Magnolia grandiflora*).

Magnolia fruit.

FIGURE 6-2

Layers in a Beech-Magnolia Forest

The trees with dark green leaves are magnolias. The trees with light green leaves, which are spread horizontally to catch the sun, are beeches. The beeches give a layered look to the forest.

In their young lives, slow-growing magnolias and beeches often have to grow in the shade. Magnolias are evergreen hardwoods and can grow in the winter sun while deciduous trees are bare. Beeches, thanks to the planar arrangement of their leaves, catch every possible ray of sunlight that filters down to them from above.

Both species can grow beneath other trees at first, and can ultimately overtop most of their competition. They create a dense shade—the magnolias, by holding their leaves all year long, and the beeches by retaining their leaves even after they have turned brown. Beech leaves fall late, mostly when the new leaves are starting to grow in early spring.

Both trees compete at the ground level, too. Magnolia leaf mulch presents a formidable barrier to other seedlings' growth, and decaying beech leaves may release chemicals that inhibit the growth of all but the beech's own seedlings. Ultimately, beeches and magnolias come to dominate the canopy, and shade out many midstory trees.

Source: Means 1994 (Temperate hardwood hammocks).

and soil organisms also repopulate gaps left where trees have fallen. The renewal process is most rapid and complete where trees fall naturally, rather than being cut. When trees fall, they raise tip-up mounds and leave pits

Puffball (a *Lycoperdon* species).

Witches' butter fungus (*Tremella mesenterica*).

Onion-stalk lepiota (*Lepiota cepaestipes*).

FIGURE 6-3

Fungi Native to Florida Forests

Each of these fungi performs chemical dismantling work in the forest. The one at right is just beginning to disintegrate a dead oak log.

behind, creating small variations in contour which foresters call pit-and-mound microtopography. Different plants colonize different small surface features around tip-up mounds. Ferns grow well on the shady side of up-raised root systems. Mosses grow best on shady, damp spots. Other plants seize other sites. In this way, the ground cover remains diverse for centuries. (In contrast, conventional forestry may prepare ground for new planting by leveling it—and not all ground-cover plants can reestablish themselves in soil exposed to sun in this fashion. Decades after clear-cut, leveled forests have been replanted, they still lack ground-cover diversity.)[2]

Key functions are performed by numerous plants in a beech-magnolia hammock. Pine grasslands have one dominant fire-starter tree (such as longleaf, slash, or cabbage palm) and one dominant fire-spreading ground-cover plant (such as wiregrass). In hardwood hammocks, many trees, shrubs, and herbs play overlapping roles. Still, whichever species are present, the hammock's structure is similar, with broad canopy boughs at the top, many branches at the midstory level, and lots of shade, moisture, leaf litter, and decay on the forest floor.

Hardwood hammocks differ from pinelands in many other ways. Pinelands are open, sunny, and spacious; hardwood forests are closed, shady, and full of branches and vines. Much of the photosynthesis in pine grasslands is conducted by the ground cover herbs, whereas in hardwood hammocks it is the canopy that captures most of the sun's energy and converts it into substance. Living ground cover is less abundant due to both low light intensity and root competition by trees and shrubs. In pine communities, fire recycles most nutrients; in hardwood forests, decay agents do. Figure 6-3 shows three fungi that perform this task in Florida forests.

Animals. Because most of the productivity of a hardwood hammock takes place in the canopy, animals that live there have to be able to fly or climb. The hammock's many compartments offer diverse habitats for animals. Thanks to the diversity of tree species, all kinds of branches offer shelter and support at every height. Diverse flowers and fruits are also available as foods for insects, birds, and other animals. Some 45 species of birds breed and nest and raise their young in north Florida's mesic hardwood forests, and 13 species breed only in such forests (see List 6-2). The deeper the forest interior, the greater the diversity of breeding birds. Many migratory birds, too, prefer large hardwood forests over small ones.

Great horned owl (*Bubo virginianus*).   Barred owl (*Strix varia*).

FIGURE 6-4

Owls of Florida's Pine and Hardwood Forests

The great horned owl is the nighttime winged predator of Florida's pine forests. The barred owl is the nighttime, winged predator of Florida's hardwood forests. Sets of species like this, which perform the same roles but in different habitats, are known as *geminate* (paired) *species.*

Because the adaptations animals must make to life in hardwood forests are different from those in pinelands, different sets of animals exist in the two habitats. Some, such as the wild turkey, use both, but most animals are more closely tied to one than to the other. As an example, animals seeking nuts and berries in a longleaf pine forest find many growing in the diverse, low ground cover, but in a hardwood forest, they will have to climb high on branches. Thus while the pine forest is home to the ground-foraging fox squirrel, the hardwood forest has the tree-climbing gray squirrel. The pine forest is home to the great horned owl, whose wingspread of 55 inches is easily accommodated by the pinelands' open spaces. Lowland hardwood forests have the smaller, barred owl, which navigates skillfully among the more closely spaced trees (see Figure 6-4). In the open pine forest, there is the red-tailed hawk (wingspread 48 inches), which soars over open spaces and dives for its prey. The hardwood forest has the red-shouldered hawk (wingspread 40 inches), which usually hunts from a perch and drops to the ground to catch its prey.

Inconspicuous plants and animals in hardwood forests are also diverse. A single fallen tree provides habitat for dozens of interacting organisms and the litter on the ground feeds and shelters many more. There may be fungi spreading through the rotting wood, lichens and mosses growing on it, bacteria and algae working within its tissues. There may be spiders in the cracks, bark beetles, boring beetles, tunneling beetles, and termites eating the wood. Preying on these are scorpions, centipedes, oribatid worms, and earwigs. Preying on these are birds, of course, and on them, others. Keen observers find hundreds of creatures living in habitats like these, where a casual glance might reveal few or none.

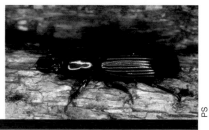

Florida native: Patent leather beetle (*Passalus cornutus*). The grub shown above develops into the beetle shown below. This is one of dozens of species of beetles and other invertebrates that eat rotten wood in Florida forests.

SELECTED HARDWOOD FORESTS AND HAMMOCKS

The beech-magnolia forest just described illustrates the principles that govern growth and competition among hardwoods, although details vary from one to the next. Moisture varies so much from the top to the bottom of a slope that, even in a single forest, many different trees and shrubs grow, each in a zone to which it is especially well adapted. People who can recognize most hardwoods can guess quite accurately how far down a slope they are from the mix of species there.

Variation also occurs along Florida's north-south axis. In general, the northern hammocks are richest in species. Most of the species of the temperate hammocks of north and central Florida do not range into the southern one-third of the peninsula. American beech, for example, does not grow east or south of the Suwannee River. In their place are tropical hardwoods from the Caribbean that have colonized south Florida.

Many forests and hammocks in Florida are prized for their special characteristics. Selected here are a few from the Panhandle and north and central peninsula.

Apalachicola Bluffs and Ravines. The Apalachicola River area is some 30 million years old, and some of its plant populations are even older because they were already established nearby before Florida emerged from the sea. Now, climate change has fragmented the ranges of some of the plants, but they still are protected within the deep, steep ravines of tributaries flowing into the Apalachicola River from the east. Many are recognizably related to plants elsewhere in the world. Several of the tree and shrub species resemble those in similar ravines in eastern Asia. A few mosses and ferns that grow in the protected understory also grow in Mexican tropical cloud forests.

Species from the Appalachian Mountains to the north are especially

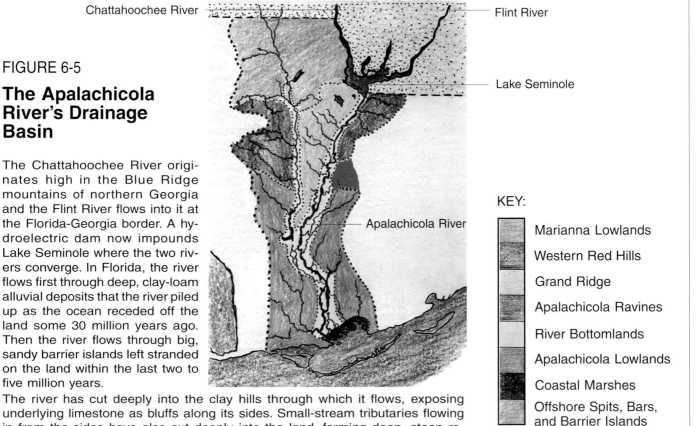

Chattahoochee River — Flint River — Lake Seminole

Apalachicola River

FIGURE 6-5

The Apalachicola River's Drainage Basin

The Chattahoochee River originates high in the Blue Ridge mountains of northern Georgia and the Flint River flows into it at the Florida-Georgia border. A hydroelectric dam now impounds Lake Seminole where the two rivers converge. In Florida, the river flows first through deep, clay-loam alluvial deposits that the river piled up as the ocean receded off the land some 30 million years ago. Then the river flows through big, sandy barrier islands left stranded on the land within the last two to five million years.

The river has cut deeply into the clay hills through which it flows, exposing underlying limestone as bluffs along its sides. Small-stream tributaries flowing in from the sides have also cut deeply into the land, forming deep, steep ravines.

Source: Adapted from Means 1977.

KEY:

Marianna Lowlands

Western Red Hills

Grand Ridge

Apalachicola Ravines

River Bottomlands

Apalachicola Lowlands

Coastal Marshes

Offshore Spits, Bars, and Barrier Islands

PRICELESS FLORIDA

abundant around the Apalachicola River. The deep ravines all along the river's watershed are shady, protected valleys into which Appalachian plants and animals were able to spread and persist. All other rivers in Florida originate on the Coastal Plain, but the Apalachicola's headwaters are in the Blue Ridge mountains of northern Georgia. No other river valley near the Gulf of Mexico, not even that of the mighty Mississippi, holds so many northern plants from times long past.

Bluffs along the Apalachicola River. The steep slopes, rich soils, and limestone outcrops along the river support diverse hardwood communities.

Figure 6-5, opposite, highlights several areas in the river's watershed that are discussed in this and several chapters to come. Forests in the ravines and in the Marianna Lowlands are described here; communities of the river bottomlands, coastal lowlands, and coastal marshes are described later.

The ravines in the younger sand hills are of particular interest because they are stable repositories of ancient species. Water flows through them steadily all year long, regardless of fluctuations in rainfall on the high land around them. In contrast, creeks in the more northern ravines in clay-lined valleys go dry between rainfalls. The sandy ravines are 50 to 100 feet deep, and their walls are among Florida's steepest slopes. Locals call them "steephead" ravines, because even the head of each valley has deep, steep walls. Views of steephead ravines are shown in Figure 6-6.

Figure 6-7, on the next two-page spread, reveals how the trees and shrubs respond to the several gradients in a steephead ravine. The soil is coarse sand, well drained throughout, but other conditions grade from top to

FIGURE 6-6

Steephead Ravines

A steephead ravine cuts into a sand hill, not because water runs over the surface and erodes it, as in clay hills, but because water runs out from beneath the hill and carries sand away. The steephead's walls slope at a 45-degree angle to the earth.

BELOW: Diagram of a single ravine showing the valley contours. Note how water runs out from beneath the sandy slopes through which it has percolated, rather than from the top of the ravine.

ABOVE: Aerial view of several steepheads. Shown are heads of three ravines that run into Sweetwater Creek, a tributary that flows through deep sand into the Apalachicola River from the east. Notice how abruptly each ravine drops off from the surrounding, level agricultural land.

Other steephead ravines are found on other stranded, ancient barrier islands: on Eglin Air Force Base in Santa Rosa and Okaloosa counties; by the Econfina River north of Panama City; by the Ochlockonee River along Lake Talquin; and along Trail Ridge in Clay and Bradford Counties.

Source: Means 1991 (Florida's Steepheads).

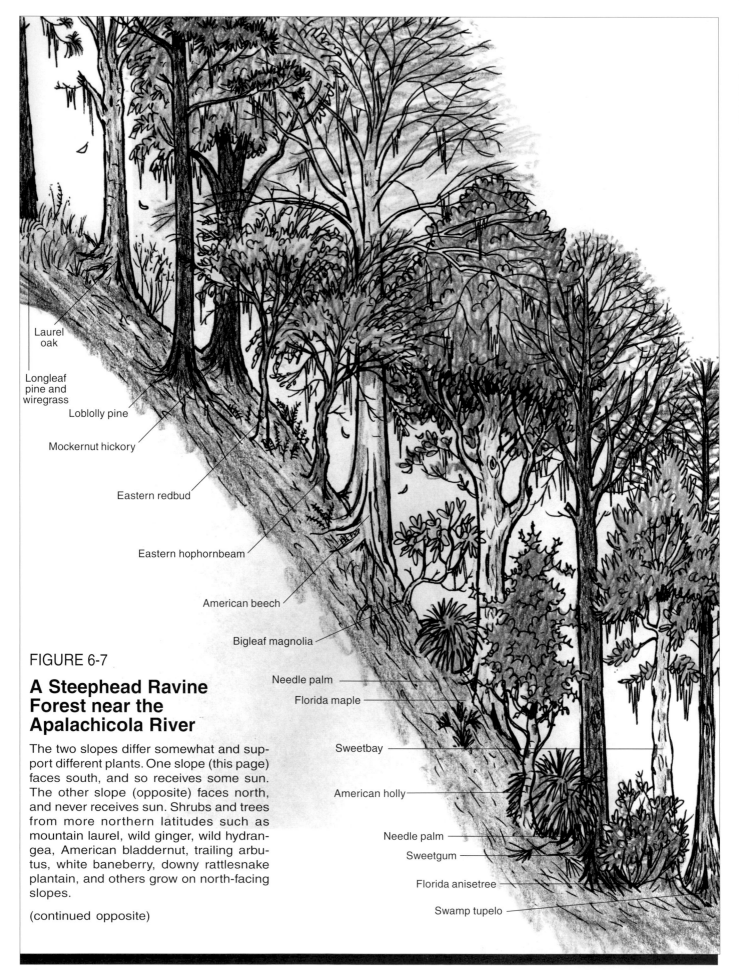

Laurel oak

Longleaf pine and wiregrass

Loblolly pine

Mockernut hickory

Eastern redbud

Eastern hophornbeam

American beech

Bigleaf magnolia

Needle palm

Florida maple

Sweetbay

American holly

Needle palm

Sweetgum

Florida anisetree

Swamp tupelo

FIGURE 6-7

A Steephead Ravine Forest near the Apalachicola River

The two slopes differ somewhat and support different plants. One slope (this page) faces south, and so receives some sun. The other slope (opposite) faces north, and never receives sun. Shrubs and trees from more northern latitudes such as mountain laurel, wild ginger, wild hydrangea, American bladdernut, trailing arbutus, white baneberry, downy rattlesnake plantain, and others grow on north-facing slopes.

(continued opposite)

Sand
live
oak

Southern
red oak

Flowering
dogwood

Pignut hickory

Red cedar (stump)

White oak

American hornbeam

Spruce pine

Southern magnolia

Silky camellia

Gopherwood (and stump)

Red buckeye

Tuliptree

Needle palm

Swamp tupelo

Mountain laurel

Florida anisetree

Virginia willow

Florida native: Pyramid magnolia (*Magnolia pyramidata*). This rare small tree was photographed skyward from deep in one of the Apalachicola ravines.

bottom, supporting great diversity among the plants. Soil temperature varies with the season at the top, but deep in the "V" of the ravine it remains relatively constant the year around. The top is sunny, hot, dry, and windy. The bottom is shady, cool, humid and still. Steephead ravines provide Florida's most dramatic examples of the ways trees and shrubs sort themselves out on slopes

To appreciate the details, imagine standing at the edge of a typical ravine. The level land behind you supports a high, dry, longleaf pine-wiregrass community (or more commonly today, stand of planted pine). In front of you, the thickly forested ravine slope plunges so steeply to the ravine floor that you are looking down into the tops of tall hardwood trees growing up from below.

Now, walk down into the ravine. Go carefully: at each step, your heels must sink deeply into the soft sand, otherwise you'll pitch forward and tumble down a dangerous incline. Grasping the stems of small oaks and some vines, you can brake your descent hand-over-hand until you are halfway down. You have already passed through a couple of bands of trees. At the rim was a very well-drained forest composed mostly of live oak or sand live oak; just below it, a mixture of laurel oak, mockernut hickory, wild olive, red cedar, and other trees adapted to dry soil. In this zone the trees are scrubby and stunted because the sandy substrate holds water so briefly after each rain.

About midway down, the forest becomes more shady and therefore more open near the ground. The slope is paved with crinkly, light-brown beech leaves or leathery, rattling, and slippery magnolia litter. The beech-magnolia combination bears testimony that the soil is more moist here. Unlike other mesic forests, though, this one has a few ancient understory trees, such as gopherwood (Figure 6-8), bigleaf magnolia, and pyramid

---

FIGURE 6-8

Florida Native: Gopherwood (*Torreya taxifolia*)

This evergreen conifer (whose genus name is pronounced *toe-RAY-uh*) is endemic to Gadsden, Jackson, and Liberty counties and to Decatur County, Georgia. Its range now is limited exclusively to the east side of the Apalachicola River north of Bristol. It is endangered in Florida: fewer than 2,000 individuals remain in the wild.

The gopherwood's habitat requirements are stringent. It grows only on wooded ravine slopes and bluffs, at elevations from 35 to 100 feet above constantly running streams, and in moist soil of moderate to rich organic content, most often on north-facing slopes.

In the past, a gopherwood could be a massive tree. A large specimen could be 60 feet tall with branches some 25 feet long. All specimens in Florida are of less than sapling size now, however, possibly due to a fungal blight.

Sources: Godfrey 1988, 59-60; Clewell 1985, 56; Schwartz and Hermann 1993, 1-3, 20.

Florida native: Wild ginger (*Asarum arifolium*). This plant, growing in the Apalachicola Bluffs and Ravines Preserve, is a member of a species population that is centered in the southern Appalachian mountains. Another common name for it is heartleaf.

magnolia. Gopherwood may be one of the oldest plant species in the ravines. It closely resembles fossil gopherwood from southern Appalachia that dates back about 100 million years—well into the age of the dinosaurs.

The floor of the ravine, where the ground water seeps out from beneath the ravine's sides, holds saturated soil and supports a true wetland dominated by sweetbay, fetterbush, and Florida anisetree. In its center flows a perennial clear-water stream that is an ecosystem all to itself—the steephead stream described in Chapter 13.

Because the Apalachicola River runs from north to south, its tributary streams run east-west. As a result, each ravine has north-facing and south-facing slopes. The path just described descended the ravine's north slope, the one that faces south and receives at least some sun each day. Look across at the opposite slope, which faces north and never receives direct sun. Many of its species are those typically seen much farther north. Wanderers may wonder in surprise whether they have suddenly been transported to the mountain cove valleys of the Appalachian Blue Ridge.

Besides the trees in the ravines, diverse shrubs, vines, herbs, fungi, and lichens contribute complexity. All these plants enable many animals to use the ravines as foraging, breeding, and nesting sites.

Most of the birds and mammals that frequent the ravines are the same as those that inhabit other stream valleys, lake and pond basins, and other lowland forests. However, in steepheads, small animals whose lives depend on continual moisture are tightly confined to the ravine bottoms because the sidewalls are so sandy, well-drained, and dry. Only at the level of the stream does much moisture and organic matter accumulate. Wherever frogs and salamanders reside in a particular steephead stream, their progeny will surely dwell nearby. Because the frog and salamander populations found in different drainage basins have long been isolated from each other, even though they are members of the same species, they possess recognizably different characteristics. A specialist may be able to tell you which steephead system a particular animal comes from, just by looking at its color pattern. Many other such differences doubtless also exist. One naturalist has observed that black widow spiders in the Apalachicola ravines live in trees and shrubs, whereas elsewhere in Florida, as far as we know, they live on the ground.[3]

The single 35-mile stretch of ravines on the east side of the Apalachicola River in Florida harbors more total plant and animal species, and more endemic species in particular, than any other area of the same size on the southeastern Coastal Plain. The ravines are home to more than 100 rare and endangered species. Despite their fame, however, the Apalachicola Bluffs and Ravines have never been exhaustively surveyed; they have many more secrets to reveal in times to come. The Florida Biodiversity Task Force has ranked the Apalachicola River basin, together with the central Florida ridge, as a "hot spot" of endemic species, a site of great value to global biodiversity. The Nature Conservancy has purchased the bluffs and ravines for ongoing preservation and now a large percentage is owned by the state and managed as Torreya State Park.[4]

The Marianna Lowlands Forests. West of the Apalachicola River around Marianna in Jackson County lie the Marianna Lowlands, centering on the

Florida native, near-endemic: Croomia (*Croomia pauciflora*), seen growing in the Apalachicola Bluffs and Ravines.

Florida native: Oakleaf hydrangea (*Hydrangea quercifolia*), growing in Torreya State Park in Liberty County.

Florida native: Wild columbine (*Aquilegia canadensis*). This plant, at the extreme southern end of its range, grows in the Marianna Lowlands forests.

Spanish moss is an **epiphyte**—that is, a plant that grows on another plant. Actually not a moss, it is a bromeliad, related to the pineapple. Chapter 7 features several other bromeliads.

Chipola River. Once continuous with the clay-loam highlands on both sides, this 450-square-mile area has extensively subsided and eroded. Today, masses of limestone are exposed at and near the surface, and only partially capped by a mixture of largely clayey sediments. This unusual combination of clay and limestone supports a hardwood forest with species not found anywhere else in Florida, even in the nearby Apalachicola Bluffs and Ravines (see List 6-3). Within the lowlands, the Florida Caverns State Park supports many plant populations that are now evolving separately from their sister populations in both northern and tropical habitats. Other ancient and isolated populations occur in the nearby Chipola floodplain.

The Marianna Lowlands' ground-cover herbs include many that are at the extreme southern ends of their ranges, such as wild columbine, Mayapple, and bloodroot. It is thought that these plants first found refuge there more than two million years ago. They have persisted there ever since, thanks to the Lowlands' deep, shaded habitats that are similar to the mountain coves of the southern Appalachians.

Other plant populations in the forests are ferns and mosses isolated from much more southern, and much more northern, groups. Because they reproduce via spores, which can float in on the air, they can grow wherever they find suitable habitat, and the limestone exposures around Marianna are ideal for them. Observer Richard S. Mitchell remarked that the mixture of species he found seemed to defy the existence of north-south zones.[5]

Xeric Oak Hammocks in Central Florida. Central Florida's slopes hold scattered hammocks of all three classes: xeric, mesic, and hydric. For reasons unknown, American beech and white oak do not occur this far south. Other trees predominate, especially oaks (other than white oak) and hickories. Some forests have abundant cabbage palms. One of the most distinc-

Florida native: Live oak (*Quercus virginiana*). Live oak hammocks are common in central Florida and hold masses of Spanish moss and resurrection fern. This was photographed at sunset, hence the orange tint.

Florida native: Resurrection fern (*Pleopeltis polypodioides* var. *michauxiana*). This is a true fern with a knack for surviving dry times. It dries to a brown, crackly remnant of itself that looks dead. Then, after rains, it turns green almost instantaneously and continues its life.

most distinctive forest types in north-central Florida is the live oak hammock, whose trees often bear festoons of Spanish moss so abundantly that naturalist Archie Carr called the forest a "moss forest." Carr asserts that "A big old live oak without its moss looks like a bishop in his underwear."[6]

Live oak, and also laurel oak, grow slowly, some to a century and a half. They ultimately reach majestic proportions of 75 feet in height and more than 100 feet in crown diameter, with trunks up to five feet across at breast height. By the time they are fully mature, these oaks form a closed-canopy forest beneath which the understory is often sparse, presenting an inviting aspect to walkers.

New oak groves may supplant pinelands wherever fire is newly excluded, halting natural fire-induced regeneration of the pines. Once the oaks are past the sapling stage, they can withstand mild surface fires. In fact, fires help to clear the undergrowth and free the oaks of competition by other, upcoming saplings. Without periodic fires, beech, dogwood, hickories, and others might in time convert an oak grove to a mesic forest type.[7]

Where there are oaks, there are acorns, of course, and animals that eat them. Acorns are especially abundant in mast years, when their major consumers, mice and deer, crowd into oak groves. Gray squirrels, wild turkeys, and wild hogs also flock to oaks to feast on their acorns. So do ducks, even on high, dry hills. Naturalist Archie Carr says the most spectacular event of a mast year is the coming of wood ducks to the live oak woods: they land in droves, even where there is no water. How they know the acorns are there remains to be discovered.[8]

Other animals visit: rat snakes, barred owls, flying squirrels, opossums, and armadillos. Many microscopic animals reside permanently on the ground in the oak forest, together with many small animals that eat them: fast-running wolf spiders, crickets, millipedes, snails, the eastern spadefoot toad, and a tiny lizard, the ground skink. Under rotting logs, Carr says, are "big, black shiny beetles (*Passalus cornutus*), wireworms (the larval stage of click beetles), centipedes, newts, a short-tailed shrew, and little burrowing or litter-inhabiting snakes including the eastern coral snake."[9]

One might never guess most of these animals are there: you have to know where to look. The spadefoot toad, for example, digs a burrow with spurs on its hind feet, then backs into it and comes out only at night. Spadefoots are so inconspicuous that even forest lovers seldom see them, but they can be numerous. One observer reported more than 2,000 toads

Florida native: Spanish moss (*Tillandsia usneoides*). The tiny gray scales of this epiphyte absorb dew, which provides water and nutrients. Spanish moss flourishes on trees with a leafy canopy, which protect the humidity on which it depends.

A **mast year** is one in which a population of trees (frequently, oaks) produces a much larger crop of seeds (acorns) than the average.

Florida native: Spadefoot toad (*Scaphiopus holbrookii*).

Spadefoot in retreat. To hide, the spadefoot digs its way backward into the ground until only its eyes are showing.

Florida native: Cope's gray treefrog (*Hyla chrysoscelis*). This treefrog calls from hammock trees and bushes on warm, rainy nights, and eats arboreal insects. Like other frogs, it seeks out ponds at mating time.

per acre of woodland hammock. Fortunately for the toad, oak hammocks grow especially well around ponds and lakes, a necessity for an animal that breeds briefly in water after heavy rains.[10]

The abundant Spanish moss in a live oak forest provides needed habitat for a variety of animals. A warbler, the northern parula, nests in clumps of living Spanish moss and apparently never builds its nests anywhere else. Three kinds of bats—the yellow, Seminole, and eastern pipistrelle—also roost in Spanish moss. This must be why rat snakes climb the oak trees, because they avidly hunt the birds, bats, and skinks that hide in the moss. For most animals, Spanish moss collects too much moisture to make suitable nesting habitat, though. The gray squirrel uses true mosses, together with strong, slender threads pulled from cabbage palms, to build its nests.

Mesic Central Florida Hammocks. Mesic hammocks in central Florida are more of a mixture of evergreen and deciduous trees. Seen from above in winter, they are a patchwork of green and gray blotches. Besides live oak, tree species in these forests include laurel oak, mockernut hickory, pignut hickory, southern magnolia, red bay, sweetgum, and two or three dozen others. As mentioned earlier, American beech, which is a signature tree in the beech-magnolia forests of north Florida, is not present.

All of Florida's hardwood hammocks are important for birds, and none is more so than the San Felasco Hammock in Alachua County. San Felasco provides indispensable food and shelter for birds that are passing through on spring and fall migrations, other birds that are staying through the winter, and still others that are breeding there in summer (see Figure 6-9). An in-depth study made in the San Felasco Hammock illuminates an age-old relationship between the hammock's fruits and fruit-eating birds. The birds benefit by eating the fruits, of course, but the plants also benefit, because birds disperse and fertilize their seeds. After birds eat fruits, they excrete the seeds, which are still viable, with their droppings. Shrubs, midstory trees, and vines with fleshy fruits, the kinds birds like and help disperse, make up half of the mesic hammock's plants—45 species in all.

The plants of San Felasco bear fruits in all seasons, but birds are especially dependent on those that ripen in winter, summer, and fall. Winter fruits are so abundant that birds can stay as residents throughout the season. Twenty species of plants, mostly evergreens, produce ripe fruits from late fall to early winter. Winter-fruiting plants include Carolina laurelcherry which, along forest edges attracts wintering robins and woodpeckers; and oak mistletoe, which grows high in treetops and attracts cedar waxwings. Cedar waxwings arrive in Florida just as the mistletoe berries are first ripening. They derive most of their nourishment from mistletoe.

Fall fruits support migratory birds that fly through on their way to winter homes farther south. The fruits ripen at times that coincide with the two peaks of the fall migration (see List 6-4). As shown, six kinds of plants bear fruits that ripen in time for the first wave of fruit-eating birds, and six other plants bear fruits that ripen in time for the second wave. These are predominantly "bird fruits"—that is, they attract mostly birds—and their diversity and timing make them dependable food sources. The migrants seem to have evolved in concert with the southern mixed hardwood forest's ability to provide them with fuel for their journey. The strategy works out well, both for them and for the plants they help disperse.

Summer breeder: Red-eyed vireo (*Vireo olivaceus*). Fruits are not abundant in summer, but the vireo eats berries, seeds, and insects and so does not run short of food.

Winter visitor: Cedar waxwing (*Bombycilla cedrorum*). The cedar waxwing, which depends largely on oak mistletoe berries, spends all winter in Florida forests.

Spring and fall migrants: At left, the wood thrush (*Hylocichla mustelina*), which finds spring and fall fruits in Florida's forests. At right, the Louisiana waterthrush (*Seiurus motacilla*), which eats insects and spiders from the ground and water.[a]

Year-round residents: Above, the blue-gray gnatcatcher (*Polioptila caerulea*). At right, a young Cooper's hawk (*Accipiter cooperii*). As its name implies, the gnatcatcher eats insects, which are available in Florida forests all year. The hawk is a top predator and finds abundant prey among birds, small mammals, snakes, and other animals in forests.

FIGURE 6-9

Birds in Florida in Different Seasons

Note: [a]Some of these thrushes spend whole summers in north Florida forests.

In summer, birds are scarce. They spend summers farther north and so are unreliable as dispersal agents for the seeds of summer fruits. As if responding to this shortage of birds to spread their seeds, the nine hammock plants that fruit in summer attract mammals, too.

Like fruit-eating birds, insect-eating birds are important to the forest. In hardwood hammocks, as in pine grasslands, scientists have found that

Florida native: Cabbage palm (*Sabal palmetto*). This is a rare, nearly pure stand of cabbage palm in Wakulla County.

trees are not overwhelmed by leaf-eating insects if patrolled by enough insect-eating birds. Wood thrushes, warblers, tanagers, grosbeaks, titmice, vireos, cuckoos, chickadees, and dozens of other species of birds control leaf-eating caterpillars and other insects so successfully that their efforts noticeably improve tree growth.[11]

The San Felasco Hammock is the largest protected stand of classic mesic hammock in Florida. It is an object lesson illustrating the principle that the larger a habitat island is, the more species it will support. All of the 45 species of birds that breed in north Florida's hardwood forests do so in the San Felasco Hammock. No other Florida forest can claim that distinction.

Highlands Hammock in Central Florida. It is often thought that forests with temperate tree species are restricted in Florida to its northern two thirds and that none exist in the southern third. It is true that the farther south the forest, the fewer temperate tree species it has, but some temperate trees are found all the way to the tip of the peninsula. Examples are red maple, live oak, Carolina ash, and sugarberry. Highlands Hammock State Park boasts a virgin forest of immense live oak and cabbage palm, including the nation's largest specimen of the latter, and there are numerous other tree species. No tropical elements are present, probably because of occasional winters with freezing and near-freezing temperatures. Highlands Hammock is Florida's southernmost large, protected, temperate forest.

Other Hammocks. This chapter has described a variety of temperate hardwood hammocks, but has not exhausted the subject. As one last example, soil that is rich in calcium supports some species of plants not found elsewhere. In much of north Florida, cabbage palm thrives on such soil, especially where it is cool in summer and not too cold in winter. If fires sweep through from time to time, wiping out competing species, these palms can become a solid stand of majestically tall trees, growing above a

Florida native: Common blue violet (*Viola sororia*). In a hardwood hammock, seedlings start up wherever light and moisture are sufficient to support them.

carpet of dried and broken fronds. Cabbage palm stands are rare, few are on public land, and only one (at the St. Marks Wildlife Refuge in north Florida's Big Bend) is protected.

Many other types of hammocks, not shown or described here, contribute to the state's ecosystem diversity. The variety is nearly endless.

Values and Maintenance of Hardwood Hammocks

Like pine grasslands, Florida's hardwood hammocks help to maintain the atmosphere, regulate the climate, and contribute to regional rainfall. They help prevent runoff and water pollution. They keep the soil soft, facilitating infiltration by rain and recharge of the aquifer, and they offer enjoyment and learning to people. They also cloak the terrain with endlessly interesting combinations of trees, shrubs, and associated wildlife. Moreover, because they are so biologically diverse, they serve as a repository of species, all adapted to the slopes of the southeastern Coastal Plain and all providing habitat for numerous other native organisms.

These diverse natural communities together possess a resilience that ensures that the land can continue to be protected by tree cover even in changing times. All that forest managers must do to ensure their continuance is to allocate land to them and protect the conditions they require in order to renew themselves. The many services they render, and their immense diversity, argue for native forests in preference to non-natural forests wherever such land uses are feasible—and for as many different native forests as possible.

Among the factors that promote a natural forest's renewal, the soil is one of the most important. Earlier, the diversity of organisms in forest soil was mentioned, and it was said that this diversity persists most reliably in undisturbed forest soil. The natural process in which large, old trees age and fall, uneven ground results, and nearby soil organisms and plants colonize this ground best guarantees the continuance of all of the native organisms. They all contribute to the ecosystem's stability and resilience: biodiversity promotes the renewal of healthy forests.[12]

Florida native, endemic: Leafy beaked ladiestresses (*Sacoila lanceolata* var. *paludicola*). This rare orchid blooms in moist woods and swamps in southern Florida.

CHAPTER SEVEN

ROCKLANDS AND
TERRESTRIAL CAVES

Limestone dominates the four natural communities treated in this chapter. The first two are strictly south Florida types. One is a pine community: pine rockland. The other is a hardwood community: rockland hammock. Both of these grow on roughly level limestone plains or on raised outcrops. The third type of community, known as a sink community, occurs all over Florida, growing on and around the walls of solution holes or sinks. The fourth type of community also occurs all over Florida, but underground: terrestrial caves.

The first two of these natural communities, pine rocklands and rockland hammocks, occupy south Florida's uplands, shown in Figure 7–1. As for sinkholes, they are scattered all over the state wherever karst activity occurs; and the locations of Florida's terrestrial caves are depicted at the end of the chapter.

Both pine rocklands and rockland hammocks are species-rich ecosystems. Both contain some temperate plants and animals from peninsular Florida and points north, mingled with numerous tropical plants from the West Indies. The next sections treat the plant communities of these two ecosystems separately; then a single section treats the animals of both.

PLANTS OF PINE ROCKLANDS

The pine rocklands were once a vast and beautiful, fire-dependent ecosystem, similar to the longleaf pine grasslands of north Florida, but with slash pine in place of the longleaf. The pines formed an open canopy, growing above a patchy but diverse understory of tropical and temperate shrubs and many herbs. They once dominated more than 186,000 acres on the Miami Rock Ridge and the lower Florida Keys. They held sway wherever fire was frequent. (Rockland hammocks grew wherever fire was excluded.)[1]

Today, clearing and development have obliterated the pine rocklands

Rockland communities grow on land with exposed limestone and little or no soil. **Pine rocklands** have an overstory (usually sparse) of slash pine. **Rockland hammocks** consist largely of tropical hardwoods.

In south Florida, the word **hammock** refers specifically to an isolated stand of broadleafed evergreens, including many West Indian species.

Florida native: Florida butterfly orchid (*Encyclia tampensis*). This common orchid grows in hammocks and swamps in central and south Florida.

OPPOSITE: A limestone sink. Water from a surface stream falls into the ground water at the bottom of this deep sink in the Devil's Millhopper State Geological Site in Alachua County.

105

drying wind. Tropical hammocks have also probably grown for hundreds of years on most of the Florida Keys and on some of the tree islands in the Everglades—raised islands of humus anchored to the underlying rock and protected from fire by marsh or prairie.[3] (Background Box 7–1 describes the water- and land-surrounded islands known as keys.)

BACKGROUND 7–1

Keys

Islands in south Florida are often called keys, whether surrounded by water or by seasonally flooded land. There are dozens of large, ocean-surrounded islands off the tip of south Florida (the Florida Keys) and thousands of land-bound keys in the Everglades. Figure 7–4 shows the Everglades keys and the Florida Keys—all islands.

The floor on which these keys rest developed as limestone accumulated under water on the flat seabed south-southwest of a then much shorter Florida peninsula. (Peninsular Florida's shoreline, at that time, lay just south of Orlando.) Then, from about 100,000 to 15,000 years before the present, more and more of the ocean's water was tied up in polar glaciers, and sea levels fell. The higher pinnacles of limestone poked above the water as bare, rocky islands and ridges in a shallow meadow of salt marshes and seagrasses.

The first plants to colonize these islands and ridges probably were carried in by birds from the West Indies. Bathed in sun and fresh rain water, they sprouted, twined their roots down into the stony labyrinths of limestone beneath them, and began to grow. As they grew, the trees dropped litter that decayed in hollows on the ground, creating a thin layer of soil and acidic water that eroded holes in the underlying limestone.

Once the first trees were there, birds flew in, landed, ate the fruits, and left droppings containing viable seeds from other trees elsewhere. Winds carried still other seeds to the islands. Communities of hardwood trees started up.

Roothold in limestone. The roots of a gumbo-limbo hang on for dear life in the rugged limestone of south Florida's rocklands.

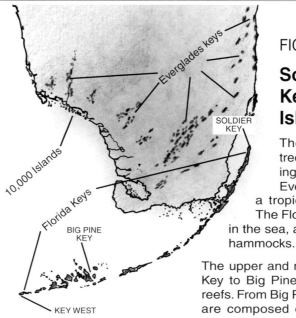

FIGURE 7–4

South Florida's Keys and Tree Islands

The Everglades "keys" are tree islands in the vast, flowing, freshwater marsh of the Everglades and each holds a tropical hardwood hammock. The Florida Keys are true islands in the sea, and they, too, hold tropical hammocks.

The upper and middle Keys, from Soldier Key to Big Pine Key, originated as coral reefs. From Big Pine Key to Key West, they are composed of limestone like that on most of the south Florida mainland.

Protected from fire by the surrounding water, the hardwoods grew taller, and a green canopy closed over each hill. A thick border, composed of dense shrubs and vines, formed around the sun-exposed edges. The interior, now that it was shady and protected from wind and cold, grew more moist and tropical. Ferns, orchids, bromeliads, and mosses took hold on limestone ledges and on the trees. Depending on local temperatures, winds, and humidity, and on what seeds might have been carried in, each island developed a character of its own.

As sea level continued to fall, the water grew shallower, more and more of the seafloor became exposed, and the shoreline crept outward. The salt water surrounding each island was increasingly diluted by rain water running off the mainland, and an expanding freshwater marsh—the Everglades—began to replace the earlier seagrass beds and salt marshes.

Florida native: Pineland passionflower (*Passiflora pallens*). A rare inhabitant of tropical hammocks, this flower blooms all year. In contradiction to its name, it does not grow in pinelands.

Mainland Rockland Hammocks. Before European alteration of south Florida, there must have been hundreds of hardwood hammocks scattered among the pinelands on the Miami rock ridge and hundreds more standing among the marshes and swamps of Big Cypress Swamp. Today, remnants testify to their rich plant assemblages and diverse animal populations. Hammocks around Miami sport some 150 species of trees and shrubs, most of which are tropical and do not grow in north Florida. Hammocks on Long Pine Key display great diversity: each has a well-known name and a distinct tree-species profile. The Wright Hammock has a preponderance of 40-foot-tall live oaks; the Osteen Hammock has tropical Asian trees known as tamarinds, also 40 feet tall. Other hammocks have larger trees. Some have false mastics, large wavy-leafed trees up to 60 feet high with trunk diameters of two to three feet at breast height. Royal Palm Hammock (also known as Paradise Key) has tall, straight Florida royal palms emerging above its canopy of broad-leafed trees.[4]

The Everglades hold thousands of tree islands, one to ten acres in size. Each is shaped like a teardrop, because marsh water flowing around each island from the north rounds it off at the upstream end and drops silt at the downstream end, forming a point. Typically, each island has a solution hole in the center and is surrounded by a moat. The hole in the center enlarges as leaf litter accumulates there, holds moisture, turns acidic, and eats down into the underlying limestone. The moat around the outside of the island forms as acidic water runs off the outside of the hill and dissolves the surrounding limestone. The island is seldom touched by either fire or flood. Dry-season fires that sweep across the Everglades cannot cross the moat, and rainy-season floods cannot cross the island's high limestone walls. As a result, the island holds a self-maintaining, closed-canopy forest that may persist for centuries.[5]

The would-be visitor to an Everglades tree island encounters a formidable, almost impenetrable barrier. One has to wade through the surrounding, deep marshy moat, struggle up the bank, and then push through a dense outer thicket of sprawling bushes and tangled vines: fierce, thorny devil's claws, muscadine, Virginia creeper, eastern poison ivy, and snowberry. One visitor complains that "It's too low to crawl under and too high to crawl over."[6] But inside, thanks to the thick canopy overhead, the tree

Florida native: Florida royal palm (*Roystonea regia*). A few tropical hammocks are graced with this rare Florida–West Indies native towering high above them. The royal palm, whose white trunk looks like a concrete pole, can be almost 100 feet tall. These are growing in the Royal Palm Hammock in Everglades National Park.

Tree islands in the Everglades. Each supports a tropical hardwood hammock.

cluster opens up. Spider webs are everywhere, and exotic, skunky odors assail the nostrils; still, the effort is rewarding. The shady, moist, cool, and quiet interior is a climate-controlled greenhouse for orchids, which bloom at eye level and higher in the trees. Bright butterflies flash in the dappled shade, and at night pale moths catch the moonlight and fireflies sparkle.

Competition for space is fierce on the limited area of a tree island. Growth piles upon growth. Epiphytes cope with the shortage of space by growing on other plants that are already there (see Figure 7–5). Each has evolved a way of capturing water: orchids and most bromeliads in funnel-like leaf whorls; mosses in specialized scales that can gather humidity from the air. The strangler fig perches in a cranny at first and absorbs nutrients from rain, then puts down twining roots that draw water from trapped litter. Finally, the roots reach the ground and the fig wraps its host in an embrace that ultimately kills it (see Figure 7–6).

Mule-ear oncidium (*Trichocentrum undulatum*). This rare orchid grows on trees in south Florida's tropical hammocks.
FIGURE 7-5

Native Epiphytes in Tropical Hammocks

Cardinal airplant (*Tillandsia fasciculata*). This is a common resident of south Florida hammocks.

The fig seed first roots in a crevice on a host tree.

FIGURE 7–6

As it begins to grow, it puts down hanging roots that tightly entwine the host. When they reach the ground, the roots rob the host of needed water and nutrients, and the host begins to decline. Finally the host dies and decays, and . . .

. . . the strangler fig stands around the empty space left by the tree it has killed, with its huge branches supported by the columns of its roots.

Strangler Fig (*Ficus aurea*)

Florida Keys Hammocks. The hardwood hammocks on the Keys differ from the mainland hammocks in several ways. They have mostly tropical species. They are drier, both because ocean breezes blow through them and because, especially toward Key West, they receive less rainfall. As a result the trees are smaller: not more than 40 feet high in the upper Keys, 25 to 30 feet high in the lower Keys, and toward Key West more like shrub thickets than trees. Where undisturbed, the ground supports coastal prairies. Hammocks on the lower Keys are fringed by well-developed mangrove swamps (see Chapter 17).

Although, as mentioned, many of the plants species populations in the Florida Keys are derived from sister populations in the West Indies, some have evolved on site. They could not have come in from elsewhere because they have no means of dispersal; they are endemic to south Florida's rockland communities. List 7–2 offers a sampling of the tree, shrub, vine, and cactus species found in Keys hammocks. Each hammock is different, all are diverse, and many plants are not mentioned here. Key Largo alone has more tree species than most *states* in the United States.[7]

Epiphytes grow less profusely than in the larger mainland hammocks but many find places where there is enough light and water to support them—some in the canopy, some below it, and some only in hammocks near the ocean. Several cactuses perch in the trees; so do several ferns: strap fern, resurrection fern, shoestring fern, and golden polypody. Mahogany mistletoe grows on West Indian mahogany trees.[8]

ANIMALS IN THE ROCKLANDS

Some of the animals in south Florida's pine rocklands are of the same species as in north and central Florida pine grasslands, notably the east-

Florida native: Key tree cactus (*Pilosocereus polygonus*). This cactus can attain a height of 20 feet, hence its name.

Florida native: Eastern king-snake (*Lampropeltis getula*, "yellow" form). Two color variants of the eastern kingsnake were depicted earlier, on page 81. This kingsnake, found near Miami, is adapted to blend in with a limestone background.

LIST 7–3
Animals of south Florida rocklands: a sampling

Invertebrates
Bartram's hairstreak
Big Pine Key conehead
Chigger mite
Florida atala
Florida leafwing
Keys short-winged conehead
Scarab beetle
Tree snail
Walking stick

Amphibians
Oak toad

Reptiles
Dusky pygmy rattlesnake
Eastern box turtle
Eastern diamondback
    rattlesnake
Eastern indigo snake
Green anole
Ground skink
Peninsula ribbon snake
Southeastern five-lined skink
Southern black racer

Birds
American kestrel
Brown-headed nuthatch
Eastern towhee
Gray kingbird
Great crested flycatcher
Northern bobwhite
Pine warbler
Red-bellied woodpecker
Red-shouldered hawk
Red-tailed hawk
Summer tanager
Wild turkey

(continued opposite)

ern gray squirrel, the red-bellied woodpecker, and the northern bobwhite. Similarly, some of the animals in the rockland hammocks are the same as in the temperate hammocks of north Florida: opossums, raccoons, and others. But some of north Florida's animal species populations are not distributed all the way to the tip of Florida. One reason is that the south end of the long, somewhat narrow Florida peninsula is less physically diverse than the north end and Panhandle; another, that the climate of south Florida is too tropical for some species. The south Florida groups occur mostly as isolated populations, cut off long ago from the main populations farther north. Some, such as the green treefrog, rough green snake, red-bellied woodpecker, cotton mouse, raccoon, and white-tailed deer, are virtually identical genetically to their northern relatives, but some south Florida populations have become endemic species or subspecies. Kingsnakes on the Miami Rock Ridge are markedly different from of north Florida populations—they are yellow like the limerock in their environment.

South Florida also has some animals whose populations don't extend to the north end of the state. Most of them are birds, bats, and flying insects such as butterflies, dragonflies, and moths, whose forebears probably flew or blew in from the West Indies. Others are small animals such as snails that were able to survive transit on floating debris. Because there was never a land connection between the West Indies and Florida, larger West Indian land animals are not present.

Altogether, Florida's rocklands contain several dozen species of animals that are today so rare that they are classed as threatened or endangered. Lists 7–3 and 7–4 present samplings from the mainland rocklands and the Keys hammocks and the next paragraphs describe a few of them.

Among invertebrates, Florida's tree snails bear witness that the Everglades keys began as true islands even though they are now parked on land. The snails arrived on floating logs from the West Indies, and whatever snails happened to lodge on a given island, only their descendants continue propagating there. Thousands of generations after they arrived, each island's snail population has evolved its own distinctive markings. Figure 7–7 shows a few of the many patterns they exhibit.[9]

Insects occupy many microhabitats among the hammocks' diverse plants. The Schaus' swallowtail butterfly lays its eggs on torchwood or wild lime trees, where its larvae find the food they need to metamorphose into butterflies. Other insects thrive, together with algae, in tiny pockets of water in wells at the bottoms of air plants. The three-inch bark mantis, camouflaged like a lichen, lies in wait for ants and beetles on tree trunks. Butterflies and moths drink the nectar of tiny, bright bromeliad blossoms and help to pollinate them. Tiny fig wasps drill holes in figs, enabling the fig seeds to be fertilized.

The relationships between birds and plants in the Keys hammocks resemble those already described for the temperate hardwood hammocks of central Florida. More than 70 percent of the woody plants in the Keys produce colorful, tasty fruits and thereby take advantage of the dispersal mechanisms offered by birds and other animals that eat them. The plants produce fruits in all seasons, guaranteeing a sure food supply for the birds. The birds spread and fertilize the seeds, ensuring reproduction of the plants.

FIGURE 7–7

Florida Tree Snails (*Liguus fasciatus*)

As if painted by children using bright watercolors, tree snails are decorated with pink, yellow, russet, green, and orange stripes and checks. Sixty distinct color patterns exist, some found on only one tree island or cluster of islands.

The snails' life cycles are tightly tied to the specialized habitats they occupy. Each snail starts life as an opalescent pink egg the size of a fat peanut, laid by the parent in the fall and remaining dormant over the winter dry season. The spring rains stimulate the buried eggs to hatch as "buttons," tiny but fast-growing snails. By the end of the rainy season, each snail has two or three twists to its shell and is ready to mate.

The snails mate on the trees, remaining intertwined for up to 48 hours. Both are female as well as male, so both become pregnant. Then they descend the trees, dig into the litter on the ground, and lay their eggs. The adults reascend the tree, glue themselves to it, and go dormant inside sealed shells until the first spring showers dissolve their seals. They feed by scraping up algae and lichens that grow on tree bark, with mouth parts that are specialized for the purpose. They will live for four to five years, attaining a length of two to three inches at maturity.

Sources: Anon. 1998 (Key Largo); Sawicki 1997; Toops 1998, 57.

List 7–3 (continued)

Mammals
Bobcat
Cotton mouse
Cottontail
Florida Key deer
Florida panther
Gray fox
Hispid cotton rat
Raccoon
Short-tailed shrew
Virginia opossum

Source: Stout and Marion 1993.

LIST 7–4
Animals of Keys hammocks (examples)

Invertebrates
Keys Cesonia spider
Keys green June beetle
Keys ochlerotatus
Keys scaly cricket
Keys short-winged conehead

Amphibians
Cuban treefrog*
Green treefrog
Greenhouse frog*

Reptiles
Bark anole*
Brown anole*
Corn snake
Eastern coral snake
Florida Keys mole skink
Green anole
Knight anole*
Reef gecko*
Rim rock crowned snake

Birds
Antillean nighthawk
Bachman's warbler
Carolina wren
Cuban yellow warbler
Gray kingbird*
Mangrove cuckoo

(continued on next page)

These mutually beneficial relationships are important in maintaining the enormous diversity of the Keys hardwood hammocks.[10]

Some special mammals also occupy Keys hammocks; two are shown in Figure 7–8. The Key Largo woodrat is endemic to Key Largo. It is a "sweet animal," says one who knows the species. "[Its] fur is soft and dense, [its] ears and eyes exceptionally large and round. Fastidious bathroom habits characterize these animals, who deposit their droppings in one place in tidy piles."[11] For reasons unknown, it hoards shiny objects, many of which, today, it finds as people's discards: belt buckles, keys, foil, even golf balls.

Note: *Some of these animals may not occur naturally in the tropical hammocks but may have been introduced by man.

Sources: Adapted from Snyder, Herndon, and Robertson 1990, 264–267; Ashton 1994 (Invertebrates), Ashton 1992 (Amphibians and Reptiles), Ashton 1996 (Birds), and Ashton 1992 (Mammals).

Key Largo cotton mouse (*Peromyscus gossypinus allapaticola*). This mouse is a tree dweller and is active at night.

Key Largo woodrat (*Neotoma floridana smalli*). The woodrat constructs a large, multi-chambered tunnel underground and usually conceals the entrance with a pile of sticks and debris. It eats the same foods as the cotton mouse, but is active by day.

FIGURE 7–8

Two Rodents Endemic to the Keys

The same habitat supports a nocturnal relative, the Key Largo cotton mouse, which occurs in several of the upper Keys. The mouse is a tree dweller; it even builds its nests in trees as squirrels do. By dwelling in different niches and being active at different clock times, the rat and mouse share the available resources without competing directly for them.[12]

The Florida Key deer is a subspecies of the white-tailed deer and is endemic to the Keys. Although descended from the same ancestral stock, it does not cross-breed with the common white-tailed deer, so it is considered a distinct species. It is a much smaller animal than the mainland deer, only about the size of a German shepherd dog. A doe deer weighs about 50 pounds, a buck about 100 pounds. Their habitat needs are well met by the Keys, where they have no natural enemies except people in vehicles.

SOLUTION HOLES AND SINKHOLES

Among the most naturally-protected environments in Florida are solution holes and sinks, which occur in karst areas all over the state. Background Box 7–2 describes the karst activity that produces these and other formations.

BACKGROUND   7–2
Karst

Wherever limestone is exposed to water, especially to acidic water, dissolution occurs. Water dissolves the stone, creating spaces into which other materials move—primarily water (below the water table), and air, sand, clay, or soil (above it). Effects on surface topography are many.

Bluffs and Limestone Outcrops. Sometimes dissolution and mechanical processes create vertical cliffs, bluffs, riverbanks, and smaller vertical rock walls in limestone. Especially along rivers and in flowing-water marshes such as the Everglades, limestone outcrops lend picturesque features to the landscape and provide nooks in which animals find homes and crannies into which plants can twine their roots.

Depression Marshes and Ponds. One typical karst feature is a depression on the ground where the land has begun to subside. This may become a marsh for a while or may deepen enough to form a pond.

Water-Filled Caves and Tunnels. Subterranean spaces are other common karst features. Water works its way down fissures and cracks in limestone, then spreads horizontally along planes between layers (bedding planes). As the limestone dissolves away, a maze of complicated, water-filled spaces forms. The underground, water-filled spaces in Florida's limestones are the subject of Chapter 14.

Terrestrial Caves and Sinks. Air-filled caves are also common in

(continued on next page)

The sinking down of land is **subsidence** (sub-SIGH-dense, or SUB-si-dense). It commonly occurs as subsurface limestone dissolves.

Sudden, major subsidence of an area, as when an underground cave's ceiling falls in, is **collapse**.

Acidic water begins to dissolve the limestone underlayer:

An underground cave forms and the land above it begins to subside:

Surface sediments fall in. The ground water is now visible from the surface:

If more sediment falls in, the pile of rubble at the bottom may grow large enough to plug the hole. It will then become a pond, lake, or wetland.

FIGURE 7–9

How a Sink Forms

A **sink**, or **sinkhole**, is a circular depression on the land surface caused by dissolution of underlying limestone.)

A **wet sink**, or **lime sink**, is a deep, conical hole in the ground that intersects the water table. The water at the bottom of a wet sink is aquifer water. The collapsing limestone fell into the ground water.

A **dry sink** is a shallower, conical hole in the ground that is dry all the way to the bottom. The collapsing limestone remained above the water table.

A **solution hole** is a hole in surface limestone that does not intersect the water table. The water in a solution hole is rain water.

DBM

A wet sink. If unpolluted, the aquifer water visible in sinks is extraordinarily blue and clear.

Interior of a terrestrial cave in the Florida Caverns State Park in Jackson County. For as long as a cave exists, rain water keeps soaking into its limestone from above and continuing to dissolve it. Here, the water is dripping from the roof of the cave, saturated with dissolved calcium carbonate. When this water evaporates, it leaves the calcium carbonate behind as a stony "icicle" (a stalactite).

Water drops that fall to the floor from stalactites, and then evaporate, form pinnacles (stalagmites) directly below the stalactites. When the two grow long enough, they finally meet, forming columns, which help to hold up the cave roof.

Columns form in air. The presence of columns is evidence that this cave has been above the water table for many years.

7–2 (continued)

karst areas and may lead to sink formation (see Figure 7–9). Having first formed below the water table as water-filled spaces, they fill with air when the ground-water level falls (usually because the sea level drops). If a cave's roof is strong enough to bear the weight of the overlying earth, it may remain an air-filled cave for centuries. If the ceiling of the cave falls in, then overlying sediment slides down into the cave, forming a sink.

Dry and Wet Sinks, Wetlands, Ponds, Lakes. When a cave ceiling collapses, together with the surface sediments that lie on top of it, the result may be a deep, conical hole in the ground that is dry all the way to the bottom—a dry sink. If the hole intersects the water table, it becomes a water-filled sink with limestone walls, a "karst window" into the aquifer. Later, more earth may slide into the sink from its surroundings and plug it closed, making a basin that holds a wetland, pond, or lake.

Rocklands. South Florida's uplands consist largely of rocky limestone plains, ridges, and hummocks. The pinelands and hammocks described in this chapter occur on these features.

Springs. Large and beautiful freshwater springs are still another feature of karst areas, and present striking evidence of the presence of fast-flowing underground rivers. Where a ground-water stream can find no way to go but up, it bursts forth as a spring that gives rise to a clear, cold, beautiful, surface stream known as a spring run. Florida has more than 700 springs, a higher concentration per land area than any other place in the world. Springs are featured in Chapter 14.

Some springs spout their water forth offshore, on the seafloor, and some are found in the bottoms of rivers that have eroded their beds down to the underlying limestone. The Suwannee River is especially notable for the many springs in its bed.

Karst Lowlands. Over time, the subsidence of small land areas over limestone add up to subsidence of whole regions. Karst features signify that the land is sinking down, slowly but relentlessly, century after century. Subsidence accounts for the lowlands around Marianna, whose temperate hardwood hammocks are described in Chapter 6.

Solution holes and sinks have high, rocky sidewalls which, like chimneys, protect the air from disturbance by wind. The limestone is moist and cool, and the water at the bottom is cold, so the air in a sink tends to stay moist and cool, too—the Greeks would have called it a grotto.

Overhanging trees also impede evaporation, and seepage keeps the surrounding walls continuously wet. In such a cool, humid environment, beautiful assemblages of delicate plants can grow around the water of solution holes and sinks. Above all, mosses, liverworts, and ferns grow luxuriantly around sinkholes, especially where the walls are steepest. Some are threatened and endangered species—Venus'-hair fern, fragrant maidenhair, southern lip fern, hairy halberd fern, and others. They are specialists. They differ from epiphytes: they grow with their roots in the moist, alkaline environment provided by the limestone rather than in the acidic water found in niches on the trees.[13]

Ground-cover plants are less numerous. Only two grass species are com-

Ferns around a sink in Paynes Prairie, Alachua County.

mon, and where the shade is darkest, nothing grows. As for animals, sinks support many species of small creatures that love moisture such as salamanders and small invertebrates.

TERRESTRIAL CAVES

Caves hold communities radically different from those of sun-exposed ecosystems because they are dark. Photosynthesis does not take place there, so there are no green plants, yet healthy caves are full of living things. This chapter deals with terrestrial caves, Chapter 14 covers aquatic caves.

Florida's uppermost limestone layers contain hundreds of caves and tunnels. Near-surface limestone is especially abundant in Florida's three so-called karst areas, which center on Marianna, Woodville, and Ocala, but as Figure 7—10 shows, there are many areas all over north and central Florida, and some even in south Florida, where the earth is honeycombed with caves and tunnels. Those shown are all terrestrial caves—that is, those that are dry and have been mapped by explorers. Divers have begun to explore aquatic cave systems as well, but have not yet completed the task.

Caves vary in size. Some are tiny pockets in the rock, some are room-sized, and some are the size of concert halls, or even bigger. The formations within them are the work of hundreds of thousands, or millions, of years: a century adds one cubic inch to each. And cave-dwelling organisms have had thousands of years in which to adapt to the limestone environment.

Life in Caves. Healthy Florida caves hold well-developed, self-perpetuating ecosystems. Living plants are present only near the entrances of caves,

Entrance to Cal's Cave in Wakulla County. This partially dry cave quickly sumps into an underwater conduit known as the Pipeline or Chip's Hole Cave system. It holds three and a half miles of mapped underground passages.

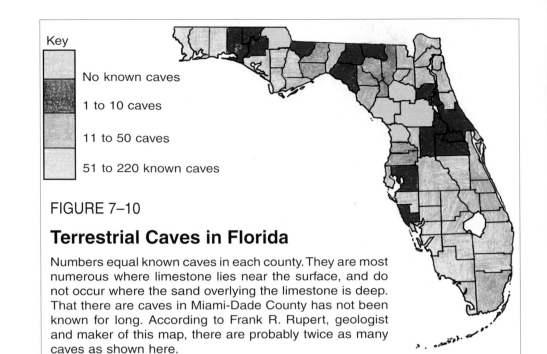

Key

No known caves

1 to 10 caves

11 to 50 caves

51 to 220 known caves

FIGURE 7–10

Terrestrial Caves in Florida

Numbers equal known caves in each county. They are most numerous where limestone lies near the surface, and do not occur where the sand overlying the limestone is deep. That there are caves in Miami-Dade County has not been known for long. According to Frank R. Rupert, geologist and maker of this map, there are probably twice as many caves as shown here.

Sources: Map from Rupert 1998 courtesy of F. Rupert; information on Miami-Dade caves from Cressler 1993.

of course, because they need light, but animals are numerous both in number and kind. Some, such as the woodrat, are casual visitors that visit caves for shelter but forage outside. Some, such as the slimy salamander, nest in caves if sites are available, but may nest elsewhere, too. Some, such as cave crickets, can live only in caves.

The foods of cave dwellers come from several sources. Some animals eat forage that they have carried in. Most subsist on energy-containing materials from outside the cave, primarily plant litter washed in by rain

FLAUSA

A coastal, partly submerged cave in Blowing Rocks Preserve, Martin County. A massive layer of limestone stands exposed at the coast as a reminder of the submarine origins of Florida.

and runoff and guano dropped by bats. And of course, some residents prey on other cave animals. Both visitors and residents also leave droppings and carcasses, and these deliver energy and nutrients to numerous decomposers—bacteria and fungi. The decomposers in turn become food for insects and other animals.

In many ways, the cave ecosystem works like any other: it has sources of energy and materials, organisms that use them, and interactions among those organisms. The system seems radically different only because the lack of light so powerfully influences all of the inhabitants.

Imagine entering a cave environment for a while, with all senses tuned and receptive. Bring a flashlight, of course: it will be totally dark just beyond the entrance. The example chosen is a small cave hidden in an upland glade in Jackson County.

The Twilight Zone. It is easy to miss the mouth of the cave, draped as it is with an overhanging fringe of grasses and ferns that are growing on its roof. Once inside the opening, it is just possible to stand upright below the cave's ceiling. The earthen floor slopes gently downward toward the back of the cave. Distinctive plants are growing on the walls and floor near the entrance, but they are microscopic algae and call no attention to themselves. What is noticeable, though, is the ambiance.

Within the cave, there is only peace and stillness. Outside, the weather may have been hot and dry, or cold and windy, or rainy—regardless, the air inside is a comfortable, mildly humid, near-70 degrees. Any organism that can do without the sunlight and greenery of the outside world gains a trade-off in the cave's stable environment.

In the dim light, in a nook in one wall, a small nest of twigs is visible where an eastern woodrat has made a home. A knowing search would turn up many other animals that divide their lives between the cave and the outside world, but except for a few spiderwebs strung across crannies in the wall and ceiling, they are not yet apparent.

Suddenly, hundreds of small, brown bats come into focus, hanging from the ceiling just a few feet inside the opening. They are asleep, and so still that they are easy to miss at first, but they are major contributors to the life of the cave. Beneath them, the cave floor is perilously slippery from an inches-thick coating of bat guano. Bats are common residents of caves, and are given attention in Background Box 7–3.

BACKGROUND   **7–3**

Florida's Bats

Sixteen bat species have occupied Florida since two million years ago or longer and are integral parts of the region's ecosystems. Twelve of them are still residents in the state (see List 7–5). Each species has its own home range, preferred foods, and habitat preferences, and they do not compete with each other. Three are cave-roosting species: the gray bat (limited to Jackson County), the southeastern bat, and the eastern pipistrelle.

(continued on next page)

LIST 7–5
Florida's native bats

CAVE-DWELLING BATS

Still present in Florida:
Big brown bat[a]
Eastern pipistrelle
Gray bat
Southeastern bat

Rare or perhaps no longer present in Florida:
Indiana bat
Keen's myotis
Little brown bat

TREE-DWELLING BATS[a]

Still present in Florida:
Brazilian free-tailed bat
Eastern red bat
Evening bat
Rafinesque's big-eared bat
Seminole bat
Wagner's mastiff bat[b]
Yellow bat

Rare and perhaps no longer present in Florida:
Hoary bat
Silver-haired bat

Notes:
[a]Today, some of these bats commonly live in buildings and under bridges.
[b]Wagner's mastiff bat occurs in the Fakahatchee Swamp.

Source: Gore 1994.

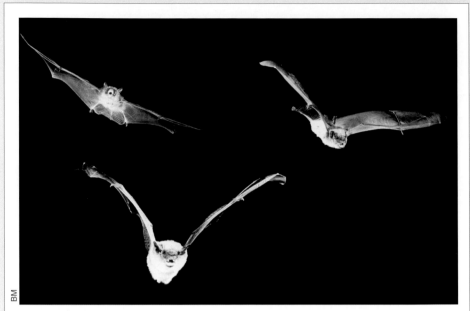

Florida native: Big brown bat (*Eptesicus fuscus*), flying at night.

Human fears of bats are based on misinformation. Bats are not blind, they do not attack humans, and they never get tangled in hair. They use the echoes of their own sharp cries to locate and avoid all obstacles in the dark. They rarely carry rabies and pose little threat when they do. (Vampire bats do bite, but none remain in Florida today, although fossils of their ancestors have been found in caves in central Florida, in Marion and Alachua counties). Bats are not ferocious animals, either. They are gentle, furry creatures that, like all mammals, suckle their young and tend them attentively. They are not rodents and do not breed explosively; they average only two offspring per mature female per year. Even as adults they are tiny. The gray bat weighs 1/3 ounce at maturity; the southeastern bat weighs even less, about 1/5 of an ounce. One of the largest of Florida's bats, the big brown bat, weighs one whole ounce when mature, well fed, and pregnant.

In some ecosystems, bats eat fruits, seeds, pollen, and other plant parts as birds do, so they are important pollinators and seed distributors.* Florida's bats are night-time insect eaters, and as such, they are powerful clearers of the air. Each one eats its own body weight in insects every night—500 insects an hour. The bat population of a single large cave can consume several tons of insects nightly. They are as effective as insecticide sprays but are nontoxic and self-renewing. Thus a swarm of bats against the moon at night is a welcome sight. Many mosquitoes bent upon sucking human blood have never raised a welt on anyone's skin but have passed through the cave food web as bat food instead. Out of appreciation for this amenity provided by bats, some resorts encourage bats to take up residence by constructing special houses for them.

Note: *Jamaican fruit bats were seen on Key West and Stock Island from 1872 to 1986, but are probably no longer present there.

Sources: Gore 1994, Wisenbaker 1991, Hartman 1992, and Beck 1996. Quantities bats eat from Tousignant 1995.

Other animals using the cave are less conspicuous, but numerous. A search might turn up any of three salamanders—the slimy, the two-lined, or the three-lined salamander. Their prey are tiny flies, spiders, snails, worms, and other invertebrates, which in turn eat organic matter dropped by the bats and others.

Without the influx of bat and rat droppings and debris to feed on, the numerous other cave dwellers would have little or no means of subsistence, but with the whole system operating naturally, a cave environment supports its inhabitants and visitors reliably from year to year. New materials are brought in at about the same rate that waste materials are disposed of. In a sense, a functioning cave ecosystem is self-cleaning, although "clean" as applied to caves is not the same as applied to homes.

The Dark Zone. Now step deeper into the cave. Ease around a corner to a space where no daylight penetrates. Be careful: the floor is so slippery with guano that it seems coated with oil. Touch a wall to steady yourself, and—are you ready? Turn out your light.

Blackness.

On keeping still for several minutes, you may sense that the pupils of your eyes have enlarged as far as they can. Or you may have closed your eyes; it makes no difference. Either way, you see nothing at all. You hear your own bated breathing and sense your heart's slightly accelerated beat, but otherwise detect only the sounds of dripping water and occasional rustling noises from the cave walls and ceiling.

The dark zone, which may consist of many caves and tunnels penetrating deep into the earth, is where the full-time cave-dwellers spend their lives. List 7–6 identifies a few of these very specialized animals. Animals to which sight is utterly useless must have other senses that are keen, instead—and they do. Many cave dwellers are designed like the familiar daddy longlegs: they have long, attenuated limbs with which they can probe and feel their way from place to place. Many have keen chemical sensors that function like our organs of taste and smell, but much more sensitive and finely tuned. Some have motion-detecting organs to pick up on disturbances of the air, or of the water where they swim, so that they can escape predators and find prey. Most are colorless, having lost the genes for skin pigments at some time in the past. In a 100-percent dark environment, no advantage comes from having any color pigment. In the light of a flashlight, cave animals show up defenselessly white and vulnerable, but in the total darkness of a cave, they go unseen. Many are not only blind,

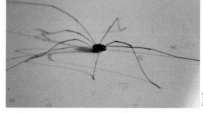

Florida native: Daddy longlegs (a *Leiobunum* species). This arthropod, which can live inside or outside of caves, serves as a janitor, cleaning up debris. It is not a spider and does not sting.

Cave millipede (*Cambala annulata*).

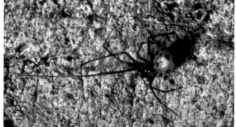

Cave spider (*Baiami teganaroides*).

Florida cave animals

LIST 7–6
Obligatory cave dwellers

Cave earwig
Cave millipedes
Cave mites
Cave spiders
Cave springtails

Source: *Guide* 1990, 57.

Florida native: Joewood (*Jacquinia keyensis*). This extremely rare plant grows, in Florida, only in coastal hammocks in Lee, Miami-Dade, and Monroe counties.

but eyeless, for the same reason. Their ancestors that lost the genes for eyes could invest resources in more useful traits and so produced more offspring.

The dark zones of Florida's caves are inhabited by many isolated, endemic populations of cave-adapted animals. Some are aquatic animals, notably crayfish, salamanders, shrimp, and snails; these are described further in Chapter 14. Each endemic population is thought to have begun evolving from ancestors that first crept into Florida's cave system a million years ago or more, when the seas last moved off the land. Future study of these animals should bring to light more interesting details about them and perhaps will reveal more species.

***

This is the last chapter describing interior Florida's uplands. The next chapters explore an altogether new realm: interior Florida's wetlands.

PART THREE

INTERIOR WETLANDS

Wetlands are areas whose soils are relatively wet, either permanently or at intervals for significant periods. They are saturated or inundated by surface water or ground water frequently enough, or for long enough, to support a predominance of plants that are adapted for life in water-saturated soils that are low in oxygen.

Florida's **interior wetlands** include bogs, marshes, and swamps with standing or flowing water.

String-lilies (*Crinum americanum*) in Big Cypress National Preserve, Collier County.

CHAPTER EIGHT

Wetlands

Having explored Florida's upland pine grasslands, scrub, forest, rockland, and cave communities, prepare now to explore the wetlands. These are realms of equally great richness and diversity. Compared with uplands, though, wetlands are less familiar, even scary, to most people. Words such as "quagmires," used to describe them in times past, evoked a fear of sinking out of sight in mushy earth. "Swamps" suggested hazards such as sickness, and predatory animals. People live and work on high land, for the most part, and tend to think of wetlands as hostile environments.

In reality, though, wetlands are worth a long look. They are justly prized as a vital resource, as habitat for diverse living things, and as places of great beauty. And today, parts of many Florida wetlands are accessible by way of boardwalks and boats, allowing visitors to explore them easily and safely, without even getting their feet wet.

But what, exactly, is a wetland? A later section, "Delineating Wetlands," explains more fully, but simply stated, wetlands are intermediate between upland and aquatic ecosystems in both wetness and position on the land, as shown in Figure 8–1. Wetlands do not stand continually under deep water, but their soil is either permanently or frequently wet (saturated or inundated with water) for long enough periods to "drown" upland plants.

Plants respire just as animals do, and for this they require oxygen. True, they generate oxygen when conducting photosynthesis, but at the same time, they use oxygen to grow and maintain themselves. However, unlike animals, plants cannot circulate oxygen throughout their tissues, and air, with its life-giving oxygen, does not penetrate far below the surface of water. If a plant is standing with its leaves in air and its roots in water, the leaves can respire, but the roots cannot. The only plants that can tolerate inundation for prolonged times are those that are adapted to it. Some go dormant during wet times. Some have roots with special adaptations for obtaining oxygen; the examples of cypress and tupelo are featured in Chapter 11. Tolerance for low-oxygen conditions distinguishes wetland plants from others.

Wetlands can thus be recognized by their plants, which are unique in being able to make the stretch between water and air and survive. Their

Boardwalk into a gum swamp. Boardwalks like these make exploring wetlands easy.

Bradwell Bay, a basin swamp in Wakulla County. Shown is the flooded footpath into the swamp. The ground is covered in water nearly the year around.

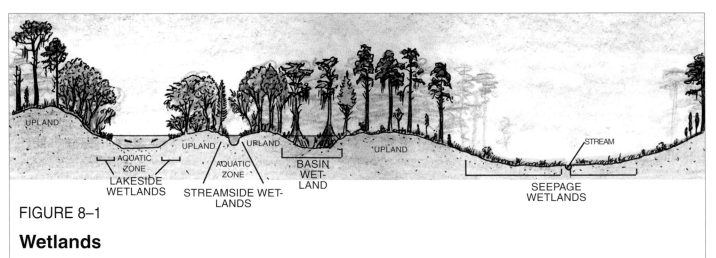

FIGURE 8–1

Wetlands

Some wetlands form borders between upland ecosystems and open water bodies such as lakes and streams. Other wetlands stand in circular or linear depressions without being adjacent to lakes or streams.

Wetlands are either *marshes*, *shrub bogs*, or *swamps*.

Marshes are dominated by grasses and forbs. Some are **seepage herb bogs**, kept wet by laterally flowing ground water. Some are **wet prairies**, **wet flatwoods**, and **bogs**, kept wet by rain and slow drainage of the soil. Some are **swales**, wide, marshy, flowing-water wetlands.

Shrub bogs are wetlands with low, usually multi-stemmed, woody plants smaller than trees, typically about 10 to 15 feet tall.

Swamps are **forested wetlands,** that is, wetlands with trees.

roots are under water or in saturated soil for much of the time, and their leaves or blades are in air.

Figure 8–2A shows Florida's wetlands as they were before the Europeans came. What a watery place it was! Swamps predominated in the north, marshes in the south. Much of south Florida was one giant swale, the Everglades, with swamps embedded in it. Figure 8–2B shows that Florida has lost most of its wetlands, but they still embrace all the lakes and streams that are in their natural state and they also still occupy large and small basins and grooves in the land. Over much of south Florida, water still flows in miles-wide sheets over marshy land in the Everglades.

Florida's wetlands are also diverse. The state has more than a dozen different major classes of marshes, bogs, and swamps. They are listed in Appendix D and treated in the next three chapters.

DIVERSITY OF FLORIDA WETLANDS

Many factors govern what kind of wetland will occupy a given area. One is, of course, the topography. Some wetlands are in basins, some are on inclined planes; and as a result, some have standing, and some have flowing, water. Another factor is the hydroperiod: the duration of saturation or flooding, the depth of flooding when it occurs, and the degree of drying between times. Wetland topography varies. Water stands deeper and for longer times on low ground than on high ground. As a result, higher-elevation areas of a wetland have shorter hydroperiods, and therefore different plant assemblages, than do lower-elevation areas.

Fire frequency is important, too. Marshes burn frequently, swamps almost never. It is because these factors vary, and the variants occur in many combinations, that Florida's wetlands are of so many different kinds. Consider the different conditions that produce a seepage wetland, a basin swamp or marsh, and a river floodplain swamp. The next three chapters provide more detail on each.

Seepage Wetlands. A seepage wetland's water oozes onto the surface

A wetland's **hydroperiod** is the length of time during which it is saturated or inundated each year. The hydroperiod is a key factor governing what will grow in a wetland.

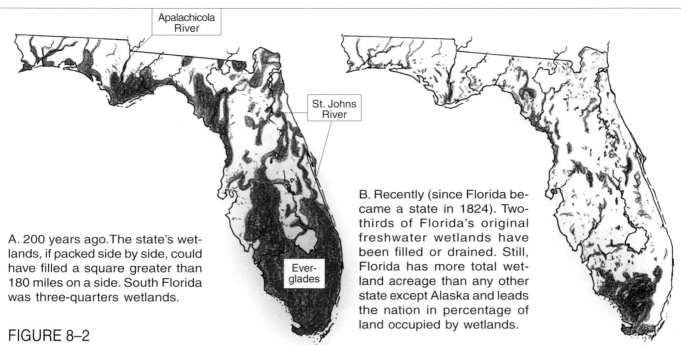

A. 200 years ago.The state's wetlands, if packed side by side, could have filled a square greater than 180 miles on a side. South Florida was three-quarters wetlands.

Apalachicola River

St. Johns River

Everglades

B. Recently (since Florida became a state in 1824). Two-thirds of Florida's original freshwater wetlands have been filled or drained. Still, Florida has more total wetland acreage than any other state except Alaska and leads the nation in percentage of land occupied by wetlands.

FIGURE 8–2

Florida's Wetlands, Yesterday and Today

Source: Adapted from figures in Fernald and Patton 1984, 63–64.

from a ground-water aquifer beneath nearby higher land and moves continually across the ground surface in sheets and small runnels, keeping the soil moist virtually all the time. Seepage wetlands never flood, because their water is always running off at the same rate as it is flowing in. The nutrients in seepage water vary somewhat, depending on the source, but most seepage wetlands on sandy soil are acidic and nutrient poor.

> Plants growing in water can be *emergent*, *submersed*, or *floating*.
>
> **Emergent** plants are rooted at depths of about five feet or less and their topmost tissues spread on or extend above the water's surface. Emergent plants identify a wetland.
>
> **Submersed** plants are rooted at depths of five to 15, or even 20, feet.Their topmost tissues do not reach the surface.
>
> **Floating** plants have roots that do not reach the bottom.

A seepage bog in Blackwater River State Forest, Santa Rosa County. Water-tolerant herbs, including pitcherplants, thrive in seepage water flowing downhill from the pine forest.

Florida native: Lizard's tail (*Saururus cernuus*). This native plant is common in freshwater marshes and swamps throughout mainland Florida.

Herbs, shrubs, and trees whose roots are able to live with low oxygen levels can all grow in seepage wetlands. What determines which will predominate is fire frequency. Frequent fires support seepage herb bogs. If fires become less frequent, shrubs take over; and with very infrequent fires, wetland trees such as bays or Atlantic white cedar come to dominate a seepage site.

Plant litter that falls into the water of a seepage wetland decomposes slowly and tends to accumulate, forming peat. Peat in a seepage wetland can become many feet thick. It holds moisture even in dry times, so it helps to keep plant roots hydrated in all but the most severe, years-long droughts.

Basin Wetlands. In a basin wetland, the land's contours trap rain. As a result, the water level does not remain constant: it rises in the wet season and falls in the dry season. Also, the water stands still or moves very slowly and is low in oxygen not far below the surface. Rain delivers few nutrients, so the water is also nutrient-poor.

Like seepage wetlands, basin wetlands tend to accumulate peat, provided that wet and dry times alternate around a constant average. In prolonged wet conditions, peat accumulates. In extended dry periods it oxidizes and disappears; and if ignited—a rare event—it burns completely. During short dry periods, peat tends to remain wet because it wicks water up from below, so it provides for tree roots a constantly moist environment.

Spatterdock-sawgrass marshes in the Everglades. Water flows slowly and constantly during the rainy season and completely dries up during nearly every dry season.

Some of the same plants grow in basin wetlands as on seepage sites. Marsh herbs can thrive in shallow basins, especially if subjected to frequent renewal by fire. Where water levels are very high or low for long times, though, plants must be able to withstand long periods of flooding, and perhaps, of drought. Herbs may not be able to survive these conditions; marsh plants thrive in rising and falling water, but only within certain limits. Certain shrubs such as titi (pronounced TYE-tye) can withstand somewhat greater water-depth extremes, and several trees are famously successful in deep basins, notably sweetbay, cypress, and swamp tupelo. The photo at the start of this chapter shows part of Bradwell Bay, a basin wetland in Wakulla County.

Floodplains. Like a basin wetland, a river floodplain may be largely dry during one season of the year and flooded during another, but each flood delivers *moving* water that sweeps loose organic matter along with

Florida native: Eastern pond-hawk (*Erythemis simplici-collis*). Dragonflies and damselflies abound around water bodies because their larvae develop there. The female of this species flies low over the water and flicks her abdomen downwards to wash off her fertilized eggs.

A floodplain swamp known as Sutton's Lake in the Apalachicola River floodplain. Deep in water during rainy seasons, this terrain is free of standing water in the dry season.

it. As a result, peat accumulates only in low places such as sloughs, backswamps, and oxbow lakes; litter left on higher ground decomposes quickly when the floodwaters recede. Fires virtually never spread into river floodplains for lack of fuel.

Floodplain soils vary. In the Panhandle, they are rich in nutrients, because floods repeatedly drop clay-rich sediments eroded from upstream headwaters and banks. Around streams that arise from limestone springs, soils may be rich in calcium, whereas around blackwater streams, soils may be rich in decaying organic matter.

A river floodplain has varied contours: mounds, ridges, basins, grooves, plains, and many other features, most of which are wetlands because they are inundated annually. Many different communities of trees, shrubs, and ground-cover plants grow on these landscape features.

With so many different factors defining the different types of wetlands, it is clear why each wetland has a somewhat different suite of plants. Still, all wetland plants have certain characteristics in common, and many are useful in distinguishing wetlands from other communities, as shown next.

DELINEATING WETLANDS

Because wetlands are so important to the health of the landscape, they are increasingly protected by law. Some activities are prohibited in and near wetlands. Landowners must seek permits before they fill, drain, or develop wetlands, and not all permissions are granted. Also, some activities that pollute wetlands or alter their water regimes are prohibited in and near wetlands. To enforce the law, however, people must know what wetlands are. If they go unrecognized, they are likely to be lost or damaged. Defining just what is, and what is not, a wetland is an important part of land-use decision making.

Hydroperiod. It might seem logical to base the definition of wetlands on their water regimes. After all, water is either continuously or periodically present in all wetlands. However, during a prolonged dry spell, a wetland may be dry for several years and during prolonged rains it may be altogether under water for a time. These extremes are not only natural; they are necessary for the perpetuation of some wetland species. Cypress seedlings can germinate only when water is absent; then they grow best if water returns in quantities enough to kill competing species. Marsh plants can survive only if water levels fall enough to permit fires to burn away encroaching shrubs. Such fluctuations, including *irregular, long-term* fluctuations, are necessary if wetlands are to be maintained. Yet land surveyors cannot wait for years to measure an area's hydroperiod. They often have to distinguish wetland from non-wetland areas within a single season. How can they recognize a wetland, no matter what the current state of wetness?

Wetland Soil. A clue can be gained by examining the soil. Surveyors look for hydric soils—soils that are wet, and therefore low in oxygen, for long times. Such soils hold accumulated organic matter that has only partially decomposed, because decomposition requires oxygen. Hydric soils

RLH

Florida native: Mexican primrosewillow (*Ludwigia octovalvis*). This plant grows in marshes throughout the state.

are found wherever the water table is near the surface or where (as in peat or clay) capillary action lifts water to the surface. The soils' chemistry also reflects the absence of oxygen. In practice this means that wetland soils are often peaty or mucky in texture, their colors are the black of slowly decomposing vegetation, and they often smell of sulfur.

Water's Signatures. Signs of water's earlier presence may also indicate a wetland. Even if an area is dry at the moment, mats of dried algae may remain on the ground; aquatic mosses may be stranded on the trunks of trees; drift lines may appear where water-borne debris has been left behind in shrubs or trees. These and other signs are summarized in List 8—1 and three indicators appear in Figure 8—3.

Wetland Plants. The clincher that identifies a wetland, however, is that certain of the plants that grow there not only tolerate saturated, oxygenless soil, but are naturally restricted to wetlands because they are un-

Florida native: Common buttonbush (*Cephalanthus occidentalis*). This wetland shrub is common in swamps, cypress ponds, and around the margins of lakes, rivers, and ponds nearly throughout Florida.

Cypress swamp during a dry season. Even though no water is in sight, the soil, hydrologic indicators, and trees tell the surveyor that this is a wetland.

Crayfish chimneys. Each chimney consists of wet soil that a crayfish has excavated from the wet earth to make its burrow. If numerous, such chimneys are evidence of prolonged saturation of the soil.

Enlarged oxygen-Intake organs. These "pimples" near the base of the tree, technically *hypertrophied lenticels*, indicate that the trunk is under water for part of the year. This is a wax myrtle growing on the margin of a lake.

Adventitious roots. Tangled, braided roots that sprout from the sides of a tree or shrub above the water line indicate a wetland. These are ogeechee and swamp tupelo trees in a swamp adjacent to the Apalachicola River.

FIGURE 8–3

Three Wetland Indicators

able to compete with upland plants. Because they can grow *only* in wetlands, these plants identify wetlands unequivocally. To identify a wetland, then, surveyors inspect not only the soil, the hydroperiod, and other signs, but also look for obligate wetland plants. A manual developed by the state identifies several hundred such plants.[1] List 8–2 and Figure 8–4 present a few of them.

In summary, to delineate a wetland, surveyors look for a combination of (1) wetland soils, (2) wetland plants, and (3) periods of saturation or inundation long enough or frequent enough to drown non-wetland plants. Using these criteria, experts can usually determine wetland boundaries accurately to within ten feet or less.

WETLAND SERVICES

Wetlands are knit into larger, interacting systems through which water, materials, and energy flow unceasingly, and they serve all of Florida's other ecosystems in many ways. They are powerful water managers, regulating water's purity, routes, and rates of flow. They also help control local climates, nourish nearby plant and animal communities, and serve as indispensable parts of the natural landscape.

Water Purification. Because under natural conditions, wetlands always lie between upland and aquatic areas, they can remove water-borne pollutants flowing out of uplands before they reach lakes, streams, and

American white waterlily (*Nymphaea odorata*).

Broadleaf cattail (*Typha latifolia*).

Chapman's butterwort (*Pinguicula planifolia*).

Pond pine (*Pinus serotina*).

FIGURE 8–4

Native Wetland Plants (Examples)

Virginia iris *(Iris virginica).*

The waterlily, cattail, butterwort, and iris can grow only in, and are diagnostic for, wetlands; they are obligate wetland species. Pond pine grows preferentially in wetlands but can grow in dryer soil as well; it is a facultative wetland species.

Three classes of plants grow in wetland areas:

Plant species that ordinarily grow *only* in wetlands are **obligate wetland species**. They are diagnostic for wetland areas.

Some plant species, called **facultative wetland species**, normally grow in wetlands but can also grow in uplands if competition is excluded. They are useful as indicators but do not conclusively identify wetlands. Examples are wax myrtle and gallberry.

Some plants grow freely anywhere. Known as **facultative species**, they are essentially "weeds," and do not help to define an area.

Some plants are *unable* to grow in saturated soil for any appreciable length of time. These are **upland species**, diagnostic for areas that are *not* wetlands.

marine waters. And they do: wetland plants extract these materials from water and deposit them in plant tissues. Then when the plants die or shed their leaves, the pollutants settle into the wetland's bottom sediments.

Until polluted recently by sugar farming and other agriculture, the water running through the Everglades sawgrass marshes carried few nutrients, which the plants were able to take up. By the time the flowing water emerged into the bays at the end of the land, it was crystal clear. Adventurer Hugh Willoughby and his men, crossing the Everglades in 1897, dipped up water from the marsh, drank it copiously, and reported that they "did not know a sick hour."[2]

A coral reef off south Florida in John Pennekamp Coral Reef State Park, Monroe County. If the Everglades does not purify the water flowing off the land, reefs like this cannot survive.

When a land area holds water so that the water sinks into the ground, this is **retention**.

When a land area slows floodwaters down before releasing them laterally, this is **detention** (or **attenuation**).

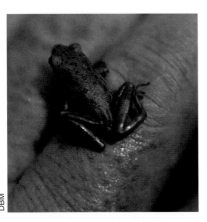

Florida native: Upland chorus frog (*Pseudacris triseriata feriarum*). This minute amphibian can use tiny wetlands, even rainwater pools, to court and breed.

Even today, the Everglades sawgrass marshes take up most of the nutrients in the water passing through them and release water that is purer—although, today, it is still unnaturally high in nitrogen and phosphorus from upstram sugar and other farming. Teeming hordes of sea creatures in Florida Bay and the coral reefs off the Keys depend on the purity of this water.

We humans have learned to use the water-filtering ability of wetlands to our own advantage. If not overburdened, wetlands can trap human-produced runoff and pollutants and thus help protect ground water, streams, lakes, and bays.

Water Capture. In contrast to the flowing-water swales of the Everglades through which rain water runs off into the ocean, some basin wetlands retain rain water. From basins underlain by clay or other water-impermeable materials, water simply evaporates, but in basins underlain by sand and limestone, newly arriving rain percolates down to become ground water. The Green Swamp system, a giant upland area in central Florida, contains 870 square miles, almost half of them retentive wetlands that recharge the Floridan aquifer system, thereby providing vital ground water to south and southwest Florida.

Slowing of Water's Flows. The vast floodplain wetlands in river bottomlands let water flow through them but they slow it down. By doing so, they reduce peak flood levels and prolong the time flood waters take to move into rivers and run off downstream. A river embraced by a broad band of floodplain wetlands rises and falls more moderately than when it is confined within man-made levees or when the adjacent land is cleared or paved. When a natural river rises and is free to wash into and out of its floodplain's swamps, the swamps retain some of the water, soak it up like a sponge, and then gradually release it. According to researchers, if wetlands occupy ten percent of a landscape, flooding is reduced by as much as 60 percent. If wetlands occupy twenty percent of the land, flood reduction is a dramatic 90 percent. Detention performs a vital service in protecting ecosystems from flood extremes.[3]

Detention also eases the stress of drought. Long after the rains have ceased to fall, water detained in wetlands continues to be available to slake the thirst of animals and plants.

The benefits of water management performed by wetlands are felt all the way to coastal estuaries whose waters oscillate between fresh and saline extremes during wet and dry seasons. Oysters and other estuarine organisms are adapted to, and in fact require, this alternation in saltiness. However, it must take place at a pace, and within limits, that they can tolerate. They cannot handle too sharp an alternation of dry, salty periods with sudden, silty floods. As nature writer Gerald Grow puts it, "Without the flood detention performed by wetlands, Florida's great oystering areas, such as Apalachicola Bay, would be devastated . . . . Wetlands sponge up floods and release them at a pace that . . . organisms [can] live with."[4]

Climate Control. Florida's highly prized muck and peat accumulate in still or slowly moving water. These organic substances tie up carbon, a global-warming gas. Muck and peat wetlands hold masses of carbon out of the atmosphere in such volumes that they help to regulate the climate. The Okefenokee Swamp and the Everglades are among Florida's big muck and peat wetlands that contribute to this effect.

The moisture in wetlands also contributes to regional humidity. Evaporation of water from leaf surfaces in the Everglades is thought to contribute to seasonal rains all over south Florida, and the draining of the Everglades in the early 1900s is thought partly to blame for the droughts of recent times.

Productivity. Wetlands are also factories that generate many products we use, not least among them, food. As wetland plants drop litter or die and decompose, they yield detritus by the ton that nourishes huge populations of animals. Wherever water flows rapidly through the floodplain swamps along high-discharge rivers such as the Chattahoochee, Apalachicola, and Ochlockonee, it sweeps the detritus downstream to nourish ecosystems where the water flows next. Most of the great salt-marsh and estuarine fish nurseries along the coast are fed primarily by detritus from river floodplain wetlands. Botanist Andre Clewell warns ocean fishermen to guard interior wetlands: "Do you fish? Your catch spent its early life in the protective and food-rich confines of a wetland—mangroves, tidal reeds, pickerelweed, maidencane . . . Take away the river swamps, and . . . there won't be any more [seafood]."[5]

Thus, wetlands generate wildlife. It has been said of freshwater wetlands that they "take in air, sunlight, water, and nutrients and turn out bass, anhingas, ducks, herons, eagles, raccoons, alligators, turtles, snakes, frogs, salamanders, mink, . . ."[6]

Links in the Landscape Mosaic. Figure 8–5 is an enlargement of Figure 8–2B with the major wetlands labeled. The map illustrates that Florida's wetlands all share the landscape with other interconnected eco-

Florida native: Purple gallinule (*Porphyrula martinica*). Floating marsh plants can be so thick that birds can walk on them.

Detritus (de-TRY-tus) is a vital material in wetland ecosystems. It consists of loose dead and shed plant and animal matter, and it serves as a source of energy for organisms that are low in the food chain.

Wetland productivity. The base of the food chain in a wetland is masses of undecomposed organic matter such as this algal muck, formed from sunlight and water beneath a bed of zigzag bladderwort (*Utricularia subulata*).

Florida native: Tiger salamander (*Ambystoma tigrinum*). This is one of many rare salamanders that depends on a type of wetland, temporary ponds, to breed.

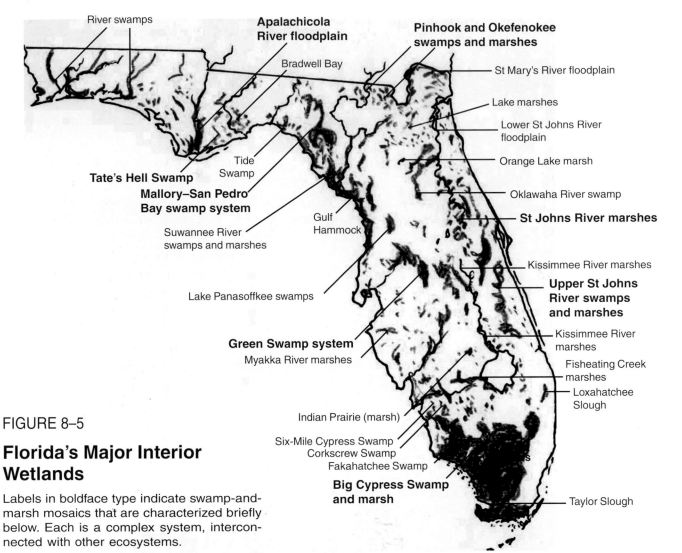

FIGURE 8–5

Florida's Major Interior Wetlands

Labels in boldface type indicate swamp-and-marsh mosaics that are characterized briefly below. Each is a complex system, interconnected with other ecosystems.

Map labels:
River swamps
Apalachicola River floodplain
Bradwell Bay
Pinhook and Okefenokee swamps and marshes
St Mary's River floodplain
Lake marshes
Lower St Johns River floodplain
Orange Lake marsh
Tide Swamp
Tate's Hell Swamp
Mallory–San Pedro Bay swamp system
Oklawaha River swamp
Gulf Hammock
St Johns River marshes
Suwannee River swamps and marshes
Kissimmee River marshes
Upper St Johns River swamps and marshes
Lake Panasoffkee swamps
Kissimmee River marshes
Fisheating Creek marshes
Green Swamp system
Myakka River marshes
Loxahatchee Slough
Indian Prairie (marsh)
Six-Mile Cypress Swamp
Corkscrew Swamp
Fakahatchee Swamp
Big Cypress Swamp and marsh
Taylor Slough

The **Apalachicola River floodplain** is described in Chapter 11. Its many pools, swamps, sloughs, oxbow lakes, and old levees harbor hundreds of species of plants and animals. Wildlife can travel the floodplain all the way from the Appalachian mountains to the Gulf of Mexico. Nearby in the Panhandle, **Tate's Hell**, a 100,000-acre tract that once was a natural wetland in Franklin County, is potentially (if restored) another such mosaic, unique in being located on the Gulf coast. The Tate's Hell swamps purify much of the water that runs into Apalachee Bay, supporting one of the northern hemisphere's great seafood nurseries. On the east side of Florida, the **St. Johns River marshes** are another great wetland system extending for several hundred miles from south to north.

The **Okefenokee-Pinhook system** of swamps and marshes in northeast Florida and southeast Georgia occupies some 600 square miles. It is one of the nation's largest, most complex, and most diverse wetland mosaics, holding an intricate network of ponds, marshes, bogs, and swamps as well as piney uplands. The Suwannee and St. Mary's rivers meander out of these swamps, one flowing to the Gulf of Mexico, the other to the Atlantic Ocean, and these rivers link the interior to the sea along hundreds of miles of wetland corridors. Wildlife can travel along these corridors all the way from the headwaters to the coast.

Central Florida's **Green Swamp system** (shared by Sumter, Lake, Polk, and Pasco counties) is another rich mosaic of habitat types. It embraces the headwaters of four major Florida rivers: the Withlacoochee, the Oklawaha, the Peace, and the Kissimmee rivers, which flow west, north, south, and east, respectively. Still another great swamp, also a mosaic, is the **Mallory–San Pedro Bay** swamp system in Dixie, Lafayette, and Taylor counties, which occupies more than 500 square miles of the Florida Big Bend. Three rivers flow from this system: the Econfina, Fenholloway, and Steinhatchee.

Finally, Florida's vast swale, the **Everglades**, and the adjoining **Big Cypress Swamp** also hold patchworks of habitats including pine grasslands, hardwood hammocks, cypress swamps, and marshes.

Sources: Adapted with the permission of the University Press of Florida from Ewel 1990, 282, fig. 9–1 and Kushlan 1990, 327, fig. 10–2.

systems, especially aquatic systems such as lakes and streams. The figure caption describes a few of the landscape mosaics within which wetlands lie.

Originally, all of Florida's wetlands were embedded in broad landscape mosaics. Today, wherever large areas have been preserved, such mosaics are still functioning, to an extent, as they did in the past. Tomorrow, perhaps, these landscapes will be more extensively restored. The Everglades, especially, is in great need of restoration, a vital task. As Marjory Stoneman Douglas said, "The Everglades is a test. If we pass it, we get to keep the planet."[7]

***

With these general remarks as introduction, the next three chapters turn to describing specific types of wetlands. The next chapter begins with a seepage bog.

RLH

Florida native: Winged loose-strife (*Lythrum alatum* var. *lanceolatum*). This loosestrife grows in marshes and around lakes throughout the Panhandle and peninsula and is very popular with butterflies

CHAPTER NINE

SEEPAGE WETLANDS

This chapter describes the four types of seepage wetland that are common in Florida: seepage herb bogs, shrub bogs, bay swamps, and Atlantic white-cedar swamps. These wetlands occur on seepage slopes—inclined areas where ground water seeps out onto the surface of the ground, keeping it cool and wet. The simplest type of seepage wetland is shown at left: water seeping from a hillside feeds directly into a stream and supports an herb bog.

Figure 9–1 depicts a gentle flatwoods slope that supports two seepage wetlands and shows the water flow that maintains them. The herb bog begins at the point where ground water first seeps out of the soil. The shrub bog takes over farther down the slope, where the soil is soggy with water-soaked peat, and water moving through the shrub bog feeds into a small, blackwater stream.

Longer sequences of seepage wetlands can also occur. Beyond the herb and shrub bogs, there may be a bay swamp, an Atlantic white-cedar swamp, or both—and beyond them, a deep-water swamp that finally gives way to a basin or channel holding open water. The transitions from one of these communities to the next may take place over a decline in elevation of only ten to twelve inches, and a horizontal distance of only 50 to 200 feet. This sequence may occur in various arrangements on slopes—in parallel bands, in adjacent arcs, or concentrically, as shown in Figure 9–2.

Seepage wetlands are diverse in appearance, but they all have several things in common. They are kept constantly wet by shallow, slowly flowing water, and it is mainly ground water; rain is a secondary source. The water tends to be nutrient poor and strongly acidic. Deeper seepage wetlands have floors of peat maintained by the constant moisture.

Seepage slopes appear in specific places in Florida. They are strung out all across the Panhandle, both on the northern highlands and on the sandy lowlands. They occur in northeast Florida along the bases of high slopes, and in central Florida along the base of the central ridge. Seepage is uncommon in south Florida where the topography is flat and wetland water comes from rain.

A **bog** is any site where the ground is wet and spongy. Wetlands known as bogs may have predominantly herbs or shrubs.

A **seepage bog** is a bog across which water is flowing from a **surficial aquifer** (one that lies just beneath the ground).

Not all Florida bogs are flowing-water seepage bogs. Herb and shrub bogs can grow in still-water wetlands, too, as described in Chapter 10.

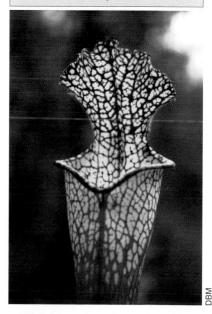

DBM

Florida native: Whitetop Pitcherplant (*Sarracenia leucophylla*). Because nitrogen and phosphorus are hard to come by in their habitats, pitcherplants are adapted to trap insects and digest them to obtain these nutrients.

OPPOSITE: A seepage herb bog on Eglin Air Force Base in Walton County. A throng of whitetop pitcherplants is growing in peat that has accumulated at the edge of a stream fed by seepage along its sides.

Here, water is flowing laterally within the ground because the underlying aquaclude blocks its downward percolation.

At this point along the slope, the flowing water emerges onto the surface of the ground and flows across it.

At the foot of the slope, the water collects into a stream.

FIGURE 9–1

A Flatwoods Seepage Slope

The underlying layer of clay or organic hardpan (aquaclude) keeps water from sinking down, creating a surficial aquifer. Water from this aquifer moves laterally through the sand, emerges onto the surface, and flows across the slope until it reaches a creek. The moving water supports one or more seepage wetlands.

Source: Means and Moler 1979.

A mosaic of seepage wetlands. In the foreground is a grass- and sedge-dominated pitcherplant bog. Farther down the slope is a shrub bog and, beyond that, a cypress swamp, with pond pine in the transition zone.

The reasons why the vegetation changes across a seepage slope—why herbs grow in one part, then shrubs, then bays and/or cedars—will become clear in the sections to come. A stroll from an upland pine community down to a stream can illustrate the array of seepage wetlands just described.

A Seepage Herb Bog

Start out from an upland pine community—say, a longleaf pine–wiregrass sandhill or flatwoods in the Panhandle. Ease down off the higher ground and you find yourself on a declining plane where the ground is moist and spongy underfoot. You walk, first, into a grassy meadow where, perhaps, a few slash pines may be mixed with the longleaf. As you walk farther downslope, the ground becomes more squishy. The wiregrass is patchy, and in its place are plants that tolerate wet conditions well: toothachegrass, other coarse grasses, and sedges. Club-moss weaves its soft tubular arms across the moist earth. Carnivorous plants are abundant: sundews dot the ground among the grasses and pitcherplants are everywhere.

This is an herb bog, and this particular bog is maintained throughout the year by seepage from a small aquifer that is perched on an imperme-

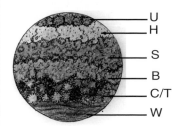

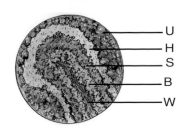

LEFT: A wetland series forms bands across a slope. Upland pine is highest. Then, where seepage water first wets the ground, there is an herb bog, beyond it a shrub bog, and then a bay swamp. Then comes a deep-water swamp of cypress and tupelo, and beyond that, open water.

RIGHT: A similar wetland series is clustered around the head of a stream. All of the wetlands might appear in the same sequence, but in the case shown here, the bay swamp borders the stream; there is no cypress-tupelo swamp.

FIGURE 9–2

Seepage Wetland Mosaics

Seepage wetlands may be arranged on the land as arcs around the head of a stream, or as bands on a slope. Not all communities appear in every array. Occasionally, Atlantic white cedar occupies sites usually occupied by bays.

A **seepage slope** is an incline where water from beneath adjacent higher land seeps out onto the surface of the ground and then runs across it. (A clay or other water-impermeable underlayer keeps the water from sinking into the soil.)

able layer of earth (a hardpan) not far below the surface. Other herb-dominated wetlands, which go dry when the rains stop coming, are classed as marshes and are described in Chapter 10. Some species of plants and animals can survive only in seepage bogs, as described here.

Herb Bog Plants. The flowering plants in an herb bog are extraordinarily diverse. Because the moisture changes gradually across such a broad expanse of land, it allows space for many different species to spread out and flourish, each in its own zone. Toward the top of the moisture gradient, 40 species of plants share the terrain. Only a few steps farther down the gradual slope, although there are still 40 species, 20 of them are different. A few steps more, and again they change. One square meter of herb bog contains the highest species richness of any such area yet inventoried. Many of the plants are also unusual (see List 9–1). Among them are rare wild orchids, rare lilies, and carnivorous plants unlike any others, anywhere on earth. Botanists travel to Florida from all over the country to see them in their season. Why have they assembled here? Acid soil and fire provide the needed support.[1]

The soil in which herb bog plants flourish is hostile to other plants. It is soggy, acidic, and infertile. Water cannot drain down because of the aquaclude below, and moves only slowly across the gradual slope. The acid comes from dead plants, which decompose slowly in the water. Acid makes it hard for living plants to obtain and retain nutrients. In acid water, plants actively lose nitrogen and phosphorus by leaching; and their uptake of mineral nutrients is inhibited. Even if fertilized, plants would grow slowly in such acidic, water-soaked sand. Aluminum, one of the minerals in the soil, reaches such high concentrations in bog soils that it, too, inhibits plant growth. Nearly everything is stunted. Even the few 100-year-old trees are scrawny and spindly, but carnivorous plants thrive in nutrient-poor, acidic soils, thanks to the insects they capture and digest.

Among Florida's carnivorous plants are four species of sundews, of which

LIST 9–1
Herb bog plants (examples)

Trees (sparse, If present)
Longleaf pine
Pond pine
Slash pine
Titi

Shrubs (if present)
Dahoon
Fetterbush
Large gallberry
Laurel greenbrier
Maleberry
Myrtle dahoon

Herbs
Beaksedge
Bladderwort (several species)
Bluestem (several species)
Butterwort (several species)
Chain fern
Cinnamon fern
Cutthroatgrass
Orchids (several species)
Pitcherplant (several species)
Sundew (several species)
Wiregrass
Yelloweyed grass (several species)

Source: Adapted from *Guide* 1990, 33.

Pink sundew (*Drosera capillaris*). These tiny rosettes spread their sparkling leaves flat on the damp earth. Gluey hairs on the leaves attract, capture, and digest insects to nourish the plant.

FIGURE 9–3

Tracy's sundew (*Drosera tracei*). LEFT: A field of Tracy's sundews. Each unfurls slender leaves up to eight inches high, and each leaf glistens with shiny, sticky hairs that attract and capture insects. CENTER: Close-up of one leaf with three trapped insects. The leaves digest their prey with powerful enzymatic juices and absorb the end products. RIGHT: Tracy's sundew flower.

Florida Natives: Two Sundews

two are shown in Figure 9–3. Florida's bogs also host six butterworts, a dozen bladderworts, and six species of pitcherplants. Butterworts have broad, tapering leaves that lie flat on the ground. The leaves are adorned with dewlike drops of a gummy substance to which tiny spiders and insects stick as if on fly paper. The plants digest them through the leaf surfaces. Bladderworts grow underground or float in the water of bogs with their stems and leaves below the surface. Their leaves are modified to serve as tiny, basketlike traps. These bladders trap swimming prey too tiny for the human eye to see—paramecia, amoebas, and other microorganisms, and these serve as food for the plants.[2]

Pitcherplants are among the oldest flowering plants on the planet. They date back to the time when dinosaurs were still roaming the earth, and in the 80-some million years since then they have evolved elaborate mechanisms for capturing insects and other small prey. The strategies of two of them are shown in Figure 9–4.

Other creatures take advantage of the attractiveness of the pitcher. A spider may spread its web across the top of a pitcher and catch bugs being drawn into the trap. A treefrog may sit inside the lip of a pitcher to escape the midday heat while catching bugs. A moth may lay its eggs inside a pitcher and then spin a layer of silk to close the top. When the larvae emerge, they eat their way out through the pitcher-plant walls. The belly of a carnivorous plant seems a hazardous place for tiny animals to live, but it is a protected place, shady, and with constant humidity and temperature. For those that can avoid its pitfalls, it provides a fine habitat. Mosquitoes, for example, breed in some pitcherplants.[3]

Besides the hapless victims of the plant and those that take advantage of their being attracted there, other creatures scavenge the parts that the plant doesn't digest. A single, giant carnivorous maggot is often found floating on the surface of the water in a pitcher; it devours whatever bugs fall in and also cannibalizes its own kind (which is why there is only one). Other scavengers include bacteria, protozoans, nematode worms, copepod

Yellow pitcherplant (*Sarracenia flava*). This tall pitcher has an umbrella that keeps rain water out. If it filled with water, it would fall over. The bait and trap are shown at right.

Yellow pitcherplant bait and trap. Bright colors and sugars glisten on the underside of the umbrella. Insects are attracted, fall into the deep, slippery trap, and then can't crawl back out.

Gulf purple pitcherplant (*Sarracenia rosea*). Having no umbrella, this pitcher fills up with water and lies on the ground. Bait and trap are at right.

Purple pitcher bait and trap. Bright colors attract insects; then when they crawl down the ramp, downward-pointing hairs prevent their escape. They die by drowning and enzymes secreted by the plant's walls digest their soft tissues.

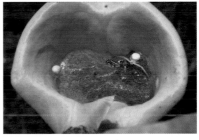

Two opportunists. Two crab spiders tend a web in the mouth of a pitcherplant, where they will easily catch insects.

Another opportunist. A pine woods treefrog conceals itself within the lip of a pitcher to catch flies lured there by the plant.

The prey. When the plant is held open, the undigested remnants of bugs, flies, grasshoppers, moths, and tiny snails are revealed. At the end of the season, when the plant itself dies down to the ground, these remains will decompose and fertilize the next generation of plants, the next season.

FIGURE 9–4

Florida Natives: Two Pitcherplants

Florida native: Panhandle lily (*Lilium iridollae*). For most of the year, the Panhandle lily is just a cluster of slender leaves growing from an underground bulb. During July and August it grows a stem four to six feet tall, with one or more showy flowers at the top.

crustaceans, mites, mosquitoes, a midge, six kinds of mosquito larvae, a gnat, a wasp, and others. Of three flies known to feed on the plant's captives, each avoids competition with the others by specializing. One feeds on new captives floating on the surface; one feeds on debris suspended within the liquid, and a third eats the corpses at the bottom. Still other small creatures, including a root borer and the moth larva mentioned earlier, eat parts of the plant. In these age-old relationships, each participant has found a niche, adapted to it, and secured a place for itself. Each kind of pitcherplant, and presumably each other carnivorous plant, attracts somewhat different insects, dividing the resource rather than competing.[4]

The plants' life cycles are synchronized with the seasons. In spring when the new pitchers first rise from the earth, they are empty, but by late fall they are full of insects. When they die, the undigested remains return to the earth. If a fire comes through, the remaining nutrients are freed as ash, and help to fertilize the plants in their next growing season. Thus these plants contend successfully with the challenges of an environment that is far too hostile for others.

Most of the herbs described here can grow without seepage water, on soil that remains boggy for most of the year due to poor drainage. Carnivorous plants appear again in the next chapter. But some plants can live *only* in seepage bogs. The Panhandle lily is an example. It grows only in Florida's four westernmost counties and in two adjacent counties in Alabama. It requires wet seepage habitats near permanent flowing water and thrives best at a clean boundary between an open sunny herb bog and a brushy, shady shrub bog. Today, the lily is found at only 20 to 30 sites in the western Panhandle, and each population is small.

Fire in Herb Bogs. Surface fires burn frequently into herb bogs from higher ground, and frequently sweep all the way across the bogs. Fine, grassy litter is readily flammable even over wet soil. The fires burn out only at the edge of the shrub bog, where the shade is dense, there is no grassy fuel, and the litter on the ground is water-soaked.

As in upland pine communities, fires stimulate the reproduction of a bog's ground-cover plants while killing woody vegetation or at least keeping it from growing tall. Herb bogs are absolutely dependent on fires, which must bite back encroaching shrubs at intervals of not more than five years.

The seepage bog's place in the landscape just downslope from a piney upland guarantees the conditions that maintain it. The slope supplies constant moisture in which even the tenderest, most water-dependent herbs can thrive. The pine community invites lightning to ignite fires, which then can burn into the bog, eliminate competition, clear away litter, and release nutrients.

Cutthroat Seeps. This description has featured a Panhandle seepage bog, but should acknowledge that other seepage bogs exist in other parts of Florida. On the central Florida ridge, there are seepage bogs in which cutthroat grass is the major ground-cover plant. Only a very few such bogs remain today; a photo of one appears in Figure 9–5.

Herb Bog Animals. Some common bog animals are shown in Figure

FIGURE 9–5

A Seepage Herb Bog in Central Florida

Dominated by cutthroatgrass (*Panicum abscissum*), a moisture- and fire-dependent endemic herb, this wetland in Polk County is known as a "cutthroat seep." The trees are pond pines.

9–6, and the Pine Barrens treefrog is singled out here for a closer look. This tiny creature, only 1-1/2 inches long, is a dazzling little jewel of a frog. Bright lime green on its back, it has a chocolate brown mask and sides fringed with bright yellow. The inside surfaces of its legs are a glowing orange with yellow spots and a lemon-yellow band surrounds each eye.

The Pine Barrens treefrog is a specialist: it requires seepage, and in Florida it lives mainly in the same bogs where the Panhandle lily grows. Outside of Florida, only two other locations in North America support Pine Barrens treefrogs: one in the Carolinas, and the other in the Pine Barrens of New Jersey. The three widely separated populations of frogs are all of the same species, indicating that they must have originated from a single population at some time in the past. There was a time, during the past two million years, when seepage bogs formed a continuous line all the way from New Jersey to Florida and even farther west. The bogs came and went over thousands of years of climate change, and the treefrog populations expanded and dwindled along with them.

Further fragmentation of the treefrogs into isolated populations occurred after the Europeans came. They built roads, dug ditches, timbered, plowed firebreaks, and thereby altered ground-water flows, wiping out thousands of acres of herb bogs and leaving only patchy remnants. Today, only a few seepage herb bogs remain in the Panhandle, and populations of Pine Barrens treefrogs are rare. You can find them only by listening, on rainy nights during the breeding season, for the males' nasal "quank, quank" calls. A typical population includes about ten males; some have only three or four; or even only one or two.

DBM

Pine Barrens treefrog (*Hyla andersonii*)

BM

Mimic glass lizard (*Ophisaurus mimicus*)

DBM

Squirrel treefrog (*Hyla squirella*)

BM

Eastern ribbon snake (*Thamnophis sauritus*)

FIGURE 9–6

Florida Natives: Herb Bog Animals

When not breeding, the adult treefrogs live in nearby shrubs or trees, either in the neighboring pine community upslope, or in the shrub bog downslope. They shelter there, defend their territories, and catch insects and spiders for food. Then in the next rainy season, they return to the herb bogs to do their calling. Pairs meet and mate in tiny seepage puddles, and the fertilized eggs develop into tadpoles there. A small puddle may be tea-cup-sized—seemingly insignificant, but without such puddles, the tadpoles cannot develop into frogs—hence the requirement for constant seepage. The water has to be cold, too, and seepage water meets this requirement. It stays at 68 to 72 degrees, like the surficial aquifer from which it emerges, even as it flows through puddles that are in full sun. The water also must be crystal clear and highly acidic, with the particular acids that characterize seepage sites. Seepage water never dries up and never stands still. It stays cold and pure, providing just the right environment for the frogs' rendezvous and tadpole nurseries.

Like the Panhandle lily, the Pine Barrens treefrog can reproduce its kind only in a bog that has a clean boundary; shrubs should not be crossing the line into the bog. The sharp dividing line is important for several reasons. If nearby shrubs began to invade upslope into the herb bog, they would suck up water, and the needed puddles would soon dry up. Frequent fires are necessary to maintain the clean boundary the lily and frog require. If you see shrubs encroaching on an herb bog, you can be sure that fire has been excluded for too long. The sun-loving bog plants will soon find themselves in full shade and will die out.

A Shrub Bog

Downslope from the green expanse of the herb bog, an evergreen shrub bog abruptly takes over, where woody shrubs form a tangled wall of greenery that rises to 25 feet in height. This is one of Florida's most sharply defined natural borders: it exactly marks the line at which the last herb-bog fire burned out.

Shrub Bog Plants. A shrub bog appears to be impenetrable, but only the sun-exposed edge is thick with greenery; shade keeps the interior open. Look for an animal trail into the shrubs: it won't be hard to find. (Wear wettable shoes that can drain, not rubber boots that might fill with water and grow heavy.) Slip into the deep shade among the shrubs and explore this miniature swamp. Each shrub is 10 to 25 feet in height and looks like a small, ancient, gnarled tree. The twisty branches are covered in prickly greenbrier vines. Birds sing and flit among them, attracted by titi fruits, holly berries, greenbrier berries, and insects.

In contrast to the bright green herb bog, this is a shady world. In places, an inch or more of cold water covers the ground, and in it, soft leaf litter is floating. The floor consists of a spongy dark soil that is rich in peat from earlier litter. Instead of tender green herbs at ground level, now the ground is covered with deep twig and leaf litter.

Plant diversity is low in a shrub bog. Two species of titi often predominate, among perhaps a dozen other species: only a few types of plants can grow in this limiting environment. Bright splashes of color indicate occasional clusters of sweet pinxter azalea. A few softwood trees stand well above the canopy: pond pine, slash pine, and perhaps Atlantic white cedar. Some young hardwood trees may be present in a dwarfed form: red maple, loblolly bay, swamp bay, and sweetbay. The water and shade prohibit the growth of ground-cover plants, except for sedges or sphagnum moss in sunny spots. List 9–2 names some common shrub bog plants.

The shrubs in the bog all exhibit similar adaptations to the environment in which they grow. They all have small, glossy, leathery leaves that are coated with wax. They resemble the leaves on desert plants, whose thick skins resist water loss—but this is a wetland. Why are the leaves drought-resistant here? Paradoxically, the answer probably has to do with dry conditions just as in the desert. Although wetland shrubs must usually cope with water-saturated soil, drought resistance is necessary, too, if they are to survive many decades.

The small leaf size may also be a function of nutrient-poor soil, which supports only slow growth at best; of acid water, which inhibits nutrient uptake; and of the aluminum compounds that accumulate in acid soil and are toxic to most plants. The leathery texture may be an adaptation to protect against water loss, because wetland shrubs are poorly equipped, in other ways, for dry times. They have shallow roots, an adaptation that helps them cope with waterlogged soil—the oxygen that roots need to grow is available only near the surface. But shallow root systems make plants especially vulnerable to droughts when water withdraws far beneath the surface.

The bog shrubs have leaves that are coated with wax, and they become a highly flammable litter on the ground. This adaptation, together with the bog's closed canopy, enables hot surface fires to burn across the shrub bog floor and leap to the canopy during severe droughts, which occur every decade or so. Just as frequent mild surface fires in herb bogs keep shrubs from invading, infrequent hot crown fires keep taller trees from overwhelming shrub bogs. Fires that burn through shrub bogs completely kill most hardwood saplings, leaving only mature trees that managed long ago to

LIST 9–2
Shrub bog plants (examples)

Trees
Atlantic white cedar (rare)
Pond pine
Slash pine

Shrubs
Black titi
Blueberry (several species)
Coastal sweetpepperbush
Coastalplain staggerbush
Dahoon
Fetterbush
Gallberry
Large gallberry
Myrtle dahoon
Odorless bayberry
Swamp titi
Wax myrtle

Vines
Greenbrier
Wild grape

Other
Ferns
Grasses
Reeds
Sedges
Sphagnum moss

Sources: Adapted from Wolfe and coauthors 1988, 152; *Guide* 1990, 33.

A **surface fire**, or **ground fire**, is one that burns through litter and undergrowth. A **crown fire** is one that burns through the tops of trees. When there is a closed canopy above a hot surface fire, an accompanying crown fire is almost inevitable.

Two species of evergreen shrubs known as **titi** (TYE-tye) grow in Florida's shrub bogs.

One of the two is **black titi** (*Cliftonia monophylla*). Black titi holds the drier territory and is the first shrub encountered when one walks downslope into a titi swamp.

The other is known simply as **titi**, or **swamp titi** (*Cyrilla racemiflora*). Swamp titi tolerates water better and holds the wetter territory.

Titis are so common in north Florida that their community is often named for them, a **titi swamp**. (Titis do not, however, grow south of Gainesville because the climate is too warm for them; fetterbush, staggerbush, and others take their place.)

DBM

A shrub bog. In this bog, the tall shrubs are black titi. Pure stands like this typically grow between herb bogs and bay swamps next to swampy creeks.

The regrowth of shrubs or trees from the bases of their stems is called **coppicing** (COPP-iss-ing).

establish themselves. But shrubs survive such fires: their above-ground stems and branches may be killed, but their thick bases are left undamaged. The shrubs soon resprout vigorously, sending up many stems from each root crown and replacing the entire thicket within a few years, much as before. The multiple stems of the shrubs is evidence that they have resprouted repeatedly after earlier fires. No wonder they look old: despite the small size of their stems, their root crowns may be several centuries old.[5]

Given periodic fire and perpetual seepage water, bog shrubs can hold their territory indefinitely. In the deep shade beneath their closed canopy and in the acidic soil produced by their litter, other species win no space.

Shrub Bog Animals. A tangled shrub thicket provides habitat with deep shade and abundant cover, radically different from the open, sunny herb bog upslope. The pine barrens treefrog and five-lined skink forage for insects among the branches. The Florida bog frog finds comfortable wetness here complete with plenty of small prey animals and many good hiding places under leaves and litter. Other frogs breed in pools beneath the shrubs. The green anole and coal skink frequent the bog margin.

Attracted by the abundance of small prey, the common garter snake and eastern ribbon snake hunt here, together with the southern black racer and eastern indigo snake. Birds find abundant berries, insects, and nesting places. Marsh rabbits and two species of shrews also use the bog. Southern mink, raccoon, and bobcat come and go.

Deer browse and bed down in the shrubby thickets; up to 65 percent of a deer's diet may be titi leaves and bark. Even the Florida black bear is at home in shrub bogs, where it can stay well hidden while dining on berries and fruits.

Florida visitor: Chestnut-sided warbler (*Dendroica pensylvanica*). This small songbird pauses in brushy habitats in Florida during its fall and spring migrations.

A BAY SWAMP

Now follow the gentle slope farther downhill, into a stand of trees that are increasingly taller than the shrubs—sometimes up to 100 feet tall. Now the ground underfoot is a springy floor of water-soaked peat. This is another seepage community, a bay swamp, and like the herb and shrub bogs upslope, it is a seepage wetland. The peat in which the trees stand may be mixed with sand, and may be anywhere from a few inches to well over six feet deep, too deep to measure with a standard soil borer. In areas where the peat is shallow, heavy rains may erode it, leaving channels or pools where the sand beneath it is visible.

Bay Swamp Trees. Usually, most of the trees in this zone of a seepage slope are bay trees—sweetbay, loblolly bay, and swamp bay (but not redbay, which grows in somewhat dryer habitats). These three bays are evergreen hardwoods adapted to life in hydric habitats that seldom burn. The leaves of these evergreens are larger, but otherwise resemble those of the evergreen shrubs described earlier: they are leathery, coated with wax, and highly flammable. Once or twice in a century, when this swamp finally burns, it will burn completely.

Bay trees grow more slowly than shrubs, but given 10 to 25 years without fire, they finally grow stout enough to withstand shrub bog fires. Then they form a high, closed canopy, shading out all but a few shrubs, ferns, and mosses beneath them. List 9—3 itemizes some of the other plants that may grow among them.

Like a shrub bog, a bay swamp hosts fewer than 30 woody species of note. The few that are adapted to these severely restrictive growing conditions have come to dominate all others. The understory is patchy and consists mostly of mid-sized offspring of the canopy trees. A few titi and other shrubs remain but are dying out. The ground is brown with fallen needles and leaves; less than one percent is covered in green, mostly sphagnum moss.

Florida native: Florida bog frog (*Rana okaloosae*). This frog is a common resident of shrub bogs.

Bay swamps can occur in basins as well as on slopes kept wet by seepage. A whole basin may then be called a **bay** (as in Bradwell Bay, in Wakulla County).

Bay swamps may form arcs around the heads of small creeks, especially in flatwoods, where they are called **bayheads** or **baygalls**.

Bay swamps may also run linearly for some distance along slender streams. In those locations they are called **bay branches**.

A bay swamp. The trees are loblolly bay (*Gordonia lasianthus*). This is Florida's largest loblolly swamp, in Ocala National Forest, Marion County.

Florida visitor: Black-and-white warbler (*Mniotilta varia*). This warbler spends its winters in Florida, foraging among the trunks and larger branches of deciduous woods.

A long-undisturbed bay swamp is an impressive place. At the center of the swamp, where the trees are tallest, towering slash pines may reach more than 100 feet in height and measure three feet across at chest level. After rains, water stands several inches deep. Even in dry weather the air is humid, thanks to the great volume of water held within the peat. The air bears a fresh, earthy aroma of decaying wood. Silence enfolds the scene.

Bay Swamp Animals. In the stillness of a bay swamp, animals are inconspicuous, but small animals including salamanders and shrews are numerous. If you see a large animal, it will probably be an opossum, rabbit, raccoon, bobcat, or bear. High in the canopy, diverse birds feast on insects, and on bay fruits in their season.

Fire and Bay Swamps. Because they are fringed by shrub bogs around their edges, bay swamps are protected from all but the most catastrophic, wind-driven fires that occur during the severest droughts at intervals longer than 50 years. Recall that in the shrub bog during the last severe drought (say, two decades ago), a hot crown fire killed most hardwood saplings (including young bay trees) outright. At the low, wet end of the shrub bog, however, the fire burned out. Beyond that point, young hardwood trees continued to grow, especially bays.

When a bay swamp burns, the peat floor may well burn, too, leaving bare ground. Then the water will return, peat will again build up, and the swamp will regrow as before.

Large bay swamps may have cypress-tupelo swamps within them and may blend into river floodplain swamps. They may also have elevated sites within them, holding temperate hardwoods.

AN ATLANTIC WHITE-CEDAR SWAMP

Walking down the seepage slope from the shrub bog, you might be lucky enough to encounter not a bay swamp but a rare Atlantic white-cedar swamp. White-cedar stands are sparsely distributed and grow in seemingly dissimilar habitats, but they all have one key feature in common: their soil is very moist and remains so in all seasons.

Florida native: Atlantic white cedar (*Chamaecyparis thyoides*). This cedar swamp lies along Boiling Creek in Santa Rosa County. The floor is of peat, the air is cool, and the understory is sparse due to the deep shade.

Overstory
Cypress
Loblolly bay
Red maple
Slash pine
Swamp bay
Swamp tupelo
Sweetbay
Sweetgum

Understory
Black titi
Coastal sweetpepperbush
Eastern poison ivy (as a climbing vine)
Fetterbush
Highbush blueberry
Muscadine
Myrtle dahoon
Odorless bayberry
Rankin's jessamine
Red chokeberry
Sweet pinxter azalea
Switchcane (as an understory plant)
Titi
Wax myrtle

Ground Cover
Cinnamon fern
Laurel greenbrier
Netted chain fern
Panicgrass (several species)
Eastern poison ivy (as a ground-cover plant)
Sedges
Sphagnum moss
Switchcane (as a ground-cover plant)
Virginia chain fern
Wild grape

Source: Adapted from Wolfe and coauthors 1988, 153.

Atlantic white cedar is also called **juniper**, and a white-cedar swamp is the same as a **juniper swamp**.

Florida native: Largeleaf grass-of-Parnassus (*Parnassia grandifolia*). This rare flower depends on continual seepage and grows well in cedar swamps. It blooms in fall along shaded stream banks in Putnam, Marion, and Franklin counties.

Atlantic White Cedar. In a stand of white cedars, the trees stand tall and perfectly straight. Their bark spirals around the trunks like candy-cane stripes, and their shallow roots spread broadly across the ground's surface. The air is cool, thanks to the cold water within the peaty floor and the trees that help to hold the coolness in. Ground-cover plants are sparse, due to the deep shade.

When you walk in an Atlantic white-cedar swamp that is supported by seepage, you are walking on water, in a sense. The ground may not look wet, but poke a deep hole in the peaty earth with a rake handle and it will

Florida native: Florida black bear (*Ursus americanus*). Swamps and forests are suitable habitat for this big animal.

Florida native: Marsh rabbit (*Sylvilagus palustris*). This rabbit frequents not only marshes but also bogs and swamps.

fill with water all the way to the top. An Atlantic white-cedar stand is a true wetland, and the tree itself is a wetland indicator species.

White cedar is remarkably disease resistant and so can outlast other trees. Other trees grow old, suffer attacks by insect pests and fungi, and eventually collapse, but cedar is immune. By the time white cedars have reached maturity, only a few other trees remain standing among them.

Not uncommonly, mature white cedars reach heights of 100 feet, girths of three feet, and ages of 200 years. A few true giants remain in Florida that measure more than *15 feet* in diameter at breast height, not unlike the redwoods of the West. The virgin white-cedar swamps in early Florida must have been extraordinary.[6]

Old cedars can be blown down, but they die more often from lightning strikes. Lightning selects the tallest cedar in a stand, shoots down the trunk, and heats the interior water to steam, blasting the tree apart and hurling slabs of wood up to 20 feet long and a foot wide all around the tree. Lightning seems to be the main agent limiting the age and size of Florida's white cedars.[7]

Other plants growing among cedars may be of many different species. Some are mid-story and ground-cover plants, unusual in Florida but typically seen farther north in cooler climates. A cedar stand at Deep Creek, in Putnam County, has specimens of fairywand, largeleaf grass-of-parnassus, and swamp thistle, all typically found further north. Tuliptree, hazel alder, and Florida willow are also found here and otherwise grow only further north.

Animals in Cedar Swamps. Animals in white cedar stands include the same ones as in bay swamps: mud and dwarf salamanders, shrews, rabbit, raccoon, opossum, bobcat, and bear. The large white-cedar stand in the Ocala National Forest supports a healthy population of Florida black bears, and the bark on many of its trees is shredded by bear claws up to a height of six and a half feet.

***

This chapter began on the green expanse of a flatwoods seepage bog, and has ended in shady bay and white-cedar swamps with towering trees. In deeper swamps, another transition occurs—to ecosystems in several feet of water, such as cypress or tupelo or bottomland hardwood swamps, described in Chapter 11.

Florida native: Raccoon (*Procyon lotor*). Raccoons are plentiful in many Florida habitats and very much at home in bay and cedar swamps.

CHAPTER TEN

MARSHES

Although Florida has plenty of topographical relief, as the previous chapters have shown, it is rightly reputed to be largely flat. Its lowlands, particularly, consist of broad, level plains holding vast marshes which, in the past, were of mind-boggling scope. Florida writer W. R. Barada, writing about rains in the marshes in the distant past, describes them this way:

When the summer rains came, lake water levels rose to spread over surrounding lowlands, and meandering rivers and streams overflowed into floodplains. . . . Marsh vegetation, fish and waterfowl thrived; water table aquifers rose to bathe the roots of thirsty upland vegetation, grazing wildlife fed on the new greenery . . . and marine life flourished in brackish coastal marshes and estuaries.[1]

Those were the days when much of Florida, and nearly all of south Florida, held water several feet deep over hundreds of square miles throughout the rainy season and for several months beyond. Those days are gone, but even now, many marshes and marsh lakes remain wherever annual rainfall exceeds annual evapotranspiration and where the land drains poorly and so holds water.[2]

Marshes occur on several types of terrain. Still-water marshes occupy well-defined basins, including lake basins, where they form the borders of open-water lakes. Flowing-water marshes may border streams all along their length. Both still- and flowing-water marshes also occupy thousands of acres of Florida's flat lands, where their boundaries are irregular. On the central ridge, basin marshes are scattered all over Alachua, Lake, Putnam, Sumter, and Marion counties. On the lowlands, flatwoods marshes are dotted all over the flats along both coasts and across the Panhandle. Florida's lowlands also have two vast, flowing-water marsh systems. Numerous marshes interconnect with the St. Johns River; and marshes abound throughout south-central Florida.

FORMATION AND MAINTENANCE OF MARSHES

Lowland marshes may exist for a long time, maintained by the constant presence of water in aquifers near the surface or in adjoining streams or

Florida native: Bandana-of-the-Everglades (*Canna flaccida*). This gauzy flower blooms in Florida swamps and marshes in spring and summer.

A **marsh** is a wetland dominated by herbs that are rooted in saturated soil on which water stands for much of the year.

Evapotranspiration is the sum of **evaporation** (in which liquid water turns to water vapor) and **transpiration** (in which plants take up liquid water through their roots and release it to the air). It is a major route of water loss from wetlands.

OPPOSITE: A marsh bordering Lake Jackson, in Leon County. The flowers are native aquatic plants, pickerelweed (*Pontederia cordata*).

155

Florida native: String-lily (*Crinum americanum*). This graceful flower blooms over vast expanses of marshes and wet prairies in summer, as shown in the photo on page 123.

lakes. Highland marshes are often relatively short-lived; they may soon become either lakes or dry land.

A highland marsh may begin to develop when a sink becomes plugged or when the land subsides, forming a depression that fills with water and wetland herbs. The plants drop litter every year and the litter decays, becoming loose, fluffy material that floats for a while, then sinks to the bottom forming muck and peat. If an area of open water remains in the middle of the basin, it is called a marsh lake. (Chapter 12 describes the aquatic zones in Florida's lakes.) If the basin fills in entirely with herbs, it is called a marsh. Later, if the water table falls and remains low for several years (as in a prolonged, severe drought), it may progress to become a shrub bog or swamp unless fire razes it and water returns. Paynes Prairie is a well-known example of transitions like this: it was once a dry grassland; then because the sink that drained it became plugged, a lake; then as the lake level fell, a marsh.[3]

In all types of marshes, even the "flat" ones, water rises and falls with the seasons, but the highs and lows vary greatly from one marsh to another. In the rainy season, one marsh may hold only a few inches of water while another may hold five feet. In the dry season, one marsh may still hold standing water, while in another, the standing water will have disappeared, although the soil remains saturated at or not far below the ground level. In north Florida, which has both winter and summer rainy seasons, only very shallow marshes normally go dry every year. In south Florida, where winter rains are rare, even the deep Everglades marshes go completely dry nearly every April. The hydroperiods of Florida marshes—that is, the spans of time during which water actually stands on the surface—range from only a few weeks to twelve months of the year, but nearly all marsh soils remain saturated year-long except during prolonged droughts.

If a marsh goes completely dry for more than a month or two, it may be invaded by shrubs and trees. Then fire, as well as water, is needed to keep it from becoming a thicket or hardwood hammock. If a marsh burns con-

Florida native: Fragrant ladiestresses (*Spiranthes odorata*). This native orchid can grow submersed with only its flowers showing above the water.

A buttonbush marsh in St. Andrews State Park, Bay County. Common buttonbush shrubs are overtaking this marsh, which has not burned in recent years.

sistently every one to three years, tree and shrub seedlings are killed before they can root effectively and all of the plants will be tender young herbs. If the fire interval is longer, perhaps as long as ten years, the marsh will contain patches of shrubs and small, sprouting trees. An occasional cypress tree may even start growing during a dry time and then survive subsequent flooding and fires. With limited water fluctuation, large stands of cypress and tupelo may form within a marsh. But if a marsh remains constantly inundated with standing water, provided that the water rises and falls with the seasons, it can exclude trees for a long time.[4]

MARSH PLANTS

The plants that grow in marshes face challenges different from those on seepage slopes, already described. Seepage wetlands never flood and rarely dry out: their soil is virtually always saturated—and with *flowing* water. Marshes, in contrast, are deep in water for part of the year and nearly or altogether dry for part—and their water flows slowly if at all. Marsh plants have seasonally timed adaptations that enable them to deal with alternating high and low water.

Most marsh herbs are perennial. They have massive root systems, and although they die back as the water falls in the dry season, they easily regrow when the rains return and the water rises again. They also have ways to keep their heads above water. Sedges, rushes, and reeds are equipped with stiffening fibers that enable them to stand tall, but to bend, if necessary, without breaking. Other marsh plants, such as waterlilies, have such buoyant leaves that they float to and fro at the ends of their pliable stems.

The thick, shallow root systems of marsh herbs monopolize the ground surface and help to exclude woody plants. Even in dry times, tree seeds can seldom find a way to put down roots among them. Tree seedlings start up slowly if at all, and the next water's rise almost invariably kills them.

It is to the benefit of marsh plants to be flammable, and they are: they can even burn over standing water. Marshes typically burn, not at the peak of the dry season, but when the first summer thunderstorms blow in. Lightning ignites the marsh and fires can burn many acres before rains become heavy and frequent. Then the seeds of marsh plants, buried earlier in the soil, have a good chance to start growing wherever fire has cleared the way for them.

If the water rises too quickly, the plant seeds cannot germinate, but the plants have another way to recover: they can resprout from their roots. Conveniently, peaty soil holds enough moisture to protect the shallow roots of the marsh plants, so they can regrow fast, fertilized by ash, just as the water is rising.

As for the variety of plants in marshes, if they appear monotonous, look again. Most likely, the marsh you are looking at has dozens of species of plants, each species occupying a subtly defined site where its own best growing conditions prevail. Spikerush, maidencane, and Jamaica swamp sawgrass grow best if inundated for 70 to 90 percent of the year. Arrowhead thrives with 85 to 95 percent inundation; American white waterlily, with

Florida native: Southeastern sunflower (*Helianthus agrestis*). This sunflower grows in peninsula marshes and blooms in summer.

Many families of plants have grasslike stems and leaves. Some are wetland plants.

A **sedge** is a grasslike plant, different from other grasslike plants in having a solid, usually three-sided stem.

A **rush** is a grasslike plant with a round, usually spongy stem.

A **reed** is a true grass with a jointed stem.

Florida native: A burrowing crayfish (*Procambarus hubbelli*). This animal burrows in flatwoods marshes in the far western Panhandle.

90 to 100 percent. These plants also require that the water levels fluctuate between certain limits. Lower the low or raise the high water level, and the plant distribution changes. Stabilize the water level and the marsh will, in time, fill in with undecomposed litter and then with shrubs and trees.

Several arrays of plants may grow in Florida marshes (see List 10–1). Which array will predominate depends partly on what plants happened to seed in to begin with, but mainly on water depth and hydroperiod, fire frequency, soil, and degree of drying between times of inundation. Which plants grow in a particular marsh seems not, however, to be influenced by Florida's range of temperate to subtropical climates. Nearly all marsh plants are temperate, and the same plants grow in marshes all over Florida.[5]

Over the seasons, different plants predominate. In spring, the shorter species reach their full height and then bloom. In summer and fall, the taller ones overtake them; then they bloom. For example, in a flag marsh (a marsh in which many of the plants have flag-shaped leaves), maidencane and big floatingheart are dominant in spring. Then when the fall dry season begins, shortbeak beaksedge and beakrush take over.

The marsh plants just described are the ones that people see, those whose tops are in the air. Other plants live wholly under water, and so are easily overlooked. Notable among them are billions of microscopic algae that coat the submerged stems and blades of every plant—periphyton (pronounced "perry-FYE-ton"). These algae dangle from the underwater parts of plants and rocks in ropes, clumps, and mats, looking for all the world like useless slime. Indeed, they are slimy, but they are not useless: they are the base of the food web, as described next.

MARSH ANIMALS

Although it may look peaceful, a marsh is an extremely busy place. Hosts of animals are present that human-sized observers hardly ever see, both below and above the water line. To begin with, nearly 300 species of tiny animals live among the periphyton. Tiny invertebrates (copepods and amphipods) eat the algae and disintegrating parts of plants. These in turn are eaten by midge larvae, water fleas, mosquito larvae, and fly larvae. Salamander larvae and tiny fish then eat these. Apple snails, millions of them, graze the marsh plants under water, then periodically glide to just below the surface to draw fresh oxygen through their strawlike breathing organs.[6]

Above the water line, innumerable insects crowd the marsh. Figure 10–1 shows just three of the dragonflies, but mayflies, mosquitoes, gnats, deerflies, horseflies, water bugs, and water beetles also thrive in marshes. To see them, beat some marsh plants over a white cloth: a horde of insects will fall out and be visible on the cloth. The air is alive with a cacophony of insect hums, clicks, and crackles. Spiders harvest a bounty of edibles in such an environment; spun among the grasses are countless webs.

In the water and on the bottom, feeding on this cornucopia of tidbits are grass shrimp and crayfish, themselves well hidden by the plants. In permanent-water marshes, fish also abound—especially small species like

Four-spotted pennant (*Brachymesia gravida*)

Comet darner (*Anax longipes*)

Royal river cruiser (*Macromia taeniolata*)

FIGURE 10–1

A Sampling of Florida Dragonflies

Florida has 112 species of dragonflies, a third of all those known in the United States and Canada combined, and also 44 species of damselflies. And no wonder: Florida has abundant fresh surface waters, the environment in which these insects breed, and offers them uncountable mosquitoes, their favorite prey.

These insects don't bite people, but they do gobble up insects that bite. The largest one, the royal river cruiser, which has a five-inch wingspan, does double duty: as a larva in the water, it gobbles up mosquito larvae, and as an adult in the air, it devours mosquito adults so avidly that it is nicknamed the mosquito hawk. Unlike other pest killers, however, it leaves behind no poisonous residue to contaminate the environment.

Source: Dunkle 1991.

topminnows and killifishes, which can survive the longest during dry periods when crowded into shrinking water bodies.

Frogs and newts are numerous; so are snakes and turtles—and alligators. Listen by a natural marsh at night: the frog serenade will be overwhelming, and in the alligator mating season, the bull gators' bellows are added to the chorus. Shine a strong flashlight into the marsh and you will see hundreds of sparkling eyes. Beneath each leaf are several tiny sapphires: these are spider eyes. Tiny pairs of rosy lights are the eyes of moths. Bigger pale ones are frogs' eyes, and—surprise! several huge pairs of bright red eyes betray several alligators closer than you would have guessed. List 10–2 names some frogs, snakes, birds, and other denizens of marshes.

The crayfish and the alligator are both key species in marsh ecosystems because they dig small and large holes. Crayfish burrows house many other tiny creatures; alligator dens accommodate big ones. Aquatic organisms can retreat to these underwater sanctuaries when the rest of the marsh is dry. Then when flood waters return, they recolonize the marshes.

Mammals are not numerous in marshes, although the round-tailed muskrat and white-tailed deer are common. Both love newly sprouting young marsh plants, which make tender eating in the early rainy season. Muskrats can swim when the water is high, and burrow in exposed peat when the water table sinks below the marsh floor. Deer can forage in shoulder-deep water; they browse contentedly on pickerelweed and waterlilies.

By far the most abundant of the animals people see in a marsh are birds, especially wading and water birds, which nest in nearby swamps at night and forage daily in the marshes. Colonies of mixed species, mostly white ibis, have in the past been estimated at 35,000 birds or more. The

Periphyton is a coating of algae on the underwater parts of rocks and plants. Literally, it means "growing around plants."

LIST 10–2
Marsh animals (examples)

Amphibians
Bullfrog
Cricket frog
Florida leopard frog
Greater siren
Green treefrog
Lesser siren
Pig frog
Two-toed amphiuma

Reptiles
American alligator
Banded water snake
Black swamp snake
Florida green water snake
Mud snake

Birds
Great blue heron
Great egret
Little blue heron
Snowy egret
Tricolored heron

Source: Guide 1990, 40.

Florida native: Wood stork (*Mycteria americana*). Common only in Florida marshes, this big bird feeds on fish, reptiles, and amphibians that become stranded in drying pools during the dry season.

red-winged blackbird, northern cardinal, Carolina wren, and purple gallinule nest in marsh vegetation at a density of several hundred birds per square mile. In several central Florida marshes, Florida sandhill cranes court and breed and raise their young. Predatory northern harriers and bald eagles find abundant small prey in the marsh and soar over it, shrieking. The eastern phoebe, belted kingfisher, and marsh wren terminate their southward migrations to stay in Florida marshes through the winter, when the bird population rises to more than 1,000 per square mile.

Land and water animals are tied together in relationships governed by marsh hydroperiods, which differ between north and south Florida. South Florida normally has a major summer rainy period, and water levels are high in summer and fall. During high water, the marshes fill with tadpoles, salamander larvae, and fish. Then in winter, when the water recedes, these animals become concentrated in small areas. Predators such as raccoons, opossums, and birds gather to feast on this easy prey, and nest in the spring, while they are well nourished and able to produce healthy young. Mosaics of marshes dotted across the landscape enable bird populations to produce thousands of new individuals in each generation.

North Florida has a second rainy season in winter, during which marshes again fill with rain and produce salamanders, tadpoles, and fish. During the normal May drought, the water falls; then in summer it rises again, and again the water animals multiply. Whenever the water rises, invertebrate eggs and cysts hatch, aquatic organisms recolonize the marsh from deep solution holes and gator holes, and the cycle begins again. Humans may disparage marsh hydroperiods, calling them "floods" and "drought," but ecologists rightly call them "the rejuvenation process."

Among Florida birds dependent on marshes are wood storks, giant birds with a five-foot wingspan that soar for up to 40 miles in a morning to find shrinking shallow ponds and marshes brimming with prey. Wood storks are most efficient at feeding when the prey are concentrated in shallow water. They grope in the mud with the sensitive tips of their bills to find their food. The dropping of winter water levels is the stimulus that prompts the woodstork to nest. It is a sign that food will be abundant enough in small areas so that the birds can efficiently feed their big, hungry chicks and rear them to fledgling size. Either too little or too much water beneath a colony site during breeding season, or too much rainfall or runoff within a foraging area, can result in nesting failure.

At the same time, wood storks require strong, stout trees in which to raise their broods. The ideal habitat for wood storks is a mosaic of ponds, cypress, and pine on the unaltered landscape.

Periodic filling with water and drying up again of marshes is thus crucial to the survival of many kinds of animals. Occasional dry winters, and especially several dry winters in a row, severely reduce animal species populations. This is a natural cycle and animals can withstand it as long as the marshlands refill with water when wet years come again—hence the importance of identifying wetlands correctly even during dry times.

Wood storks nesting in a cypress swamp in the Florida Panhandle.

TYPES OF MARSHES

The Florida Natural Areas Inventory (FNAI) names marshes, not by their predominant plants as in List 10–1, but by characteristic biological and physical features (see List 10–3). FNAI counts nine natural communities as marshes, of which six are given attention in this chapter (the other three are treated in other chapters). The next sections begin with the smallest and shallowest of Florida's marshes and end with its greatest marsh system, the Everglades.

Depression Marshes. A variety of wetlands and water bodies occur in depressions. Depression marshes and basin marshes are among them. Basin wetlands holding a predominance of trees are, of course, basin swamps. Basins with open water in the center are ponds and lakes, and these may be surrounded by either swamps or marshes. No clear dividing line separates basin and depression wetlands in nature, but it is convenient to classify them, and treat them, as separate types for discussion purposes.

A depression marsh is so subtle that one can easily fail to notice it at all. It may be very small—only fifty feet across, and it may dry up completely each dry season, or even for several years at a time, yet at other times it may fill with so much water that it seems more like a true pond. It is classed as a wetland but often called a temporary pond, and it is important precisely because for *a lot* of the time, it is *not* filled with water. This feature makes it a habitat that is indispensable to some creatures.

Depression marshes occur all over Florida, in all environments from deep hardwood forests to open flatwoods. They form wherever small, shallow depressions are found, and they are ringed with various plant assemblages—some with cypress and tupelo, others with oaks, some simply with wiregrass and pines. While walking on upland terrain in the dry season, if you come upon a circular depression that has dried-up organic sediments, especially algae, in the center, you are standing in a depression marsh.

The plants in a depression marsh are similar to those in other shallow marshes, but some of the animals are unique. Small, egg-laying animals such as certain crustaceans, salamanders, and frogs, can reproduce successfully in wetlands like this, because there are no fish to eat them. Fish can't survive in temporary ponds, because they die when the ponds go dry, but the other small animals can go dormant and wait for the next rainy season to come alive again.

LIST 10–3
Marsh types

According to FNAI, marshes include:

Basin marshes*
Bogs (flatwoods herb bogs, except seepage bogs)
Depression marshes (very shallow basin marshes)
Floodplain marshes*
Marl prairies
Seepage slopes*
Sloughs
Swales
Wet prairies

***Note:** *Basin marshes are included with lakes in Chapter 12. Floodplain marshes are included with streams in Chapter 13. Seepage slopes are treated in Chapter 9.

Source: *Guide* 1990, 40–43.

A **depression marsh** (or **temporary pond**, or **ephemeral pond**) is a flat depression that holds water during rainy seasons and dries up completely between times.

Temporary pond in the dry season. Because it completely dries up each year, the pond supports no fish.

Same pond during a winter rainy season. The pond now measures about 75 feet across.

Same pond in 100-year flood. Now 500 feet across, it is still free of fish, and safe for small aquatic animals.

Florida native: Little grass frog (*Pseudacris ocularis*). This is Florida's smallest frog. It depends on fish-free ponds for breeding and nursery grounds.

Free of predation by fish, the crustaceans, salamanders, and frogs can live out their life cycles and reproduce most successfully in depression marshes. Some of these animals are found *only* in the depression wetlands of north Florida. They breed in the winter rainy season and winter rains are rare south of Ocala.

The fruitful time in a depression marsh begins after the first winter rains, when tiny animals come alive from their dormant forms and swarm in the shallow water: fairy shrimps, clam shrimps, grass shrimps, crayfishes, isopods, amphipods, and decapods. Thanks to the great numbers of these small prey, the larvae of a special set of salamanders can find food there. Panhandle Florida has four of these: the marbled salamander (in hardwood bottomlands), the flatwoods salamander (in flatwoods), the tiger salamander (in ponds in loamy soils), and the mole salamander (which lives in all types of ponds). Other salamanders, called newts, breed in the ponds and spend a few months to about a year in them as larvae, then metamorphose into terrestrial juveniles (called efts) and depart to take up their lives in adjacent woodlands. They metamorphose a second time, becoming adult salamanders, when they return to water to breed.

Even more dependent on depression marshes are frogs—specifically for egg-laying, fertilization, and the early life of their young. Different species breed in different seasons, as if they were "time-sharing" the ponds. During winter rains, from November to February, the spring peeper, ornate chorus frog, and Florida chorus frog breed in the ponds. From February to April, the southern toad and southern leopard frog breed. The dusky gopher frog, a strictly terrestrial frog, requires the ponds between October and March, just for breeding. Beginning in May or June and continuing until September, the oak toad's tadpoles, together with those of the narrow-mouth toad, the pine woods treefrog, the barking treefrog, the squirrel treefrog, the little grass frog, and the cricket frog, are teeming in Panhandle depression marshes. Other frogs also use the ponds, but can survive where there are fish as well: the green treefrog and Cope's gray treefrog.[7]

The chicken turtle also depends on depression marshes. It wanders from pond to pond in flatwoods and pine woods. When one pond dries up it finds the next wet one, and so continues its life. When it cannot find a pond, it buries itself in leaf litter in longleaf pine woods and keeps still until rains come again.

Florida natives: Turtle hatchlings. At left is the chicken turtle (*Dierochelys reticularia*). At right is the pond slider (*Trachemys scripta*). These turtles use marshes and ponds as their habitats.

Because they are small and the law protects only large wetlands, depression marshes require special attention. Size alone is not an adequate indicator of the value of a wetland.

Flatwoods Bogs and Wet Prairies. Besides temporary ponds, many broad, flat parts of Florida hold standing water for long enough times each year to be classed with marshes. Their water depths may be minimal, to be sure, and they may form mosaics with mesic and dry terrain interspersed, but they fall within the shallow end of the marsh spectrum. They have prosaic, water-soaked names such as "herb bogs," "boggy flatwoods," "seasonally flooded flatwoods," "wet flats," and the like, but they can be extraordinarily beautiful. Here, they are called *wet flats*.

Wet flats communities are also remarkably biologically diverse, due to subtle differences in small patches of terrain. Water leaves flat, poorly drained land primarily by evaporation, and it stands for longer or shorter times depending on differences in elevation of only an inch or two. That inch or two, by affecting hydroperiod, makes the difference between one suite of plants and another. As a consequence, floral diversity in wet flats is as great as in seepage herb bogs, with dozens of species per square meter.[8]

Given such a great variety of plants, immense numbers and kinds of insects and other small invertebrates thrive, and where those animals are, others follow. To give but one example, twenty different species of burrowing crayfishes have been observed in different parts of Florida's wet, flat lands. Crayfishes can tolerate alternately wet and dry conditions and can use both plants and insects for food. Each crayfish species creates its own distinctive type of burrow and each conducts its affairs in its own style. Some prefer mud, some sand. Some occupy areas where the water table is within inches of the surface, others where it is several feet down. Some roam the flats and dig burrows only when they need water in dry seasons. Some live in burrows and move into open water during rainy seasons. Some always stay in their burrows. That so many crayfish species live on Florida's flat terrain speaks volumes about the diversity of available habitats in what looks like featureless level ground.[9]

Crayfish species in wet flats may well be keystone species there, similar to the gopher tortoise in pine grasslands. Where crayfish are numerous, the ground is riddled with their tunnels. By heaping up piles of earth wherever they dig, crayfish expose earth where seeds can start to grow even when fires have not burned through and bared the soil. Many plants of wet flats release their seeds at the end of the winter dry season, the season of a crayfish's most intense burrow digging activity. The crayfish serves as another example of the general principle that the plants of a community depend on animals, just as the animals depend on plants. Also, for small

Florida natives: Pine lily (*Lilium catesbaei*) with eastern tiger swallowtail (*Papilio glaucus*). The pine lily grows best in flatwoods bogs. The butterfly is widespread across the eastern United States.

A wet flat (herb bog) in the Florida Panhandle with October flowers blooming. The diversity is evident from the colors in the foreground. This wetland may well support 150 species of herbs.

Florida native: Bogbutton (a *Lachnocaulon* species). Bogbutton can thrive in flatwoods bogs and wet prairies, but can also survive dry times, as seen here.

animals, crayfish burrows probably serve somewhat as gopher tortoise burrows do on higher ground: they are places to escape weather extremes and predation.

Marl Prairies. Marl prairies occur only in south Florida, notably in the Everglades National Park and in Big Cypress National Preserve, where they lie interspersed with sawgrass and other marshes. Marl forms in marshy areas where limestone is near the surface. Marl is derived from periphyton, masses of algae and other minute organisms that grow floating or loosely attached to bottom limestone and to underwater plant parts. While growing under water, the algae take up calcium carbonate; then when the water level drops and they dry up, they form a crumbly, highly alkaline, thin soil that resembles fine gray or white mud (see Figure 10–2).

Walking on a marl prairie is an altogether different experience from a stroll across a wet flat. The ground may be rock-hard in the winter dry season, but wet, soft, and slimy during summer rains. As in wet flats, the soils are seasonally flooded but for extremely variable periods of time. Marl prairies are unlikely to burn; there is not enough dry vegetation to carry fire.

Some of the plants found in marl prairies are itemized in List 10–4. Many are endemic. A stunted form of cypress may be present, known as dwarf, toy, or hat-rack cypress. Dwarf cypresses may be hundreds of years old as evidenced by the huge boles at their bases, but they seldom grow to more than ten feet in height. Animals are shared with neighboring ecosystems; see "The Everglades," next.[10]

Swales and Sloughs. Swales and sloughs are unlike the other marshes described so far in that their water flows perceptibly during much of the year. During wet seasons, swales are "rivers of grass," and they can be many miles wide.

Swales are found predominantly in east, central, and south Florida where the land is nearly flat. Along the coast, rims of sand, clay, or limestone

FIGURE 10–2

Periphyton's Various Aspects

Periphyton forms as loose, half-floating, half-sinking vegetation at or beneath the water's surface (A) and (B). When it dries, it remains draped over living vegetation (C) or forms a fissured mat of crumbly organic matter—that is, marl (D).

Florida native: Cypress (*Taxodium distichum*), dwarfed form. These tiny, ancient cypress trees grow scattered across a wetland that is marshy, but flooded only for short periods. Dwarf-cypress marshes may burn, but fires seldom kill the trees, both because fuel on the ground is sparse and because cypress is very resistant to fire. This marsh has not burned in recent memory. (For more about cypress, see the first pages of Chapter 11.)

contain them so that their water does not run out to sea. Many marshes of this type are found along the St. Johns River, which is confined behind an old coastal dune and barrier-island system. Rain falling in the river basin meanders north for several hundred miles before finding its way to the sea at Jacksonville. Central Florida's Kissimmee River flows similarly, but in the opposite direction—south—and it, too, is a marshy, flowing-water system. In fact, from the Kissimmee's headwaters all the way to the tip of Florida, all water flows south down the almost imperceptible slope of two inches per mile. South Florida, too, has a rim—in this case, of limestone—along its coastlines; it is shaped like a broad, shallow spoon. The result is a grand, slow flow of water over land for hundreds of miles from the Kissimmee's headwaters through Lake Okeechobee and the Everglades, finally reaching the sea at the extreme southern tip of Florida. The Everglades is Florida's premier example of a swale-slough system.[11]

> A **swale** is a low-lying stretch of land, a depression between ridges. In the Florida interior, the Everglades lies in a giant swale. At the coast, the hollows between dunes are called swales.
>
> The word **slough** (SLOO) is used to describe any minor drainage channel in a wetland. A slough may be a deep-water channel in a *swamp* (Chapter 11). It may be a channel in a *floodplain* that drains into a river (Chapter 13). Here, it means a deep-water channel in a *marsh*.

THE EVERGLADES

Whatever words are used to describe the Everglades, *vast* is always among them. Before engineering projects altered the region's drainage, the whole Everglades system encompassed 3,500 square miles (see Figure 10–3). Even today, although much reduced in size, the Everglades system is bigger than the whole state of Delaware. Everglades National Park alone occupies 1.5 million acres and the adjoining Big Cypress preserve holds another 728,000 acres.

The Everglades is an endless plain of Jamaica swamp sawgrass in shallow, slowly moving water that extends to the horizon. It is almost featureless. Here and there is a hummock (a tree island, described in Chapter 7); here and there a mound of earth in the marsh that represents an alligator's den-digging work. Except for the scattered, dark humps and an occasional trail of open water (a slough), the marsh seems utterly monotonous. Even the seasons, although intense, are repetitive. The pattern of sun, rain, and storms varies little from year to year. Dry and wet seasons follow each other predictably and the marsh dries and refills in concert with them.

No ground water flows into the Everglades from subterranean springs or seepage slopes as in north Florida. The Floridan aquifer, which is relatively close to the surface in central and north Florida, is a thousand feet below the surface in south Florida. Immediately beneath the Everglades is a sand-and-gravel aquifer that fills only with rain and empties by evapotranspiration (and pumping). All of the Everglades' water is from rain, some falling directly on the Glades, and some flowing overland from hundreds of miles to the north.

Chapter 7 describes how the Everglades keys first emerged above sea level between 100,000 and 15,000 years ago. The ocean withdrew off the land of south Florida, first exposing the highest limestone islands, then exposing the slightly lower limestone ridges that connected them, and finally draining off altogether, leaving the grooves between the ridges filled with fresh rain water. Flowing rain water has moved through the Everglades ever since, eroding the grooves ever deeper. The marshy waterways are the youngest parts of today's Everglades: they began to develop only

Everglades marsh. Jamaica swamp sawgrass (*Cladium jamaicense*) stretches for miles under an ever-changing sky, broken only here and there by a tree island or a slough.

some 5,000 to 6,000 years ago or less. Most of the plants and animals that came to colonize the marsh, however, have been around for thousands of years. The predominant plant is Jamaica swamp sawgrass (Figure 10–4).

Where the water is too deep for sawgrass, spikerush and other rushes take over. Bladderworts by the millions float among them, their yellow blossoms lighting up broad expanses of water with a golden glow. Hundreds of other, less conspicuous plants also live in the Everglades.[12]

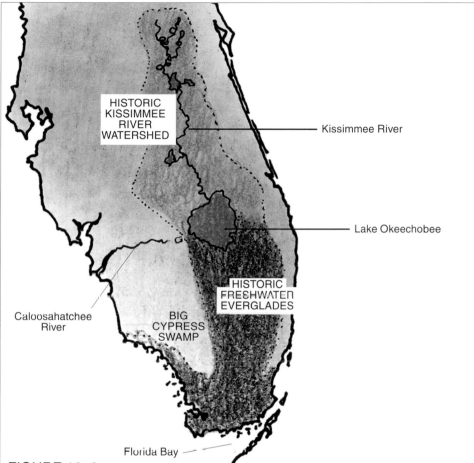

FIGURE 10–3

The Everglades (before Alteration)

South Florida tilts gently southwest between coastlines that are slightly higher than its center, so that from the Kissimmee River's headwaters all the way down the peninsula, all surface waters flow south over land. Before the system was altered, rain coalesced into lakes, the lakes overflowed into streams and swales, and these meandered down the Kissimmee River watershed into Lake Okeechobee, a lake basin 50 miles wide. Then the lake overflowed at its southern end as a sheet of moving water 40 to 70 miles wide—the great swale known as the Everglades. The Everglades' water eased down the slanted Florida peninsula and, finding few outlets east or west, finally slid off the land's southern and southwestern end through mangrove swamps into Florida Bay.

The dark blue area represents the original Kissimmee River/Everglades ecosystem watershed, which covered much of Palm Beach, Collier, Broward, Monroe, and Miami-Dade Counties. Today, to redirect the original flows of water, much of the Kissimmee has been straightened, canals now carry much of the system's water to coastal cities, and Lake Okeechobee is diked.

Sources: Strahl 1997; Douglas 1947, 5. Drawings are adapted from figures in Hinrichsen 1995.

Today's Everglades. Note the greatly reduced area of flowing water from the Kissimmee's headwaters all the way down the peninsula.

Florida native: White peacock (*Anartia jatrophae*). This daintily decorated butterfly occurs throughout southern Florida and much of tropical America.

In the water beneath the plants is a mass of mucky periphyton. When first measured 100 years ago south of Lake Okeechobee, the muck was 15 feet deep. Today, people periodically drain the land and let oxidation proceed, but the muck is still nearly five feet deep. Recently formed layers are loose and fluffy, partly floating and partly sinking in the water. The oldest layers are compactly stacked on the underlying rock.

The sawgrass and rushes support an immense marsh community. Hundreds of species of fish and other animals thrive in this environment, and in enormous numbers. Microscopic plants and prey become food for a myriad of flashing silver minnowlike fish through whose tiny forms most of the marsh's material repeatedly passes. The tiny fish become food for larger fish and water snakes.

The enormous algae-and-insect base of the food web supports impressive numbers of larger animals as well—otters, raccoons, and opossums; hawks and alligators. Most conspicuous are the wading birds that breed in the Everglades—shallow water makes a perfect fishing environment for a wader. In the 1800s, egrets, herons, roseate spoonbills, and white ibises nested there in rookeries of more than 100,000 birds. Bird numbers have declined but can still number in the tens of thousands.[13]

Other birds that prosper on the marsh's abundant fare include the Florida sandhill crane, limpkin, purple gallinule, and common moorhen. These support top predators—bald eagle, barred owl, osprey, and two kinds of kites (Figure 10–5). The predators among the birds reflect the richness of the ecosystem. Each has its hunting specialty and there is enough for all.

Although birds are most conspicuous, other animals are present, too: panthers, bobcats, rats, marsh rabbits, bears, cottonmouths, and alligators. Altogether, more than 600 species of terrestrial vertebrates live in all

FIGURE 10–4

Jamaica Swamp Sawgrass (*Cladium jamaicense*)

In the foreground is the flowerhead; in the background is the "grass"— actually, a sedge. This is the dominant plant in most of Florida's flowing, freshwater marshes. Native to the U.S. Southeast, the West Indies, and Central and South America, it is well adapted to the nearly level, marshy terrain it occupies. It may reach some ten feet in height and every blade has fierce serrations made of silica that can cruelly cut flesh and tear clothing, so the plants form a forbidding, impenetrable mass. They also store nutrients beyond their own needs, depriving other species of the chance to grow in the nutrient-poor marsh.

Jamaica swamp sawgrass tolerates cold weather, and it loves fire. The bud grows buried in the soil wrapped in its own living leaves, which protect it from the flames. The higher, older leaves, in contrast, are so flammable that they will burn even above standing water. Then they rapidly regrow, fed by the nutrients that the burning sawgrass released as ash into the water. If floods rise too high too fast, newly growing sawgrass will die back, and even die out, but it can usually outpace rising water levels. It can grow more than a foot in two weeks.

PRICELESS FLORIDA

the park's different habitats. For these reasons, the Everglades National Park today is a United Nations World Heritage Site.[14]

Seasons in the Everglades. Water is king in the Everglades, definer of the seasons. No spring, summer, fall, or winter occurs here; rather, there are two seasons: wet and dry. The marsh is green in the rainy season, brown in the dry. Depths and seasons of inundation define the habitats and determine the life cycles of all living things in the marsh.

The rains begin in May, sometimes late May. They may go on all summer, or they may hold off at times, but storms set in, in earnest, in late July through September. Sometimes a single storm will drop 10 to 12 inches on part of the marsh in an hour. Says an observer:

Thunder rocks the earth and lightning splits the sky. All the silence becomes rattling, booming noise. All the emptiness is filled with blasts . . . of light. The whole wide horizon is a sound and light show, and the windows of Heaven open. The Everglades rise up a foot and the desert turns back into a river.[15]

More than half the year's rainfall may arrive in just two months and evapotranspiration can't keep up. After a heavy rain, water runs over the land at a steady pace. And rains are also falling on central Florida around Lake Okeechobee. That water, too, runs south over the land and arrives in the Everglades days later.

In October or November, daily thundershowers end and the marsh begins to dry. For months, it rains only a little or not at all. Evapotranspiration exceeds rainfall, the water level falls, the flow ceases, and parts of the marsh dry up altogether. Dew forms and drips in the mornings, but sun soon dominates in cloudless skies. Some 60 percent of the water that falls

Snail kite (*Rostrhamus sociabilis*). A rare bird, this kite nests in shrubs a few feet above the water, hovers low in the air, and drops to snatch up apple snails when they approach the surface to breathe. It holds the snails in its talons and uses its curved bill to extract them from their shells. The kite depends absolutely on these snails: its talons are too weak to hold other prey, such as turtles, and its beak is the wrong shape to pull them out. When rearing its young, the kite may catch 60 large snails a day.

Swallow-tailed kite (*Elanoides forficatus*). This kite nests in pinelands, where it uses Spanish moss or hairy lichens to weave its nests together. Unlike the low-flying snail kite, the swallow-tailed kite flies gracefully, high in the sky. It hunts for snakes, lizards, and bird nestlings over both upland habitats and marshes. Nature writer Susan Cerulean says, "The challenge swallow-tailed kites offer us [is] to hold as much as possible of the state's remaining natural landscape intact."

FIGURE 10–5

Two Native Kites

Sources: Robbins and coauthors 1983, 69; Toops 1998, 23; nest behavior and quote from Cerulean 1994.

The wettest part of the Everglades. This is Taylor Slough, one of many marshy waterways through the Everglades. Even when the Glades are dry, as here, the slough holds water.

The rock beneath the Glades. Low water and recent fire have exposed the underlying rock, whose sharp contours have given it the name *pinnacle rock.*

Florida native: Green anole (*Anolis carolinensis*). On green vegetation, this animal is green, so in the Everglades' rainy season, it is well camouflaged.

On brown vegetation, the anole has the knack of turning brown, which comes in handy in the dry season. Here, the animal is shedding its skin. The new skin underneath is even more like the twig on which it is perched.

in the rainy season rises again in the dry season—it literally dries *up*. The water level sinks below the soil; the soil becomes parched marl with deep crevasses in it; and the slimy periphyton that was suspended in the water dries on the ground to crispy curls like cracked paint.

From December through April, puddles become fewer and smaller. Fish and bird populations decline. Lightning, when it strikes, starts fires that burn for miles across the marsh, renew the marsh plants, and kill young shrubs and trees. Fish crowd into shrinking, soupy pools, struggling to stay under water. Big fish—bass, bream, and gar—die and float, belly up, to the edges of the pools, attracting millions of buzzing flies. Vultures circle and land to strip the decomposing carcasses.[16]

Unlike the big fish, the tiny fish keep going. Some, known as topminnows, can swim in the very top inch of water, where oxygen is still available. Their protruding lower jaws and up-tilted mouths are adapted for surface feeding. Many species of these tiny fish live in Florida's marshes; four are shown in Figure 10–6.

Then the rains resume, perhaps in late May, renewing life in the glades. Sawgrass and other sedges and grasses spring back to life and regrow. Frogs, which have lain dormant, emerge to join in a noisy chorus and mating frenzy that will soon produce millions of glistening fertilized eggs. Small fish mobilize from their expanding ponds, swim into rivulets, and energetically work their way upstream against the runoff coming down from the north. The glades are a swale again, and these fish are adapted to hold their ground against its moving water. They have to be: to keep the glades occupied, they must replenish their upstream populations or they will be washed away.

Most of the glades animals, from mosquitoes and snails to fish and alligators, breed just before or during the rainy season. Then their young

Water rules the Everglades. Ear-splitting claps of thunder and blinding lightning flashes dominate the Glades throughout the rainy season.

Fire takes control. This portion of the Glades has burned, but in the foreground, sawgrass is resprouting and in the background, the cypress trees are making a comeback.

are born as the rains are bringing forth lush new vegetation that will support the rapid growth of plant eaters and, in turn, their predators. By the time the dry season begins, all are prepared to withstand its harsh rigors. Eggs, larvae, and adults are buried in the mud or lingering in alligator pools, awaiting the next year's rains.

The alternation of wet and dry seasons places demands on all the marsh inhabitants. To keep going, they have to be adapted to both extremes. They may thrive in one season, but unless they can at least survive in the other, they will be unable to perpetuate their kind.

The alternating seasons mostly favor native species, which are equipped to cope with the extremes. Most recent arrivals succumb. Unfortunately, however, some invader trees such as the Brazilian pepper and punktree possess adaptations similar to those of the native plants, and are shouldering aside native vegetation in much of the Everglades.

The Glades Engineer: The Alligator. A keystone species in the glades ecosystem is the alligator, whose deep dens hold water during dry times and help hundreds of animals hang on from one wet season to the next. This important ecological role is crucial in the Everglades.

When the marsh has gone dry, the alligator goes on the move, looking for water beneath the caked mud and marl. When it finds such a place, the

Mosquitofish (*Gambusia holbrooki*).

Mosquitofish (*Gambusia affinis*, melanistic variant).

Seminole killifish (*Fundulus seminolis*).

Golden topminnow (*Fundulus chrysotus*).

FIGURE 10–6

Native Topminnows

These tiny fishes and their many relatives thrive by the billions in the Everglades marsh, even when the water is low.

CHAPTER ELEVEN

FLOWING-WATER SWAMPS

Swamps are distinguished from other wetlands in being forested. Three seepage swamps were described in Chapter 9, swamps in which titis, bays, or white cedars grow and in which water moves slowly through the ground. This chapter covers strands and river floodplain swamps, which are flooded every year with flowing water. The swamps that are the subjects of this chapter are named in List 11–1. Notice that two of them are called *forests*, but contradicting the definition of forests given earlier, these two are considered wetlands.

Swamps occupy thousands of acres all over the state, and taken all together, they contain well over 300 species of plants, all adapted to wetland habitats. This chapter begins by describing swamp trees, whose adaptations give them predominance on land where water rises and falls. Then it turns to the swamps in which they grow.

SWAMP TREES

Swamp trees grow in environments where most marsh herbs would decline and die. Water levels vary much more widely in swamps: the highs are higher, and the lows lower, than in marshes. Swamps sometimes hold deep water for longer periods than marsh plants can tolerate; and swamps also may go completely dry, whereas marshes nearly always hold some water. Once swamp trees have claimed a wetland, no natural force can displace them except catastrophic fire—an unlikely event among hardwoods on swampy land.

Cypress and tupelo trees are Florida's foremost swamp trees. They serve as prime examples of the adaptations that enable trees to grow in wetlands.

Cypress Trees. Cypress trees have long been thought to exist as two separate species, pond-cypress and bald-cypress. Recent genetic studies suggest they may be merely variants of one species, however, much as north and south Florida slash pine are variants of slash. In this book, cypress is given its generic name only (see Figure 11–1).[1]

WBSP

Florida native: Red maple (*Acer rubrum*). This common native tree, which can grow both on uplands and in wetlands, makes the splashes of red one often sees in swamps in spring and fall.

LIST 11–1
Flowing-water swamps

The swamps and floodplain forests treated in this chapter are:

Bottomland Forest
Floodplain Forest
Floodplain Swamp
Freshwater Tidal Swamp
Hydric Hammock
Strand Swamp

Dome swamps and basin swamps are treated with lakes in Chapter 12.

OPPOSITE: A floodplain swamp along the Apalachicola River. The trees are young water tupelo.

Rolled-up leaves. Trees with this appearance may be called pond-cypress, but it may be that in dry times, any cypress tree rolls its leaves around the branchlets to reduce water loss.

Flattened-out leaves. Trees that look like this may be called bald-cypress, but it may be that wherever cypress stands in permanent water or saturated soil, it spreads its leaves at right angles to the branchlets to enhance release of water.

FIGURE 11–1

Cypress (*Taxodium* species)

The question whether cypress exists as two species is still investigation. It may be one species whose leaves vary in appearance.

Cypress can grow on high land that never floods, but meets with tough competition from other trees there and seldom wins a place for long under natural conditions. In many wetlands, though, cypress competes more successfully than any other tree. It is adapted to grow where water stands several feet high around its trunk for months at a time. It easily survives droughts, and it is fire tolerant, as well.

The base of a cypress tree is a masterpiece of biological architecture. The trunk grows strongest and broadest exactly where strength is needed most, producing a swollen, extra-stable base, or buttress. This is an adaptive response to the inundation that kills other trees.

Trees require oxygen for the activities of their root systems as all plants do, and they need more oxygen than marsh herbs do, because they are more massive. Oxygen is poorly available under water and in saturated soil, so no tree growing in a wet site can grow a taproot downward. But cypress can grow knees and adventitious roots (already shown in Figure 8-3, page 132). These structures help to obtain oxygen as well as to anchor the tree and feed it nutrients and water at times of different water levels. Figure 11–2 shows a cypress knee and the legend describes the functions of adventitious roots. Because cypress copes so well with rising and falling water, it can grow not only in basins but along rivers and around oxbow lakes in river floodplains.[2]

The events of a cypress's reproduction coincide with events in the wet and dry seasons. In fall, the tree goes dormant and therefore, as winter flood waters rise and deprive the roots of air, its oxygen needs simultaneously become minimal. In winter, the needles (leaves) turn brown and begin to fall, but the cones produced the previous summer remain closed. In spring, as the new, green leaves are coming in, the cones open and drop their seeds. Thanks to this timing, the seeds land in retreating water, and it is not long before they lodge in a mulch of the tree's own needles, on wet shores. Thus the seedlings begin to grow at a favorable time, when floodwaters are receding, and in moist soil kept moist by the tree's own litter.

A buttress. This cypress tree is in a basin where the water rises to about the same level every year. Below the water line, downward transport of hormones and nutrients halts and they accumulate there, stimulating the trunk to thicken and form a buttress.

PRICELESS FLORIDA

FIGURE 11–2

Cypress Knees

Knees grow up from deep cypress roots where floods deprive the main roots of oxygen for long times. The tops of the knees just reach the average seasonal high-water level, which in this case is higher than a man is tall. The function of knees is unknown, but they respire fastest when the water is highest, so it seems they may serve to capture oxygen from the air.

Cypresses may also have adventitious roots as shown for tupelo on page 132. Adventitious roots may be many feet long. They may sprout from buttresses or from the main roots of a tree. During floods, they float out around the tree on the surface of the water, so they can obtain oxygen when the drowned, lower roots cannot.

Sources: Clewell 1981/1996, 134; Ewel 1990.

Up to the age of two years (up to 1-1/2 feet in height), a cypress seedling must keep at least a few leaves above water, or else ten days of summer flooding will kill it. (In a colder season, when a seedling is dormant, it may be able to survive inundation longer.) The seedling also needs sun—exposure for 80 percent of the day is optimal. Cypress seedlings in deep shade may grow vigorously at first, but they won't survive for long without sunlight.

The seedling's requirements of moisture (without flooding) and abundant sun are not often met. Most cypress seedlings drown, in most years. In a dry year, they grow fast, but then they have to compete with hardwoods that are also springing up. The final step in the successful launching of cypress trees is a properly timed flood that kills all the competing hardwood trees, leaving only the cypresses alive. Given the optimal timing of high and low water, cypress can claim and hold wetland territory where most other trees are excluded.

Fires are rare in cypress territory, but they are not unknown, and cypresses are equipped to withstand them. The trees are usually spaced with an open canopy, so their crowns are unlikely to carry fire. Also, at least along streams where floodwaters frequently wash away litter, there is little fuel on the ground. When attacked by fire, a cypress's corky bark insulates its inner stem, so the tree still can remain alive. Small cypresses, even if severely burned, quickly resprout from their roots.

Old cypresses are huge. They achieve heights of 100 to 120 feet, even 150 feet. They continue growing in girth throughout their lives, so they can become massive around the base. Not uncommonly, once a giant falls over, the hollow trunk forms a spacious tunnel through which an adult can walk upright. Most of Florida's original cypress trees have been cut

Senator Cypress, Big Tree Park, Longwood, Orange County. This famous, giant tree is 118 feet tall.

down, but a few still loom above Panhandle rivers, evoking the majestic river swamps of times past.[3]

Cypress, like Atlantic white cedar (described in Chapter 9), is resistant to insect pests and fungal attack, so it can survive when other trees are aging and dying. As other trees disappear, these paragons of longevity come to dominate large areas, sometimes living to 1,000 years of age.[4]

Canoeing among giant cypress trees can be an exalting experience. They stand in stately ranks, with their swollen buttresses making them look like enormous, broad-based bottles. Cypress existed as a species millions of years before Florida first appeared as dry land, and they were among the first trees to colonize this land. When you enter among ranks of old-growth cypress trees, you are part of an ancient world.

Tupelo Trees. Tupelo trees exhibit their own adaptations to the demands of life in deep-water swamps. Like cypress, tupelos are deciduous, but they drop their leaves in fall. They stand dormant during deep winter floods; then in summer they leaf out, gather energy, and grow. Also like cypresses, tupelos have trunks that are flared at the base. Unlike cypresses, though, tupelos split up wetland territory among three species: swamp, water, and ogeechee tupelo.

In other ways, the adaptations of tupelos differ from those of cypress. For example, the ogeechee tupelo sprouts many stems, even near its base, and then doesn't grow very tall. Typically it reaches only 25 to 30 feet in height; rarely 50. It grows as a massive, hollow barrel, open at the top, with branches sprouting from around the rim, and the branches are hollow, too (see Figure 11–3). The branches age and break, leaving short, open tunnels into the tree at all levels. Animals living in these tunnels and in the barrel drop feces which become fertilizer for the tree. At flood stage, water and litter flow into the tree through its hollow, broken-off branches and are trapped inside the trunk. Later, when microorganisms decompose the litter, roots growing from the *inner* wall of the trunk absorb the nutrients. Botanist Andre Clewell says "the barrel-shaped trunk essentially attracts its own private compost heap."[5] The trapped compost provides much-needed fertilizer, because the only other influxes of water ordinarily are from rain. In short, although a mature ogeechee tupelo looks aged and half rotten, it is remarkably well equipped for the life it leads.

Like cypress, tupelos seem practically immortal. They can live for hundreds of years and they keep replacing their stems, so they need not reproduce frequently. Still, new tupelos have to get started sometimes, or the species would finally perish. They do so in a variety of ways. A seed may start its life on a hummock of peat where it won't be inundated at first. As it grows, the seedling surrounds the peat with fine roots that hold it together, and because they are above water nearly all the time, the roots continue to have access to oxygen throughout the life of the tree.[6]

During droughts, seeds may also fall on dry ground, where ordinarily they would quickly die out due to competition by upland trees. But when the habitat becomes wet again, then the tupelo seedlings will be the survivors, because they can tolerate inundation while the other hardwood species cannot.

To minimize confusion over names, it is helpful to remember that **swamp tupelo** is often called **gum** (or **blackgum**), and that a *tupelo swamp* is a *gum swamp*.

The two other tupelo species, **water tupelo** and **ogeechee tupelo** (o-GEE-chee) have no "gum" names.

Swamp tupelo usually grows in basins; water tupelo in floodplains. Ogeechee tupelo may join them in transition zones and may grow along river banks.

Another tree called a gum is **sweetgum**, but it is an altogether different tree, not closely related to the tupelos.

Young thicket in winter. These saplings must have started up in peat at a time of low water, then survived high water in which competing species perished.

FIGURE 11–3

Ogeechee Tupelo (*Nyssa ogeche*)

Old ogeechee tupelo, winter. This giant, barrel-shaped tree stands in Sutton's Lake along the Apalachicola River.

Champion ogeechee tupelo. This is a rare example, because it has only a single stem. One of the biggest tupelos in Florida, it stands in the heart of Bradwell Bay, Wakulla County.

Tupelos are well equipped to survive both droughts and fires. During a severe drought, a tupelo swamp may dry and its peaty soil may even become deeply fissured, but because undergrowth is sparse, fires are rare and seldom damage tree trunks enough to kill them.

STRAND SWAMPS

A strand is a water-filled channel in which trees are growing. Water flows slowly through a strand, often from a source such as an overflowing basin swamp or swamp lake, and often terminating in a blackwater stream.

When a strand supports a cypress swamp, it is known as a cypress strand. But strands often accommodate many other plants. In south Florida, where sunlight is especially abundant, Jamaica swamp sawgrass may dominate the undergrowth, or a lush understory of shrubs, ferns, and mosses may grow. Pond apple is often found alongside the deeper channels in cypress strands. Willow thickets grow along stretches where water or fire create the conditions for them. (Willow thickets grow where water levels fluctuate too much for herbs but not enough to exclude shrubs. They also grow where fires occur too often for trees but too seldom for herbs.)

The Big Cypress region was named for its cypress strands, which are everywhere. Virgin stands of cypress remain in Corkscrew Swamp, where the trees are 400 to 700 years old and 100 feet tall. Unaltered strands have luxuriant bromeliads and orchids on the trees and lush, soft ferns beneath. Bromeliads are so common in cypress strands that they are given attention in Figure 11–4.

Ancient cypress strands house major animal populations. On fallen logs decked with ferns, Florida redbelly turtles sun themselves while wood storks

> A **strand** is a water-filled channel in which trees are growing.

--------------------SAME LEVEL----------------

DBM

DBM

FIGURE 11–6

Flood Volume in a High-Discharge River

These two pictures were taken of the same tree in the Apalachicola River flood-plain. The photo at left shows the height of a normal flood. Flood stage that year was more than 18 feet, not unusual for a river of this magnitude. The photo at right was taken during the dry season.

The river is now dammed where the Chattahoochee and Flint rivers come together just north of Florida, but it still remains in a partially natural state. It possesses the largest forested floodplain in Florida and it holds one of the last unbroken bottomland hardwood communities in the United States. It provides a protected north-south corridor along which plants can disperse and land and water animals can travel back and forth, all the way from the southern Appalachian mountains to the Gulf of Mexico.

FLOODPLAIN FEATURES

The next sections describe an imaginary tour of parts of a floodplain, with attention to the vegetation. The first stops are the banks of a curve in the river: an outer bank, with its high levee, and an inner bank, with a point bar. Then, in the interior of the floodplain, a point of interest is an oxbow lake surrounded by a swamp, and a great bottomland hardwood forest. On higher ground within the floodplain and on the river banks are several other types of forests. Down the river towards the coast, these forests give way to a hydric hammock and then a freshwater tidal swamp. The last section of the chapter presents the animals.

Levees and Point Bars. A levee is the highest part of the floodplain and may bear a high, dry forest of oak, hickory, and spruce pine like high xeric hammocks elsewhere. Directly opposite a levee, a point bar is a gently sloping bank often with bare sand at the water's edge indicating its recent deposition. Such new land is colonized by pioneer species. The point bar has a zoned array of plants: herbs by the water, a thicket of shrubs behind them, and trees farther back on the land. New sand just laid down by the water is first colonized by pioneer herbs such as panicgrasses or dotted smartweed. These plants are native weeds: they appear promptly on disturbed sites where newly exposed earth affords opportunities to move in. They spread easily and grow fast, and advancing point bars are always offering newly exposed earth.

Behind the herbs, most commonly, is a thicket of fast-growing black willows. As the willows elongate, they lean toward the water, finally toppling onto the sand. Wherever they touch the sand, they put down new roots, grow new branches, and become new willows, a mode of propagating that gives them an advantage over competitors in claiming new ground in the sun. At high-water times, the growing willow thicket slows the water down so that it tends to drop more sand and further add to the point bar. Thus, in a sense, the willows keep creating new habitat for themselves: the point bar and willow thicket advance together. Willows can grow five feet in a single growing season, so they rapidly colonize newly-laid-down sand.

As the black willows march forward, they overtop the panicgrasses and other herbs, but new herbs keep springing up in front of them. Meanwhile, the oldest willows, farthest from the river, grow old and die, giving way to bottomland hardwood trees. On the Apalachicola River, some point bars have several recognizable zones of trees: willows on new ground, swamp cottonwood behind them, and American sycamore still farther back. Other species may accompany these, such as planertree and river birch.

Some day, of course, the river will cut across the point bar and leave it behind as an island between two river channels. Later still, levees will close off the old river meander, creating an oxbow lake, the story that is told next. Between the lake and the river old levees will remain as forested humps on the floodplain.

Florida native: Swamp thistle (*Cirsium muticum*). This common native weed grows freely in swamps and its blossoms provide nectar for many insects.

A point bar. The river constantly adds sand to this little beach as it rounds the bend in the Blackwater River. As the point bar grows, pioneer herbs, then shrubs, and finally trees march onto it.

Oxbow Lake and Floodplain Swamp. Figure 11–7 shows the genesis of an oxbow lake. As soon as it is closed off from the main river, an oxbow lake becomes a place of standing water, but unlike other lakes, it is frequently swept clear of debris by river floods, setting back its progress towards becoming a swamp. Its deep standing water is ideal for cypress and water tupelo, which may have started up when the land lay along the main river channel. At the center of an old oxbow lake, giant cypress and water tupelo may have been standing for centuries.

With repeated inundations by sediment-laden river water, the oxbow lake slowly fills in and supports the growth of trees, becoming a floodplain swamp. As the lake grows shallower, the hydroperiod grows shorter, and different hardwood trees succeed those that grew there at first. Undergrowth is sparse in the deep shade, but shrubs and ferns grow here and there beneath the trees, including royal fern, swamp titi, wax myrtle, and others.

Bottomland and Floodplain Forests. Behind the river's levees, and protected by them for most of the year, are the bottomland hardwood forests. Here, tremendous trees stand on massive root systems in muddy earth. Some have toppled over, leaving behind deep holes in the mud.

Bottomland hardwood forests are America's most towering forests. Many tree species grow taller, faster, in river bottomlands than anywhere else. The alternation of floods and retreat of flood waters speeds up the decay and release of nutrients on the forest floor. Continual flooding would deprive decomposer organisms of oxygen and so inhibit decay, but alternating wet and dry conditions give decomposers both the water and the oxy-

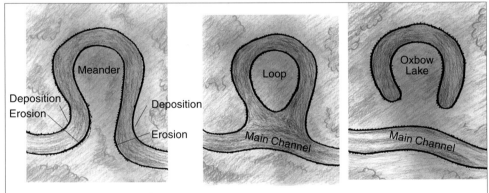

1. A meander. The neck of the loop grows narrower as the river erodes sediment from the outside bend of each curve. At the same time, the river deposits sediment on the point bar of each curve.

2. A meander loop. The river cuts across the neck at its narrowest point. Now there is a new main channel, in which the current flows fast, while it flows slowly around the loop.

3. An oxbow lake. Sediment accumulates alongside the river's main channel and closes off the loop, forming an oxbow lake.

FIGURE 11–7

Origin of an Oxbow Lake

These drawings show one stretch of a river at three successive intervals in time:

1. The river has doubled back on itself, forming a meander.

2. The river has cut across the narrow neck of the meander, forming a loop. Now, most of the river's flowing water follows its new main channel, bypassing the loop, in which the water flows sluggishly if at all.

3. The river has dropped sediment across the ends of the loop and cut it off, forming an oxbow lake. Later still, the lake will fill in, becoming a swamp.

An **oxbow lake** is a U-shaped lake formed when a river cuts off one of its meanders at both ends and leaves it behind.

An oxbow lake or other depression in a floodplain ultimately fills in and becomes a **floodplain swamp** or **backswamp**.

PRICELESS FLORIDA

gen they need to do their work. Soil moisture, held in place between floods by leaf mulch, feeds needed water to growing trees throughout dry times.

On higher ground within the bottomland forest, you will encounter several different assemblages of trees. On the lower parts of slopes are green ash and sweetbay, trees that can tolerate quite moist conditions. Higher on slopes, overcup oak and red maple become prevalent, then American sycamore and water oak. Finally, on the tops of ridges in the floodplain, other oaks, hickories, and spruce pine are growing.

Just by looking at the tall, dark trunks of the trees, one can hardly tell that each species grows best at a certain level above the river. However, when the members of each tree species are systematically counted, it is found that in general, different trees predominate at elevations that differ by only a foot or two.

How do trees "know" to grow along a strip of land at a certain elevation? One part of the answer is the hydroperiod at each level. The lower a tree lives in a floodplain, the longer its roots will be in saturated soil. Another part of the answer is that a floodplain's soil texture varies with elevation. Rivers deposit sand on high places while still running fast. They deposit silt and clay lower down when running more slowly. Soil organisms are different, too. Long periods under water create airless conditions in which one set of organisms grows. Shorter periods permit more penetration by oxygen and a different set of soil organisms.[10]

As a result of all these variations, and thanks to the annual delivery of fresh nutrients by floods, many more species of trees can grow in bottomland and floodplain forests than in acid swamps or mesic hardwood hammocks. Altogether, some 52 canopy tree species flourish in the Apalachicola River floodplain, and there are even more species of small trees, shrubs, and ground-cover plants than canopy trees. List 11—2 displays a sample of the prodigious variety found in a well-developed bottomland hardwood forest.

Trees themselves work variations on floodplain topography, thereby promoting further diversity. Because their roots are shallow for lack of oxygen, they tend to fall over when they die, raising giant, muddy tip-up mounds and leaving hollows filled with water where their roots have been pulled from the ground. Even long after a tree has completely decomposed, a mound and pool remain where it stood, and present an array of microhabitats for various floodplain plants and animals. Red maple, for example, can grow in a bottomland forest only because its seedlings can get started on mounds where the hydroperiod permits their growth. When attempting to restore bottomland forests and swamps that have been clearcut and leveled, ecologists have learned that if they recreate mounds and pools, then more species will return spontaneously and the community will develop more naturally.[11]

Forests and Swamps Upriver. Forest composition changes along, as well as across, the river floodplain. Certain trees, such as sweetgum and water oak, are more prevalent in the upstream valley than in the coastal floodplain. Other trees, including water tupelo and cypress, are more abundant near the coast than in the interior. However, because the river shifts

The Florida Natural Areas Inventory identifies three forested communities growing in river floodplains:

Bottomland forests grow on the floodplain floor.

Floodplain forests grow at slightly higher elevations within the floodplain.

Floodplain swamps grow along stream channels and in the lowest spots in floodplains.

Florida native: Atamasco lily (*Zephyranthes atamasca*). This lily is an obligate wetland plant. Here, it is growing in a cypress swamp.

Cambarus pyronotus. This crayfish digs its burrows in seeps along small streams that feed into the Apalachicola River.

DBM

BM

Cambarus diogenes. This crayfish burrows in marshy areas near stream banks.

FIGURE 11–9

Swamp Crayfishes

Of Florida's several dozen species of crayfish, eleven find habitats in stream floodplains—among them, these three.

BM

Procambarus peninsularis. This crayfish is widespread in wet sites across Florida.

LIST 11–3
Floodplain swamp amphibians and reptiles: a sampling

Salamanders
Amphiuma (two species)
Dwarf salamander
Marbled salamander
Rusty mud salamander
Slimy salamander
Southern dusky salamander
Two-lined salamander
Three-lined salamander

Frogs and Toads
Bird-voiced treefrog
Bullfrog
Cope's gray treefrog
Cricket frog
River frog
Southern leopard frog
Southern toad

Snakes
Black swamp snake
Brown water snake
Cottonmouth
Glossy crayfish snake
Mud snake
Rainbow snake
Redbelly water snake
    (continued on page 192)

Many fish species depend absolutely on flooded swamp habitats for parts of their lives. As a flood is ending, they feed in quiet backwaters and breed. Their offspring grow for several weeks there, then swim out to the main stream.

Amphibians and Reptiles. List 11–3 itemizes the amphibians and reptiles commonly found in Florida's floodplain swamps. One of them, the one-toed amphiuma, requires a habitat of brown muck that is derived from a mixture of hardwood and cypress litter. Perpetual seepage into creek and river floodplains creates pools of this muck up to about 20 feet above the Apalachicola River, and narrowly inland from the coast. The amphiuma occurs only in these few places.

Figure 11–10 illustrates that salamanders find numerous living spaces in riverside wetlands, spaces with different soil types, degrees of steepness, and hydroperiods. In the Ochlockonee River basin alone, 17 different salamander species live and share resources.

Where litter, muck, insects, and other small invertebrates abound, frogs can also thrive, of course. The standing water in sloughs is home to the green treefrog, the southern chorus frog, the spring peeper, the southern cricket frog, and the southern toad. Other frogs live along the river bank and the trees house treefrogs. The tiny, one-inch-long bird-voiced treefrog lives in the canopy of swamp trees. The bullfrog, a giant among frogs, readily eats fish, other frogs, and even baby ducks. Each frog has its own mating season, so the several species found in river floodplains do not compete for pools in which to mate. And each has its own call. The chorus in a large river floodplain at night is a cacophony of different songs, each singer straining to outshout all the others. Figure 11–11 presents a small sample of the many frogs that thrive in Florida's river floodplains.

Floodplain turtles include the Suwannee River cooter and alligator snapping turtle, the latter once rare due to overexploitation. Even more rare, and inhabiting only the Apalachicola River drainage basin, the very timid Barbour's map turtle lives in the water and suns along the shore (see Figure 11–12). Several species of snakes also occur in large populations. Among them are the cottonmouth and non-venomous snakes such as the brown water snake and redbelly water snake.

Apalachicola dusky salamander (*Desmognathus apalachicolae*), male and female

Two-lined salamander (*Eurycea cirrigera*), male and female

Red salamander (*Pseudotriton ruber*)

UPSTREAM SALAMANDERS These three species of salamanders live near the origins of streams (that is, along first- and second-order streams) in the Apalachicola and Ochlockonee river watersheds. These are raviny areas where the water is shallow and well oxygenated. The three species coexist relatively peacefully because they are of different sizes and eat different-sized prey. The Apalachicola dusky salamander eats only animals in leaf litter above the water line—spiders, sow bugs, and dozens of others. The little two-lined salamander, which has a tiny mouth and a long tongue, eats a suite of things that are beneath the dusky's notice. The red salamander is so big that it eats all kinds of animals that are too big for the other two, including the other two salamanders themselves.

Southern dusky salamander (*Desmognathus auriculatus*)

Three-lined salamander (*Eurycea guttolineata*)

Gulf coast mud salamander (*Pseudotriton montanus*)

DOWNSTREAM SALAMANDERS Each of these three species of salamanders in the second row is a very close relative of the one above it—a member of the same genus (the names, *Desmognathus*, *Eurycea*, and *Pseudotriton*, are the genus names). These three live in swampy floodplains along the same streams, but in their fourth-order or greater reaches, where the water is less well oxygenated. The two sets of salamanders are sets of geminate species, like the owls shown in Figure 6–4 (page 91).

FIGURE 11–10

Six Riverside Salamanders

Salamanders began evolving in eastern North America since 150 to 200 million years ago and later colonized many other parts of the world. No mammal has existed over comparable time spans. Six of the eight salamander families alive today have populations in Florida.

Source: Means 2000.

Birds. Birds also flourish on the abundant fare in floodplains. According to researchers Larry D. Harris and others, more than nine-tenths of all bird species of eastern North America use bottomlands at one time or another. The deep, protected canopy attracts songbirds not seen elsewhere. Warblers and thrushes, passing through, rest and feed in the swamps and forests and fill them with song. American woodcocks make their homes deep in the forest. Birds of prey, including the osprey, bald eagle, barred owl, red-shouldered hawk, and the swallow-tailed and Mississippi kites, find abundant prey in floodplains.[14]

Many bird families are represented by several species in floodplains. For example, because there are so many insects devouring wood and litter on the floodplain, several species of woodpeckers can share the resource. The red-bellied woodpecker feeds on beetle larvae in live trees. The ivory-billed woodpecker (now presumed extinct) fed on grubs under the bark in newly

Florida native: One-toed amphiuma (*Amphiuma pholeter*). Although shown here against a contrasting background, this animal actually lives in black muck, where it blends in almost invisibly. It thrives in muck pools in the Apalachicola ravines.

Spring peeper *(Pseud-acris crucifer)*. This treefrog emits a sweet, drawn-out whistle in the spring.

River frog *(Rana heck-sheri)*. This big, shy frog makes a snoring sound at night from the margins of streams and rivers.

Bullfrog *(Rana cates-beiana)*. This big frog's calls create a deep, bass drumbeat in Florida swamps at night.

Pig frog *(Rana grylio)*. This frog utters a loud, grunting call both day and night around ponds and floodplain lakes.

Eastern narrowmouth toad *(Gastrophryne caro-linensis)*. This animal goes "ba-a-a" like a sheep.

Northern cricket frog *(Acris crepitans)*. The call of the cricket frog is a treble "gick-gick-gick."

FIGURE 11–11

Six Native Florida Frogs

All of these frog species and many more thrive in Florida's river floodplains.

dead trees. The pileated woodpecker drills for grubs from deep in the wood of older, dead trees. List 11—4 identifies a few of the birds of Florida's river floodplain swamps.

Shelter in a floodplain is as diverse and available as food. Tree branches at different levels offer a surfeit of nesting sites. Floodplain hardwoods develop cavities, rotted by many kinds of fungi. These make needed homes for the wood duck, eastern screech-owl, and others, as well as for small invertebrates, lizards, and mammals.

Eastern river cooter *(Pseudemys concinna)*. This turtle spends its life in and near streams except when nesting.

Common musk turtle *(Sternotherus odoratus)*. This small turtle spends nearly all of its time in the water.

Barbour's map turtle *(Graptemys barbouri)*. This turtle occurs only in the Apalachicola River drainage basin, where it must have evolved.

FIGURE 11–12

Three River-Swamp Turtles

Two summer natives. Both of these birds spend their summers and breed in Florida. LEFT: A flycatcher (*Myiarchus* species) seeks out moist woodlands and floodplain forests, where it spends most of its time in the canopy. RIGHT: The prothonotary warbler (*Protonotaria citrea*) nests over water in Florida swamps.

Florida native: Cottonmouth (*Agkistrodon piscivorus*). Widespread across the southeastern United States, this venomous snake is found wherever there is water.

On the lower river near the coast, waterfowl abound: mallards, pintails, red-breasted mergansers, American black ducks, and gadwalls. Wading birds nesting over sloughs need only drop down and hop back up to feed their broods a bounty of nutritious morsels.[15]

Mammals. Finally, mammals, too, find ample fare in floodplains. Thanks to their bundant hickory nuts, mulberries, pecans, acorns, and other nourishing nuts and seeds (mast), floodplain forests support many more deer than do upland forests. The opossum and two species of squirrels also live on mast. Beavers work the trees and create pools in small tributaries where the sound of flowing water stimulates them to build dams. The eastern woodrat and two species of mice have habitats in the floodplain; so do the Florida black bear, the raccoon, the long-tailed weasel, the striped skunk, the river otter, the gray fox, and the bobcat.

In sum, natural, forested river floodplains are rich habitats that are home to many animals. The invertebrate animals have never been completely inventoried for any floodplain, but the vertebrates around some rivers have been counted. The Suwannee River floodplain, for example, supports 385 species: 39 species of amphibians, 72 species and subspecies of reptiles, 232 species of birds, and 42 of mammals. How many snails, worms, spiders, insects, and soil organisms must there be? How many fungi? lichens? bacteria?[16]

* * *

Floodplain wetlands are only one among several natural ecosystems associated with streams and rivers. Streams themselves, and their estuaries, contain major aquatic habitats that are described in Chapters 13, 14, and 18. The floodwaters that sweep muck and litter from floodplains feed these systems with the materials on which they run. Floodplain wetland organic matter by the tens of millions of tons fuels tremendous production of seafood in Florida's streams and offshore waters.

In concluding these four chapters on Florida's freshwater wetlands, it is important to bring the focus back to their intrinsic value. Wetlands, when functioning naturally, play crucial roles on the landscape. Florida is a flat land, and millions of acres are soggy for much of the year. Given this topography and climate, what will grow best over all those acres? The answer is native wetland plants. Each is adapted to a specific wetland envi-

LIST 11–4
Floodplain swamp birds (examples)

Acadian flycatcher
American woodcock
Barred owl
Carolina wren
Chimney swift
Hairy woodpecker
Hooded warbler
Mississippi kite
Northern cardinal
Northern parula
Pileated woodpecker
Prothonotary warbler
Red-eyed vireo
Red-shouldered hawk
Swainson's warbler
Swallow-tailed kite
Veery
White-eyed vireo
Wood duck
Yellow-crowned night-heron

Source: Adapted from *Guide* 1990, 37.

Florida native: Golden mouse (*Ochrotomys nuttalli*). Seldom seen, this little rodent lives in many Florida environments from high, dry scrub to low, wet swamps. Because it nests in shrubs and trees, it survives periodic floods.

Florida native: Bobcat (*Lynx rufus*). These wild cats thrive in swamps and their kittens are born skilled at climbing trees.

ronment, and native animals are, in turn, adapted to use the plants. For these reasons, the genetic information in native wetland species is valuable. After all, what other plants or animals are better adapted to occupy these lands and deliver their products and services, than those that have evolved here?

PART FOUR

INTERIOR WATERS

Florida's **interior waters** include freshwater lakes, ponds, streams, rivers, and underground waters.

Bluegill (*Lepomis macrochirus*) in Ginnie Springs, Gilchrist County.

WS/KP

Florida native: Halloween pennant dragonfly (*Celithemis eponina*). Like all dragonflies, this animal spends its larval life in lakes.

small organisms living on the plants and in bottom sediments. It is also a lake in which fish can live everywhere, where there are no pockets of "dead water." That means a lake whose water is well aerated, for fish breathe oxygen just as land animals do. Because most of Florida's lakes tend to be shallow, wind can create turbulence that reaches to the bottom and aerates all of the water. Typically, Florida's natural lakes and ponds range from seven to about twenty feet in depth. Most formed as limestone dissolved and caverns collapsed beneath them, so if viewed from above they are circular, and in cross section they are shallow bowls.

A few of Florida's shallow lakes are other shapes. Some lakes formed originally as lagoons between sand bars or barrier islands and then were left on land and filled with fresh water. Some are oxbows left behind by rivers. An occasional odd-shaped upland swamp lake may have been scooped out by a meteor impact. And a few lakes are circular but deep: true sinkholes, whose bottoms have fallen into limestone caverns underground.

Within the basin of a shallow lake, aquatic zones offering different habitats are arranged concentrically (see Figure 12–1). Each zone supports a different kind of life. The shoreline zone, from zero to five feet of depth, has the greatest variety of life forms. (From Chapter 10, recall that this zone is a marsh with open water in the center.) Aquatic and wetland plants spread their leaves in and above the water, offering food and hiding places for insects, other invertebrates, and small fish. These small animals serve as food sources for larger fish and birds. The debris shed by living things around the lake's edge supports decomposer organisms, and they generate acid as they work. As a result, the shoreline zone is the most acidic part of the lake.

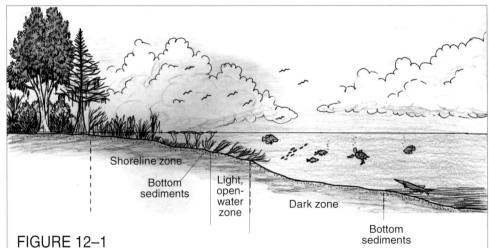

Shoreline zone

Bottom sediments

Light, open-water zone

Dark zone

Bottom sediments

FIGURE 12–1

Zones in a Productive, Wild Lake

The border of plants around the edge to a depth of five feet is the shoreline zone. Their robust growth signifies that this must be a productive lake. Animals' eggs can develop among the plants, and their young can hide, feed, and grow there. Among users of this zone are healthy, reproducing native reptiles, amphibians, and fish. Water birds—both waders and swimmers—feed on the aquatic plants and lake-dwelling animals.

Inside the shoreline zone is a deeper zone, the light, open-water zone, where underwater plants grow. Less animal life is present here. In the center of the lake, which is deeper still, sunlight cannot reach the bottom. This is the dark zone, but the surface is light and floating plants can thrive here. Finally, the bottom sediments support an indwelling community.

Reminder: Wetland plants are *emergent* plants, rooted on the bottom and protruding above the water.

Most *aquatic* plants are *submersed* plants (plants that are rooted on the bottom and growing wholly under water). A few, such as duckweed, are *floating* plants (plants that float free in the water).

In the light, open-water zone, rooted, emergent plants drop out, but in clear water sunlight can still penetrate to the bottom and support the growth of rooted *underwater* plants. (Notice, the next time you boat into a natural lake with shoreline vegetation, that underwater plants are visible on the bottom just beyond the zone where the emergent plants end.) Microscopic floating algae and other plants (plankton) can grow here, too, because they don't have to compete for light with taller plants. Microscopic animals eat these plants and each other. These tiny living things may drift into the shoreline zone and become other creatures' lunches, or they may spread into deeper parts of the lake, where they will finally drop out of the sunlight, die, and feed the cycle of decomposition at the bottom. Few fish inhabit the light, open-water zone, because it offers them little food or shelter.

In the deepest part of the lake, the dark zone, bacteria and fungi feed on organic matter that drifts down to them. In the process, they free nutrients into the water, which become available again for life along the shore. Decomposition depletes the oxygen supply, but deep water offers a temperature-stable environment to a few specially adapted fish such as catfishes and chubsuckers, which feed on decomposing materials.

The sediments on and in the floor of a lake contain a community of microorganisms, worms, mollusks, crustaceans, insect larvae, and other larvae. When healthy and well developed, this community (known as the *benthos*) is highly organized. Distinct groups of organisms live at different depths in the bottom sediments and different species are abundant at different seasons. The benthic community is especially rich beneath beds of rooted aquatic plants where oxygen and plant litter are available.

For simplicity's sake, this description was of a smoothly conical lake, but of course real lakes may be much more complex. Where a lake has a deep hole near the edge, a dark zone may interrupt the border of water

> **Plankton** are passively floating or weakly swimming, tiny (mostly microscopic) plants and animals in a body of water.
>
> The plant members of this community are the **phytoplankton** (FYE-to-plankton); the animals are **zooplankton** (ZOH-o-plank-ton).

> The sediments on the floor of a water body and the organisms that live among them are known as the **benthos**, or **benthic community**.

MS

Florida native: Largemouth bass (*Micropterus salmoides*). Fish like to hang suspended in underwater meadows like this where, as predators, they are concealed from their prey, and as prey, they are hidden from their predators.

Florida native: Canada goose (*Branta canadensis*). Most Canada geese seen in Florida do not migrate north in summer but reside continuously and rear their goslings within the state.

weeds. Where a peninsula extends into the middle, or where an island stands, another shoreline zone will be present. Lakes may have bumpy bottoms and hold interesting structures—rocky outcrops, springs, fallen trees, stream beds. From the point of view of a fish, a lake with a complex bottom makes an excellent homesite.

DIVERSITY OF FLORIDA LAKES

Fly over Florida in a wet season and you will see the mirrorlike reflections of thousands of lakes and ponds. Lakes and ponds occur wherever there are basins that, at least in rainy seasons, hold open water in their centers (water typically more than five feet deep). Figure 12–2 shows that lakes are especially numerous on level terrain that is deep in sand. The high, sandy uplands of the central peninsula are outstanding: four counties (Lake, Orange, Polk, and Osceola) account for more than a third of all of Florida's lakes. Lakes are also associated with the St. Johns River region, which is mostly low, sandy, and flat with many wetlands. Smaller lakes and ponds lie in sandy basins across all of Florida's flatwoods and nestle behind coastal dunes near the shore.

Although lakes are more numerous in some regions than in others,

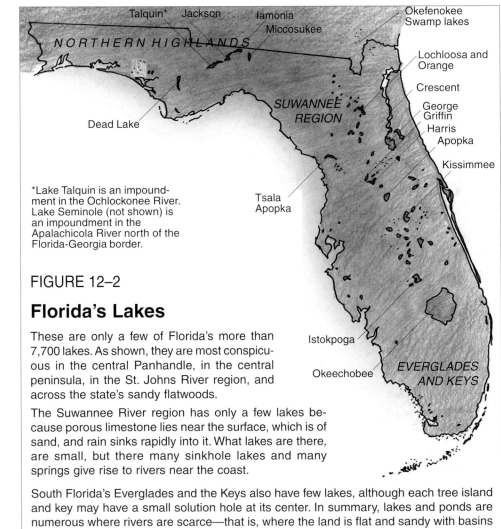

Talquin* Jackson   Iamonia Okefenokee
                    Miccosukee Swamp lakes
NORTHERN HIGHLANDS

                              Lochloosa and
                              Orange

                    SUWANNEE   Crescent
                    REGION   George
Dead Lake   Griffin
                               Harris
                               Apopka

                               Kissimmee

*Lake Talquin is an impound-
ment in the Ochlockonee River.   Tsala
Lake Seminole (not shown) is   Apopka
an impoundment in the
Apalachicola River north of the
Florida-Georgia border.

FIGURE 12–2

Florida's Lakes

These are only a few of Florida's more than 7,700 lakes. As shown, they are most conspicuous in the central Panhandle, in the central peninsula, in the St. Johns River region, and across the state's sandy flatwoods.

Istokpoga

Okeechobee   EVERGLADES
             AND KEYS

The Suwannee River region has only a few lakes because porous limestone lies near the surface, which is of sand, and rain sinks rapidly into it. What lakes are there, are small, but there many sinkhole lakes and many springs give rise to rivers near the coast.

South Florida's Everglades and the Keys also have few lakes, although each tree island and key may have a small solution hole at its center. In summary, lakes and ponds are numerous where rivers are scarce—that is, where the land is flat and sandy with basins that hold water, and where the water does not flow overland to the sea.

every region has at least some permanent, still-water bodies, and each has a character of its own. A given Florida lake or pond may be alkaline or acid; rich in nutrients or poor; clear or colored. It may have a sandy, silty, clayey, or organic bottom. It may have streams running into it, or out of it, or both, or neither. Some lakes and ponds hold water permanently, some temporarily; and the permanent ones all have fluctuating water levels. The surrounding soils and ecosystems vary greatly and affect the water's chemistry. Water temperature varies with the amount of shade afforded by encircling forests and swamps as well as with the climate from north to south Florida.

Lakes can be classified in a variety of ways.[1] The Florida Natural Areas Inventory (FNAI) recognizes seven idealized types of lakes (see List 12–1). This chapter treats six of them, leaving sinkhole lakes to Chapter 14. The next few paragraphs briefly describe each type of lake and Figure 12–3 shows diagrams of the first two types.

Clay-Hill Lake. Clay-hill lakes of varying shapes, depths, and slopes are scattered in depressions all across north Florida's highlands. Their shorelines vary greatly; some are grassy, some shrubby, some forested with swamp trees. Conspicuous within the water are about a dozen species of aquatic plants, 25 species of fish, and about 30 other kinds of animals (see end of chapter).

Pristine clay-hill lakes are rare today. The lakes that most travelers see from highways have been greatly altered by human settlement, destruction of lakeside vegetation, and pollution.

Sandhill Lake. Sandhill lakes lie on ancient sand formations that now lie inland. The typical sandhill lake is a big, round lake with a smooth, sandy bottom, underlain by limestone and bathed in sun. The water is slightly acid from vegetation in the surrounding sand, low in minerals, and supports sparse populations of water plants and fish. Even so, it is an important breeding place for local amphibians and insects.

LIST 12–1
Six categories of Florida lakes[a]

- Clay-hill lakes
- Sandhill lakes
- Swamp lakes[b]
- Flatlands lakes[c]
- Coastal dune lakes
- Coastal rockland lakes

Notes:
[a]Sinkhole, or karst, lakes are treated as parts of the ground-water system, in Chapter 14.
[b]Basin swamps are treated with swamp lakes here.
[c]Basin marshes are treated with marsh lakes here.

Source: Adapted from *Guide* 1990, 45–51.

Clay-hill lake. The lake is perched on a bed of clay, through which water does not sink down to the limestone aquifer below. It receives its water from rain and runoff, and perhaps from a tributary or two. It loses water by evaporation.

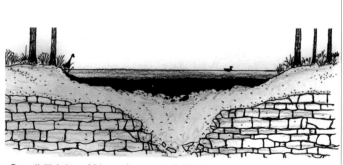

Sandhill lake. Although a sandhill lake has a floor of sand, accumulated organic matter largely prevents downward percolation of the water. Its clear, shallow water comes only from rain and ground-water seepage, and is lost by evaporation. As it evaporates, the water may become very dark, but still translucent (like tea).

FIGURE 12–3

Clay-Hill and Sandhill Lakes

A sandhill lake on Caladesi Island, Pinellas County.

Because they are so shallow, usually less than ten feet deep, sandhill lakes expand and shrink dramatically with seasonal rains and droughts. They even dry out completely every 20 or 30 years.

A pristine sandhill lake typically has much of its surface occupied by emergent plants—in fact, but for the open water zone in the center, it might as well be considered a wetland. Insects are swarming everywhere, feeding on the plants and using them as perching, hiding, and breeding places. Many of the insects are unusual; some are endemic. Because insects thrive, animals that eat them are numerous, too. Sandhill lakes are indispensable breeding and nursery places for frogs and salamanders. The gopher frog, now threatened, makes its home near sandhill lakes.

A number of long-legged birds wade along the shore, dipping for fish, frogs, and salamanders. Ducks visit to eat the aquatic vegetation, and many birds court, breed, feed, and nest nearby.

Swamp Lake or Basin Swamp. Swamp lakes are forested lakes in swamps. Swamp trees around the edges of these lakes produce a deep shade in the shoreline zone; this and the acidic water prevent the growth of most aquatic vegetation. As a result, although swamp lakes support some plant and animal life, they are not highly productive.

The formation of one type of swamp lake, an oxbow lake in a river floodplain, was described in Chapter 11; others lie in basins on higher ground. Swamp water washes into such lakes from the surrounding swamps and forests carrying disintegrating organic matter, so they are somewhat acidic, low in minerals, and dark brown or black in color. These characteristics give them the alternative name, blackwater lakes. If a swamp lake has no open water in the center, it is called a basin swamp. The distinction may be seasonal: what is a swamp lake during the rainy season may be a swamp during the dry season.

The water level in all basins rises and falls with the rainy and dry seasons. In the case of basin lakes in river floodplains, the water also flows.

Swamp lakes, also called **blackwater lakes**, include:

- **River floodplain lakes**
- **Oxbow lakes**

PRICELESS FLORIDA

A swamp lake in Leon County.

Shoreline of a swamp lake.

Lakes in the floodplains of high-discharge rivers may have their contents completely swept away once or twice a year, but life quickly returns from bottom mud and from upstream waters.

Flatlands Lakes and Ponds. Lakes in Florida's flatwoods and prairies may be surrounded by sparse prairie plants or by dense shrubs. Often, saw palmetto encircles them at the high-water line. Within the water are numerous emergent plants: varieties of yelloweyed grass and spikerushes along with maidencane and creeping primrosewillow. Many animals live in flatlands lakes, and many others visit them to feed or breed. As with swamps, flatlands lakes in the wet season may become basin marshes in the dry season.

Among flatlands ponds are cypress ponds, which lie in those small, roughly circular depressions that have formed all over Florida's coastal

Flatlands lakes include:

- **Cypress ponds/domes/ donuts**
- **Prairie lakes**
- **Marsh lakes (basin marshes)**

Two flatlands lakes. At left is a prairie lake, Lake Okeechobee, which is Florida's largest lake. Lake Okeechobee occupies 690 square miles in Glades, Hendry, Highlands, Martin, Okeechobee, and Palm Beach counties. At right is a marsh lake in Paynes Prairie, Alachua County. During wet times, a sinkhole that drains the prairie becomes plugged, rainwater accumulates, and parts of Paynes Prairie become marsh lakes.

Chapter 12—Lakes and Ponds

Florida lakes fluctuate by five feet or more in a year, and a five-foot change in depth makes for a much greater change in area. A foot or two added to the depth of a shallow lake may expand its margin by 100 feet or more. A foot or two lost may shrink it drastically.[5]

Expansion and contraction of a lake's area profoundly affect shoreline life. When a lake expands, terrestrial vegetation becomes submerged and dies back around the newly flooded margin. Then rotting plant matter (muck) accumulates in the water. Aquatic plants thrive. Fish and other water-loving wildlife expand their ranges and population sizes. Each rain increases the acidity of a lake temporarily by flushing dead plant matter into it.

When a lake dries down, it becomes cleansed—that is, its plant matter oxidizes, and disappears. Fires often burn through surrounding marshes and the organic lake bottom itself may even burn. While the shoreline and bottom are dry, the seeds of many nearby plants quickly colonize the newly available soil, and plants whose seeds have been lying dormant under water grow and flower and reseed quickly. Unusual species may bloom in profusion in dry lake beds, then go dormant again for many years while under water. An observer of lake cycles in the Panhandle reports seeing two such flowers only a few times in 50 years: Kral's yelloweyed grass and Panhandle meadowbeauty, both endemic to north Florida's sandhill lakes. When the rains return, low-oxygen conditions keep the fresh crop of seeds dormant until once again the lake dries and they become able to sprout.[6]

Once cleansed by drying, a lake basin can refill with clear water. Then it becomes habitable again by aquatic plants and animals. When refilled by rain, a lake may regain its fish from eggs that have lain dormant in the mud or fry that have survived in deep holes. Some lakes receive overflow from other water bodies that brings in animals, and occasionally wading birds fly in, bringing fish or salamander eggs stuck to their legs.

Just as terrestrial plants seem to disappear when a lake bed is full of water, lake animals go out of sight when the lake goes dry. Then during the next rainy season, lake animals grow and mature quickly, breed, and lay their eggs or produce their young. Anyone who has heard the sudden calling of frogs after the first rain that terminates a drought knows how promptly these animals respond to the arrival of water. They seem to have come out of nowhere.

When lakes shrink, the adult aquatic animals that have lived in them become more and more crowded, making them easy prey for other animals. Predators devour them and thrive in their place—a population shift in response to the lake's seasonal changes. Lakes vary year by year, too. About once every ten years, a third less rain falls than the average.[7]

Cycling is indispensable to lake health. If a lake is not permitted to dry down and cleanse itself periodically, it will increasingly fill with muck and become so low in oxygen that its fish and other animals die. Some Florida lakes have been created by damming streams, and if they are not allowed to dry down periodically, they become unproductive, unhealthy places.

Over centuries, lakes are born, live, and die. They are born when the land slumps down, either gradually or suddenly, due to subsidence or col-

Cypress pond, nearly dry. When the pond is full, its margin extends all the way to the hardwood trees in the background.

The pond cycles repeatedly from wet to dry to wet again. When wet, it accumulates peat. When dry, it loses its peat to oxidation.

lapse underground. They fill with water and life, but they last only as long as they can cleanse themselves as described. Otherwise, they fill with undecomposed organic matter and become solid ground.

Importantly, except for floodplain lakes, most of Florida's larger lakes retain their water for years. In contrast, most lakes in mountainous regions like those along New England's rivers have inlets and outlets, and all of their water is replaced about every three to six months. Whatever sinks to the bottom of a Florida lake tends to stay there indefinitely.

That Florida lakes retain whatever runs into them makes them especially vulnerable to pollution. Even excess mud and silt, sediments normally found in lakes, can be harmful if allowed to run in and settle on the bottom. When sediments slide into a lake, they smother the benthic inhabitants. Oxygen is consumed by decomposition of overly abundant nutritive material, and the benthos becomes impoverished. Whereas the bottom sediments once housed many species of thriving worms, mollusks, crustaceans, insect larvae, and others, only a few worm species and midge larvae may remain. An impoverished benthos signifies a lake's lost health and a need for restoration. Guidelines for the monitoring of Florida lakes are available.[8]

PLANTS IN LAKES AND PONDS

The still water of lakes and ponds permits floating and suspended, as well as attached, vegetation to develop well. Microscopic algae and photosynthetic bacteria may flourish in a healthy lake in full sun, both floating and attached to aquatic plants as periphyton. Microscopic plants are the basic foodstuffs in the lake food web. To illustrate their diversity, Figure 12–4 depicts a few of the microscopic plants found in Florida lakes.

Exactly which species of microscopic plants will grow in a given lake or pond depends on the nature of the bottom sediments, on the water chemistry, and on the other plants and plant eaters present. A single lake may have a dozen or more species, and all of Florida's lakes and ponds taken together hold hundreds of species of floating and attached algae and cyanobacteria.[9]

All plants, including aquatic plants and algae, require nitrogen, phosphorus, and other nutrients to grow, and up to a point, the more nutrients are present, the more abundant the plants will be. However, a superabundance of nutrients running into a lake can lead to the growth of so many algae that the lake cannot support its other aquatic organisms. As algae pile up in the water, some sink out of the sunlight and die; this leads to overgrowth of bacteria, oxygen depletion of the water, and death of the lake's fish and other animals. The progression of this process to the extreme, known as eutrophication, presents a major hazard to Florida's lakes. Broadleaf cattail, which is native to Florida, becomes an invasive weed when a lake becomes eutrophic.

Large aquatic plants are also diverse in Florida lakes, thanks to the long growing season and the mostly shallow water. List 12–2 itemizes some of them and Figure 12–5 illustrates a few.

Reminder: **Periphyton** is an assemblage of algae growing under water attached to plant stems, tree trunks, and rocks and layered on bottom sediments.

Lake waters may be ***dystrophic***, ***oligotrophic***, or ***eutrophic***.

Dystrophic waters are low in nutrients and/or acidic and support little growth of plants.

Oligotrophic waters have moderate nutrient levels.

Eutrophic (YEW-tro-fic) waters have abundant nutrients and produce abundant plant life.

Eutrophication (YEW-tro-fi-CAY-shun) is the process by which a lake becomes more eutrophic, and can lead to death of the lake's living things.

Florida native: Ebony jewelwing (*Calopteryx maculata*), a damselfly. Damselflies have swarmed around lakes every spring since 300 to 350 million years ago. Their eggs and larvae develop in the bottom sediments and are important links in the aquatic food web. The decline of a lake's damselflies signifies decline of the lake's water quality.

water. Key members of this water-filtering community are lake mussels, clams, and filter-feeding worms, which are numerous in the bottom sediments of a healthy lake. Filter-feeding fish may also be plentiful in lakes; one is the threadfin shad, which cruises the open water in the center of a lake and filters plankton from it.

The number and variety of creatures that dwell in the bottom sediments give a reliable indication of a lake's health. A productive lake's sediments include the tiny larvae of mollusks, crustaceans (isopods and amphipods), insects, and filter-feeding worms. The fecal pellets dropped by these animals and by fish serve as fertilizer for rooted plants and as food for decomposer organisms.

Given high-quality water, freedom from infestation by exotic weeds and fish, a minimum of shoreline disturbance, and healthy bottom sediments, a Florida lake may offer a bounty of healthy foods to many native game fish. The ever-so-popular largemouth and other bass species find feasts in healthy Florida lakes.

Many lake fish are of the several kinds called bream (or sunfish, because their discoid, slim bodies coated with metallic scales flash in the sun). Many others are bottom feeders such as catfish. List 12–3 names some of the 40-odd fish species that are native to Florida lakes and ponds.

A fisherman can sing you a litany of the game fish to be found along a healthy lake shore, and fishermen who know the fishes' habitat preferences can cast their lures in just the right places. Are you fishing for bream? Fish in five-foot deep water bordering the shallow zone and cast your flies near underwater hiding places such as fallen trees or aquatic plants. Are you fishing for bass? Cast your flies on the shore, wiggle them into the shoreline vegetation, and pull them through the thickest cover.

LIST 12–3
Fish found in Florida lakes (examples)

Florida's fish include two families of cartilage-skeleton (holost) fishes (gars and bowfins) and eleven families of bony-skeleton (teleost) fishes.

<u>Gars and bowfins</u>
Bowfin
Florida gar
Shortnose gar

<u>Minnows</u>
Coastal shiner
Florida pugnose minnow
Golden shiner
Taillight shiner

(continued on next page)

Florida native: Fishing spider (a *Dolomedes* species). A dolomedes spider spends most of its time on top of the water, hunting surface insects, or diving to catch swimming prey. Hairs on its legs trap oxygen, so the spider can breathe under water. It may also perch on the shore and stir the water with its front legs to attract tiny fish that are looking for insects.

PRICELESS FLORIDA

Bass (*a Micropterus* species) by a submerged tree stump. Bass thrive in lakes that offer them abundant food and shelter. Foods popular with newly hatched bass include plankton and tiny aquatic insects. Larger bass eat snails, crayfish, salamanders, frogs, fish, small water birds, and even (if they are so unfortunate as to fall into the water) mice.

Figure 12–6 displays a few of the other animals found in lakes. Besides these, the aquatic realm beneath the surface is full of microorganisms, insect larvae, benthic worms, clams, mussels, young fish, tadpoles, and numerous other living things.

Lakes are important to visiting animals, too. Water birds prey on fish, and the more fish are in a lake, the more water birds will be there. To sustain the life of a lake, keep its shoreline wetlands natural and undisturbed so that fish and fish-eating birds will be numerous. Visiting birds might include bitterns, herons, eagles, rails, wrens, and grackles, among others. Animals also come to lakes from the surrounding land for fresh drinking water. Wild turkeys gobble at dawn from the swamps around lakes, and barred owls hoot at dusk. Deer, bear, raccoon, opossum, bobcat, and other animals visit lakes from nearby wildlands.

Lakes are stopping places for migratory birds. Early in the last century, thanks to the abundant undisturbed lakes in the area, ducks and geese used to come to Florida's lakes by the thousands. Fall and winter saw vast influxes of ring-necked ducks, wood ducks, mallards, American black ducks, pintails, gadwalls, American wigeons, teals, lesser scaups, common goldeneyes, buffleheads, red-breasted mergansers, and ruddy ducks. These birds need healthy lake vegetation and fish to feed on, and they in turn enrich the variety of life there.

✳✳✳

Lakes are highly valued by people who love to swim, boat, and fish in them, but they also offer services, as all ecosystems do. They are reservoirs

CHAPTER THIRTEEN

ALLUVIAL, BLACKWATER, AND SEEPAGE STREAMS

In 1867, while traveling through Florida, John Muir commented that "Most streams appear to travel through a country with thoughts and plans for something beyond. But those of Florida are at home, do not appear to be traveling at all, and seem to know nothing of the sea. . . . No stream that I crossed today appeared to have the least idea where it was going."[1] And indeed, Florida has some of the flattest streams on the continent. To be sure, across north Florida, some streams run down slopes out of clayey uplands, but most Florida streams arise in lowlands where the land is nearly level. These waterways make hairpin turns and reverse them, traveling as much sideways as forward. A walker on the bank of a typical, flat Florida stream can easily keep up with it—the average flow is only 1 mile per hour. The very flat St. Johns River, 150 miles from its mouth flows at only a third of a mile per hour. And while the St. Johns River flows north, the Kissimmee, which is parallel to it, flows south through essentially the same terrain.[2]

Still, streams do flow, however slowly, and this makes them altogether different from the lakes and ponds of the last chapter. Anything that lives in a stream, plant or animal, has to be able to cope with and take advantage of a constantly moving environment. Plants must be rooted on the bottom, or, if floating, must circulate in eddies off the main stream. Animals must anchor themselves in bottom sediments, seek quiet backwaters, swim constantly against the current, or alternately swim upstream and drift downstream, perhaps at different times in their life cycles. Mullet jump repeatedly on their way upstream.

Streams in Florida are of many kinds; no two are alike. However, they are grouped here into four idealized types, of which three are treated in this chapter: alluvial, blackwater, and seepage streams. (The fourth type, spring-run streams, are treated in Chapter 14, together with the aquatic caves with which they are connected.) Finally, to demonstrate the complexity that exists in an actual stream, one of them, the Suwannee, is described in more detail.

OPPOSITE: The Santa Fe River, a major tributary to the Suwannee River. The photo was taken in Gilchrist County, looking toward Columbia County.

All natural, flowing, freshwater systems are **streams**. Thus all **rivers** are streams, as are **creeks** and **strands**. A canal, because it is not flowing, is not a stream.

Florida native: Blackbanded sunfish (*Enneacanthus chaetodon*). This small fish, shown here in an aquarium, swims in lowland streams all along the U.S. east and Gulf coasts.

An **alluvial stream** carries sediments eroded from the land and detritus from streamside wetlands. At times of flood, an alluvial stream is nutrient rich and muddy.

A **spring-fed stream** (**spring run**) arises from one or more limestone springs, from which it receives mineral-rich water. Spring runs typically run clear.

A **blackwater stream** seeps out of sandy soil underlying swamps or marshes and carries few nutrients. Dissolved compounds from decaying vegetation make its water acid and dark in color.

A **seepage stream** receives most or all of its water via ground-water seepage from its banks. Seepage streams are usually nutrient poor and clear. (A **steephead stream** is a special kind of seepage stream. It seeps from beneath the walls of a steep-sided, sandy ravine.)

Figure 13–1 identifies several dozen Florida streams. None is large in comparison with the world's greatest rivers. Compared with the Amazon, for instance, even the Mississippi is small, and Florida's largest stream, the Apalachicola, is smaller still. But Florida's 1,700-odd streams move water across thousands of miles of terrain. Taken together, they carry an enormous volume of water because they drain regions of such high rainfall. Florida has more than a dozen major rivers—that is, rivers that drain areas greater than 500 square miles each—and the 55 to 70 inches of rain that fall each year on 500 square miles make an awesome pile of water.[3]

DIVERSITY OF FLORIDA'S STREAMS

As indicated by the definitions in the margin, Florida's streams carry water from various sources in all kinds of terrain. Some streams are muddy, some colored with organics, some clear. Some are acid and rich in iron, some are alkaline and rich in calcium and magnesium, some carry other salts or sulfur, and some are nutrient poor. Some vary much more in volume and temperature than others. Some are long and have many tributaries, others are short and have none.

Streams vary in volume over the year, some much more than others. In areas that are impervious to water such as the clayey highlands of north Florida, rain or drought in the upper reaches of a stream may greatly amplify or deplete its downstream flows. The alluvial streams that flow through these areas depend almost entirely on rainfall and so vary manyfold in volume from wet to dry seasons. In contrast, on land that is permeable to water such as Florida's sand hills and karst lowlands, ground water exchanges readily with stream water, replenishing it during droughts and accepting it after rains. As a result, the seepage streams and spring runs found in these areas vary only a little in volume. Some small blackwater streams may dry up altogether during dry times, but some are maintained by base flow from seepage along their banks, but can vary more than a hundredfold over the year.

Two seasons on the Alapaha River. The Alapaha is a mostly blackwater river that flows out of swamps in northeast Florida. At left, the river is in flood, at right, it is dry. Both phases of the river are normal.

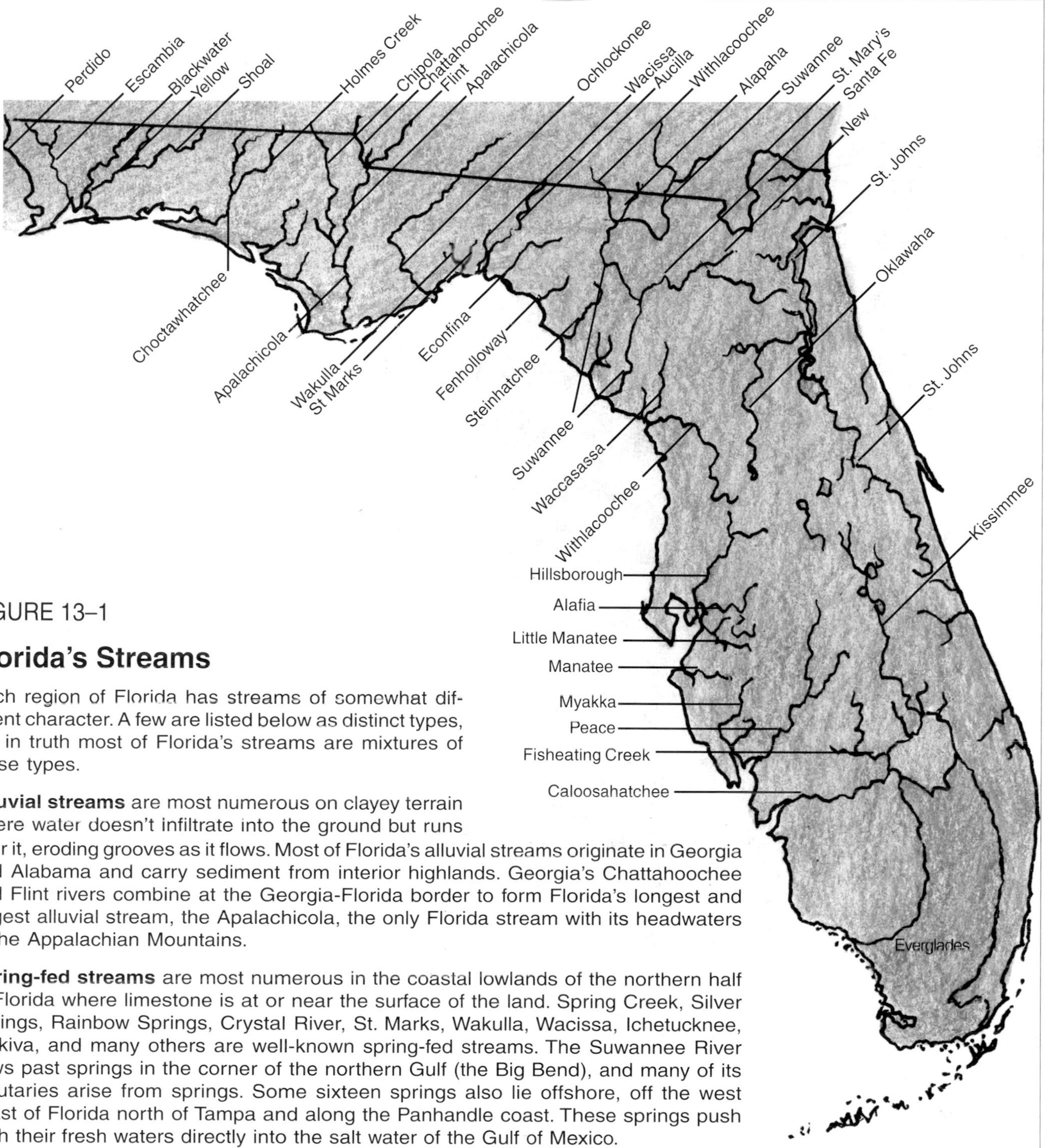

FIGURE 13–1

Florida's Streams

Each region of Florida has streams of somewhat different character. A few are listed below as distinct types, but in truth most of Florida's streams are mixtures of these types.

Alluvial streams are most numerous on clayey terrain where water doesn't infiltrate into the ground but runs over it, eroding grooves as it flows. Most of Florida's alluvial streams originate in Georgia and Alabama and carry sediment from interior highlands. Georgia's Chattahoochee and Flint rivers combine at the Georgia-Florida border to form Florida's longest and largest alluvial stream, the Apalachicola, the only Florida stream with its headwaters in the Appalachian Mountains.

Spring-fed streams are most numerous in the coastal lowlands of the northern half of Florida where limestone is at or near the surface of the land. Spring Creek, Silver Springs, Rainbow Springs, Crystal River, St. Marks, Wakulla, Wacissa, Ichetucknee, Wekiva, and many others are well-known spring-fed streams. The Suwannee River flows past springs in the corner of the northern Gulf (the Big Bend), and many of its tributaries arise from springs. Some sixteen springs also lie offshore, off the west coast of Florida north of Tampa and along the Panhandle coast. These springs push forth their fresh waters directly into the salt water of the Gulf of Mexico.

Blackwater streams predominate in the Panhandle and in northeast Florida wherever water seeps out of swamps and marshes. Some arise in swampy uplands, but many arise in the coastal lowlands and flow only a few miles to the sea. The Blackwater River in the Panhandle is a blackwater stream, and the St. Johns and St. Mary's on the east coast originate as blackwater streams.

Seepage streams arise from ground water that seeps out of the banks of sand hills. Valley wall seepage also is an important source of the water in the Choctawhatchee, Ochlockonee, Suwannee, and Apalachicola rivers.

Steephead streams are a special class of seepage streams. They seep from the bases of steep-sided, sandy ravines as described in Chapter 6.

Sources: Heath and Conover 1981, 115; Johnson 1996.

Ocean fish such as sturgeon, striped bass, and shad, which come into freshwater streams to spawn, are known as **anadromous** (an-ADD-ro-mus) fish.

American eels and hogchokers, which leave freshwater environments to spawn in salt water, are **catadromous** (ca-TAD-ro-mus) fish.

MS

Florida native: Striped bass (*Morone saxatilis*). The striped bass is one of several fish species that enter streams on both the Atlantic and Gulf coasts to spawn.

The fluctuations (or lack of them) in a stream's total volume affect the life it supports. Because alluvial streams receive large influxes of fresh nutrients and water twice a year, the ecosystems they support are the most diverse. Because spring runs vary only twofold or less in volume, they offer a nearly constant environment for a rich assemblage of plants and animals. Blackwater streams, which often almost dry up, place stringent limits on the plants and animals that can grow there. Variations in stream volume also affect estuaries as described in Chapter 18.

Florida's streams also vary in the topography of their banks and bottoms and in the amounts of light and shade they receive. Thanks to these variations, different streams are home to different sets of algae, rooted plants, bottom microorganisms, fish, and other animals. Florida's streams provide habitat for some 100 species of strictly freshwater fishes and another 100 species of saltwater fishes that swim off the coast but make their way up freshwater streams to forage or to spawn (release their eggs and sperm for fertilization). Eels and hogchokers do the reverse: they spawn in the ocean and live out the rest of their lives in interior waters.

To complicate matters further still, the various reaches of a single stream differ from each other. The first- and second-order streams near the headwaters offer habitat for different assortments of plants and fish than do the downstream reaches (recall that Figure 11–10, page 191, illustrated this same point for salamanders). List 13–1 shows how the fish populations of the Yellow River in the western Panhandle change along its course. The many variations seen in streams make them tremendous reservoirs of wetland and aquatic biodiversity.

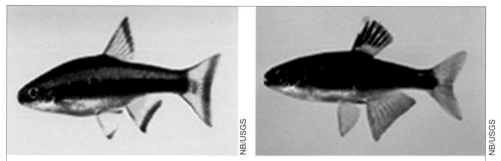

Two native shiners. At left is the Flagfin shiner (*Pteronotropis signipinnis*), from Alligator Creek. At right is the Sailfin shiner (*Pteronotropis hypselopterus*), from Tenmile Creek. Both creeks are in Okaloosa County.

ANIMAL SPECIES ENDEMIC TO STREAM SYSTEMS

Just as an island or continent may be the permanent home of endemic land plants and animals, so a stream system may be the permanent refuge of endemic freshwater species. Land animals can wander long distances, and saltwater plants and animals can swim all over the ocean. But freshwater animals, especially small ones, may be restricted to only one stream system. If a freshwater animal travels too far from the waterways it inhabits, it will die of exposure to dry conditions, and if it is washed out to sea, it will die of salt exposure.

Moreover, compared with lakes, which typically persist for only a few thousand years, streams can be millions of years old, allowing time for new species to evolve in them. Not surprisingly, then, streams flowing out of Florida's northern highlands, which have been above sea level for tens of millions of years, have distinct species of animals. These animals are not endemic to Florida alone, but to whole stream systems that flow through Florida to the sea.

In some cases, neighboring stream systems share endemic species. This reveals that when sea level was lower, the streams ran far out on the land and joined before they reached the ocean. Then the marine waters rose and swamped the junction. Now each stream runs directly into salt water, but may still have endemic species in common with a neighboring stream. Box 13–1 describes animals endemic to distinct stream systems.

BACKGROUND

13–1

Fishes and Turtles Endemic to Distinct Stream Systems

Many of Florida's stream systems, those that run from origins in the peninsula, are relatively young: they have been completely drowned by the ocean within the last two million years. However, those streams that flow out of the northern highlands are very old, and their upper reaches in the Appalachian Mountains have been above sea level for up to 70 million years. The old stream systems are, in a sense, like islands. They have been isolated from other stream systems for so long that some of the animal populations in them have evolved to become distinct species. All of Florida's 100 species of freshwater fishes occur in the old stream systems. So do many turtle species—and for that matter, snails, mussels, clams, and other aquatic animals.

(continued on next page)

LIST 13–1
Fish habitats along the Yellow River

Small creeks at the river's origins are home to small darters and shiners. Larger creeks downstream hold other mixes of fish, each different.

Numbers represent percentages of total fish in each habitat. For example, of the fish in small creeks, 18 percent were sailfin shiners. Only the five most numerous fish are listed for each habitat.

Small creeks
Sailfin shiner	(18)
Weed shiner	(15)
Flagfin shiner	(14)
Blackbanded darter	(9)
Longnose shiner	(6)

Larger creeks
Weed shiner	(23)
Blacktail shiner	(18)
Longnose shiner	(16)
Blackbanded darter	(9)
Blackspotted topminnow	(7)

Main channel
Weed shiner	(30)
Blacktail shiner	(21)
Blacktail redhorse	(6)
Bluegill	(6)
Longear sunfish	(4)

Swamps and backwaters
Pirate perch	(15)
Weed shiner	(14)
Warmouth	(8)
Brook silverside	(8)
Bluegill	(7)

Tidal swamp and marsh
Bluespotted sunfish	(22)
Coastal shiner	(13)
Ironcolor shiner	(12)
Redear sunfish	(10)
Spotted sunfish	(10)

River mouth and bay
Rainwater killifish	(34)
Coastal shiner	(24)
Gulf pipefish	(13)
Redear sunfish	(7)
Inland silverside	(6)

Source: Adapted from Bass (D. G.) 1991.

Florida native: Suwannee River cooter (*Pseudemys concinna suwanniensis*). These turtles often bask in groups on logs along streams. This population is distinct from other river cooter populations and is endemic to the Suwannee River drainage system.

13–1 (continued)

The most diverse array of freshwater fishes occurs in the Apalachicola River: 83 species. It and the Suwannee and Escambia rivers harbor most of those that are endangered, threatened, and rare. Three bass species are localized to the Suwannee and neighboring rivers west of it.

Many of Florida's diverse species of freshwater turtles, too, are restricted to one or a few stream systems. Two of them do not range to the north at all but are endemic to Florida alone—Barbour's map turtle and the Suwannee River cooter.

Following is a list of the fish and turtles endemic to river systems that flow through Florida to the ocean or Gulf of Mexico. The accompanying map shows the locations of the rivers.

A Escambia River: Crystal and Harlequin darters.

B Santa Rosa County: Blackmouth shiner.

C Okaloosa and Walton counties: Okaloosa darter.

D Apalachicola-Chipola River: Shoal bass, Bluestripe shiner, and Barbour's map turtle.

KEY

E Ochlockonee to Suwannee rivers: Suwannee bass.

F Ochlockonee to Suwannee rivers: Suwannee River cooter.

G St. Johns River: Shortnose sturgeon and Southern tesselated darter.

Sources: Bass (Gray) 1987; Suwannee River Task Force 1989; Mount 1975.

LIFE IN ALLUVIAL STREAMS

Reminder: **Detritus** (de-TRY-tus) is the finely divided material that results from the decomposition of organic matter and is an important source of energy in wetland and aquatic ecosystems.

Alluvial streams are turbid. They carry clay, silt, sand, and detritus, especially during floods. Water volumes and temperatures vary with seasons and rainfalls. Stream beds are muddy; berms and levees are sandy. These factors define the aquatic habitats.

Plants are not numerous in an alluvial stream. Too little sunlight penetrates the cloudy water to support growth below the surface, and any plants that manage to take hold in the loose bottom sediments are uprooted and swept away by each year's powerful floods. Most of the plant food that animals consume, therefore, is detritus—litter and muck washed in from the floodplain, an energy and nutrient source for filter feeders parked on plants and rocks and buried in bottom mud. Quiet backwaters in large, wide, slow-moving streams may harbor some plankton. One reference names 22 different genera of plankton found in alluvial streams—not many compared with the hundreds of species in lakes, but more than might be expected.[4]

Alluvial streams offer diverse habitats to animals that can cope with their varying flows and temperatures. They have sandy bottoms and banks, fast-flowing currents around outer bends, and quiet inner curves. And every year when streams overflow their banks, their floodplains, too, become aquatic habitats. Then fish, frogs, turtles, water snakes, and others swarm all over the floodplain feeding, breeding, and leaving behind eggs and young fish to start their next generations. During droughts, the ponds that remain are important reservoirs for many species.

An important habitat in flatter portions of the floodplain is the slough, a quiet part of a former channel off the main channel. Sloughs are sometimes water-filled, sometimes muddy. They are an ideal place for tiny animals to be born and live their early lives. Because they are connected to the stream, sloughs do not trap fish as ponds do. They permit maturing fish to move out into the larger aquatic world when they are ready.

The bottom sediments of alluvial streams house a rich and varied, little known, life. Who but a specialist in bottom-dwelling organisms would know that Florida's alluvial stream bottoms include at least 60 species of clams? Some 40 of these are in the Apalachicola River, and three are endemic to just that river's watershed. The Apalachicola also has 20 snail species, of which five are endemic. Only if you were to scoop up buckets of sediments from the bottom and sift them would you find these mollusks, with their distinctive sizes and shapes and variously colored shells. Yet they have dwelt there for hundreds of thousands of years, evolving adaptations to the particular chemistry and biology in the sand, silt, and clay in that particular stream. The same is true of other bottom dwellers even more obscure: worms, crustaceans, and others. They, too, are numerous, varied, and specialized.[5]

Mussels in stream-bottom sediments are of special interest to biologists because they are indicators of the health of an aquatic ecosystem. Mussels clean the water they are in; they have been called "silent sweepers of the

Florida native: Mud snake (*Farancia abacura*). This large snake lives in most of Florida's freshwater habitats. By day, it burrows in the mud; at night it prowls.

The Ochlockonee River, a mixed blackwater and alluvial river in the mid-Panhandle. Viewed during the fall dry season, the river is relatively clear.

COOL, SWIFT, AND CONSTANT: This unnamed stream in Torreya State Park, Liberty County is a tributary to the Apalachicola River.

SUNNY AND ACIDIC: This is the Waccasassa River, a 29-mile-long, blackwater stream in Levy County.

LIMY, SHADY, AND COOL:The Aucilla is a 69-mile-long, part black-water and part spring-fed river in Jefferson County.

BROAD AND PEACEFUL: The Tolomato is a tributary to the Guana River near the estuary in St. Johns County.

FIGURE 13–4

Diverse Florida Streams

FAST-FLOWING WITH RAPIDS: This is a stretch of the Hillsborough River in Pasco County.

other Panhandle streams possess their own animal species, and doubtless many remain to be discovered and described, for hardly anyone has even looked in the bottom sediments to see what burrowing animals are there.[9]

The foregoing descriptions help to distinguish among general types of streams, but to reiterate, no Florida stream conforms strictly to any of these types; each is a unique mixture with its own special character. To help convey the true diversity and uniqueness of Florida's flowing waterways,

Figure 13—4, opposite, presents a sampling of streams not previously described. The next section provides a closer look at a single famous, long, and varied Florida river, the Suwannee.

THE SUWANNEE, A RIVER OF MANY PARTS

The Suwannee changes character many times along its length. Each segment of the river supports a different balance of aquatic life. The river originates as black water overflowing from the Okefenokee/Pinhook Swamp, the huge blackwater swamp in southeast Georgia/northeast Florida from which the St. Marys River also takes its blackwater character. Meandering 165 miles before it empties into the Gulf, the Suwannee cuts in places between 40-foot-high banks, crosses numerous flowing springs and sinks, develops rapids on slopes, and winds across massive floodplain forests before reaching the ocean. It drains 10,000 square miles, half in Georgia, half in Florida, but it is relatively small at the Florida-Georgia border. It picks up an enormous volume from springs and tributaries thereafter and is second only to the Apalachicola in discharge at its mouth. At least 50 springs feed it, nine of them first-magnitude springs, so even during droughts when the swamps are dry, the Suwannee keeps flowing.

The upper Suwannee River, above White Springs, flows fast and narrow between high, steep banks over densely packed sand and limestone. With every major rainstorm, the water rises rapidly and sweeps the sandy banks clean, preventing debris accumulation and limiting diversity. Because it is acid at this point, with water low in minerals, it supports few clams and mussels, which depend on alkaline, mineral-rich water for formation of their shells. The fast flow keeps plants from rooting and supports no significant masses of algae, so there are few alligators, wading birds, or water mammals. Few large bottom animals can withstand the swift current, but small burrowing types such as crayfish thrive in the bottom sediments, and these serve as food for 41 species of fish.[10]

Fish in the upper Suwannee must be able to make use of the few plants

Suwannee River: Big Shoals at low water. The limerock outcrops over which the river flows are clearly visible. The river is impassable for boaters and moss on the rocks makes footing treacherous, but the plants and animals in and around the river are adapted to both high and low water.

Big Shoals on the Suwannee River in Columbia County. Big Shoals is the largest rapids in Florida.

Native to Florida waters: Common snook (*Centropomus undecimalis*). Ordinarily considered ocean fish, snook sometimes swim up freshwater streams. These were photographed at Homosassa Springs in Citrus County.

rooted plants such as lilies ("bonnets"), wild rice, fanwort, and others. These offer hiding, feeding, and egg-laying sites for snails, other invertebrates, and fish. At high tides, salt water moves up the river beneath the fresh, and several species of ocean animals move in and out of this environment, including eel, pipefish, mullet, red drum, and sea trout as well as sturgeon and manatee. Finally, at its mouth, the Suwannee broadens into a vast, productive tidal estuary.

STREAM VALUE AND QUALITY

Florida's 1,700-odd streams offer unparalleled recreation opportunities. To travel them is to know why people have always been drawn to them. Thousands of years ago, the Indians, attracted by freshwater shellfish, snails, and other mollusks, rode the rivers in canoes and walked the banks and floodplains silently on foot. For the past several hundred years, Europeans and other settlers have used faster modes of travel, but canoers and hikers can, if they wish, still see many of the rivers as the Indians did.

Florida's streams are also diverse and important wildlife habitats. They serve as corridors, permitting aquatic animals to travel between interior waters and the ocean. They are important as biological refuges for many different constellations of aquatic life. They are a valuable economic resource. They serve as conveyor belts, constantly carrying water and nutrients from upstream to estuaries. And wherever the surrounding floodplain wetlands remain unaltered, they help control both floods and droughts, cleanse and purify both surface and ground waters, serve as storage reservoirs for fresh water, help recharge aquifers, and prevent saltwater intrusion into underground water supplies.

Survival of the aquatic communities in streams depends on two factors: the quality of stream waters, and the preservation of intact ecosystems around them. It is true that streams can purify themselves to some extent, unlike lakes and ponds, which accumulate all materials that flow into them. Fresh influxes of rain and ground water dilute streams, and their sediments gradually work their way downstream. But most of Florida's streams are nearly flat and the cleansing process is slow. They can't handle much contamination without suffering a decline in their quality.[13]

Streams are also sensitive to runoff. If only ten percent of a stream's watershed becomes impervious (paved or built upon), measurable changes occur in the stream's high and low water levels, together with declines in habitat structure, biodiversity, and water quality.

To keep track of the health of streams, it has long been the custom to monitor their water chemistry. However, the shape of a stream, its shoreline vegetation, and its bottom sediments also are keys to its health. The best indicator of all is the actual presence of normal, living plants and animals. The healthy growth of water-cleansing microorganisms and the presence of robust aquatic vegetation reveal what researchers call a stream's biological integrity. Thanks to the structural complexity of their sides and bottoms, and to the abundant food in them, natural rivers associated with undisturbed floodplains hold more fish than other rivers that are *stocked* with fish.[14]

Florida native: Osprey (*Pandion haliaetus*). Ospreys build their big, messy nests in the tops of trees near lakes, rivers, and the coast. They perch or hover high over the water, then plunge for fish to eat.

According to James Karr, stream quality assessor, the question to ask is not "How clean is the water?" but "How well does the stream support aquatic life?" The best way to answer that question is to evaluate the aquatic habitat with its plants and animals. How many surfaces are available for organisms to cling to? (When a stream bottom is dredged or choked with sediment, the number of "clingers" declines markedly.) How much food is there (leaves, twigs, detritus, and mineral nutrients) for organisms in the stream? How many native organisms, and of what kinds, live in the stream (phytoplankton, zooplankton, mussels, crayfish, fish, and rooted plants)? Chemical indicators used alone fail to reflect about half of the biological integrity of streams. To keep streams healthy, then, we must observe them intelligently and dedicate resources to their protection and management.[15]

Florida native: Tricolored heron (*Egretta tricolor*), juvenile. This young wading bird finds plenty of fish in a healthy river with flourishing native vegetation.

* * *

Although this chapter has touched on a wide variety of streams, it has omitted Florida's most notable, and in many ways most beautiful, waterways. Underground streams, springs, and spring-run rivers are the subjects of the next chapter.

A wet sink. This diagram, repeated from Figure 7–9 (page 115), shows that aquifer water is visible at the bottom of this round hole in the ground. The land that fell in to form the hole lies on the bottom in a cone-shaped pile.

Sinkhole pond in Jefferson County, near Wacissa.

vast, regional ground-water system, a sink's water is not much affected by local droughts and rains. Only long-term region-wide depletion of the ground water lowers the water table enough to deplete a sink; and only repeated deluges (for example, by tropical storm rains) will raise its level.

Over centuries and millenia, of course, regional water levels do shift, as the ocean's level does. Some sinks that once held water in the distant past now lie well above the water table, and rain water rushes down into them as into a funnel. Other openings into the aquifer lie much farther down in level and propel water upwards: they are springs, described in the next section.

LIFE IN SPRINGS AND SPRING-RUN STREAMS

Some of Florida's springs are polluted today, and some are naturally dark with swamp water, but many of them still produce pure, hard water that is perfectly transparent. These crystal-clear springs and the streams that flow from them are in the spotlight for the rest of this chapter. They are swirling gems of crystal, emerald, and sapphire set in stone, and each one is different. Their bottoms are of gleaming white sand and limestone, partly free of vegetation, partly decorated with greenery like underwater gardens. Box 14–1 presents background information about springs.

BACKGROUND

14–1

Florida's Springs

Florida's springs are most numerous in the northern half of the state, wherever the underlying limestone is near the ground surface. The force behind springs is produced by pressure created by the great height of water within the clay hills; the water bursts forth after working its way down to openings near the coast. (Figure 14–2 shows the source of a spring.)

Florida has more freshwater springs than any other area of comparable size in the world. In addition to more than 300 previously mapped, a survey in 1997 discovered about 300 more, and there are now known to be at least 700 springs in the state. Some springs give rise to clear, cold streams (spring runs) such as the springs at the origins of the Wakulla, Wacissa, Ichetucknee, Silver, Crystal, and Rainbow rivers. Other springs well forth from the bottoms of streams such as the Suwannee, Santa Fe, and St. Johns rivers, which have eroded their beds down to the limestone underlayer. And at least sixteen springs occur offshore, in the Gulf of Mexico north of Tampa and along the Panhandle coast. The largest known spring in the world is just offshore at Spring Creek.

Springs are classified by their discharge. The very largest, first-magnitude springs include Wakulla Springs, Spring Creek, and Silver Springs, each of which may flow hundreds of millions of gallons a day. Collectively, the springs of Florida discharge more than 7 billion gallons of fresh water each day, exceeding all current human water consumption in the state. List 14–2 names Florida's largest-discharge springs by county.

The flow of a spring is determined by the amount of rainfall in its watershed and the permeability of the surrounding limestone. Since a spring may lie in a watershed covering hundreds of square miles, a local shower or dry spell has little or no effect on its volume of flow. A spring will cease to expel ground water only if droughts or withdrawals are so extreme and widespread that the ground-water level falls drastically all over a region. Then the spring becomes a funnel sink, through which newly falling rain may flow *into* the aquifer.

For the most part, the chemical composition of any single spring's water is extremely stable, but from one spring to the next, chemistry varies. Springs discharge ground water from both shallow and deeper layers of the aquifer. Some spring waters are high in calcium, magnesium, and bicarbonate ions. Some produce water that has been filtered by its passage through sand and limestone and is colorless and clean. A few

(continued on next page)

> A **funnel sink** is a sink in which water is swirling down to a lowered water table below the ground.
>
> If the water table were to rise, suddenly, that same sink would become a spring, and would propel water upward with similar force.

> A **first-magnitude spring** is one that produces 100 cubic feet of water per second (about 65 million gallons a day) or more.

The spring boil at the origin of the Silver River (Silver Spring, Marion County). Concentric ripples on the surface (foreground) show the force with which the water shoots forth from the spring. Loose sand "boils" vigorously in the water around the spring opening.

FIGURE 14–2

The Source of a Spring

Water is piled up in channels in the limestone.

At a break in the earth, it is visible in a sinkhole.

At a lower elevation, it gives rise to a spring, from which . . .

a stream flows over land to the ocean.

The height of a great volume of water in the hill gives rise to a fast-flowing spring-run river down below.

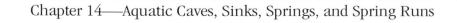

14–1 (continued)

springs, after rains, discharge brown, tannic acid-stained water that has only recently left the surface. Some spring waters are low in dissolved oxygen or high in sulfur, iron, or other minerals. Some springs that come from the deepest layers of ground water, discharge briny salt water.

The water temperature of a given spring is constant and matches the temperature of the ground through which it has flowed, which in turn reflects the average annual air temperature. Panhandle springs have temperatures of 66 to 75 degrees Fahrenheit. South Florida springs are at 75 degrees or more.

The time that has elapsed since water flowing out of a spring was last on the surface can be determined by measuring several different chemical isotopes. The residence time can range from years to decades, with larger springs tending to discharge older water than smaller springs. In most cases, water that, today, is flowing out of the limestone cave or tunnel at the bottom of a spring has spent several decades underground since it was last at the surface.

Source: Anon. 1995 (*Springs of Florida*); Scott and coauthors 2002.

To gaze into a spring is to view a world of deep blue, so vivid that it almost has texture. The sky-blue color results from the pure water's absorption of all but blue wavelengths of light. Once in the water, though, a diver floating above the bottom sees no color and seems to hang, suspended and weightless, in nothing at all.

Because Florida's springs provide constant water flow, temperature, and chemistry, and admit an abundance of sunlight, they hold some of the most lush, biologically productive ecosystems in the world. They are home to the smallest species of fish in North America—the least killifish; to the largest reptile—the alligator; to endangered manatees; and to a spectacular assortment of wading birds and ducks. And not only does life abound in Florida springs, it is also clearly visible.

Spring floors vary. They may be covered by submerged flowering plants like pondweed, by mats of bright green algae, or by coontail and stonewort—large, dark green algae that look like complex plants. Some springs, largely free of vegetation, are floored with golden-white sand bars, a miniature landscape of hills and valleys, sparsely carpeted with green mats of stonewort and tiny black snails tracing spiral trails in the sand.

Meadows of tapegrass carpet many Florida springs and spring runs, undulating and swaying endlessly in the current, their delicate white flowers blooming just below the surface. The water flows swiftly, so there are few floating plankton, but the tapegrass and other aquatic plants offer vast surfaces on which strands of green algae can attach and photosynthesize, forming the base of a complex and rich food web. Rocky outcrops present a coarse, multifaceted surface to which algae and other tiny plants also cling. A wader or swimmer in a clear, fast-flowing spring-run stream might expect the limestone rocks, having rough surfaces, to be easy to walk on, but if coated with algae, they may be extraordinarily slippery.

Strands of green algae also trail from logs, the shells of live turtles, and other hard surfaces. Altogether, more than 25 species of algae grow in springs. One study, done in Silver Springs, showed that the biological productivity of its algae and submerged aquatic plants exceeded that of many agricultural crops. The algae support huge populations of freshwater shrimp and their relatives, as well as freshwater snails, minnows, and aquatic insect larvae. Mayflies, caddisflies, dragonflies, damselflies, waterstriders, and diving beetles are everywhere, and their aquatic larvae serve as a major food source for larger animals.

Snails are extraordinarily numerous in springs because the water is rich in shell-making calcium, algae on which they graze are abundant, and there are luxuriant aquatic plants on which they can lay their eggs. Of Florida's 100-odd species of freshwater snails, most occur in spring runs. Many species are restricted to ranges of only 20 to 30 feet around the mouth of a single spring. They must have diverged from other, related species at a distant time in the past and the springs where they reside have flowed constantly ever since. Many other snails are nearly as restricted in distribution, occurring only within single drainage systems.

Apple snails may be abundant on submerged spring-run stream vegetation. Apple snails are the largest freshwater snails in North America, growing to more than three inches in diameter. They occur only in Florida and along the southern edge of Georgia. Canoeists find clumps of their pinkish-white eggs glued above the water line on streamside vegetation. Apple snails can take in oxygen both in and out of the water—using their gills when submerged and their siphons when at the surface. Apple snails are the primary food of the limpkin, one of Florida's rarest birds. They are also a major food of otters, mink, and raccoons.

The numerous insects and larvae of a spring run feed a host of small fish such as redbreast sunfish, which in turn are hunted by larger fish. Schools of fish roam all over Florida's springs. Freshwater fishes include sturgeon, eels, gizzard shad, redeye chub, spotted suckers, white catfish, sunfish, killifish, and largemouth bass. Large crayfish make an excellent food for large fish such as bass. In fact, bass are adapted especially to eat shelled creatures: they have teeth not only along their jaws but also all over their tongues and the roofs of their mouths.

More saltwater fishes enter freshwater streams in Florida than in any other state. Springs near the Gulf coast, especially Homosassa Springs, are visited by schools of saltwater species such as gray snapper, jacks, mullet, sheepshead, tarpon, pipefish, snook, mojarra, and many other marine fish. Juvenile hogchokers use springs as their main nursery areas, where they feed on insect larvae. Some scientists believe that marine fish are attracted to hard water that is rich in ions, where they can most easily regulate their own body salts. Many fish are also drawn to the warmth of spring water, especially in winter when marine waters and surface fresh waters have cooled.

Some fish travel in and out often, others stay in springs all winter. Diver Doug Stamm says it is a thrill to view these fish at night:

Shining silver caravans of schooling mullet twist and swerve through the beam of a diver's light . . . Prolific blue crabs . . . wander spring floors . . . Unpredictable

Snail eggs. The apple snail (*Pomacea palludosus*) lays its big, shiny, pink, pea-sized eggs on streamside vegetation and cypress trees just above the water line. When young snails emerge, they go under water promptly to escape predation. They graze on algae and other periphyton, visit the surface periodically, and during the dry season bury themselves in the mud and go dormant.

Florida native: Limpkin (*Aramus guarauna*). The limpkin subsists largely on a diet of apple snails. Its curved bill probes deep into the snail shell to pull out the creature's soft body.

needlefish, elusive and skittish by day, become fleeting lances . . . American eels . . . arise from caverns and burrows and swim snakelike in search of prey . . . Gars' . . . archaic forms glide by . . .[4]

Some marine fishes access springs by circuitous routes. The St. Johns River is primarily a blackwater stream, but some of its tributaries are spring runs. American shad travel 100 miles up the St. Johns to the Oklawaha and then follow the Silver River, to arrive finally at Silver Springs. Soon after spawning, they reverse their journey and travel 100 miles back downstream to the ocean again. Mullet and needlefish travel even farther to other springs. Stingrays visit springs, and juvenile flounder, only an inch long, may hide in bottom sediments. An important ecological principle is illustrated here. As with the fox squirrel and the wild turkey, a mosaic of different habitats is needed for the survival of these fish. Many require several aquatic environments, including a stream's headwaters, middle reaches, and estuary, as well as the open ocean.

Two "living fossils," the gar and the bowfin, are common throughout southern waters but are most easily seen in springs. Gar are survivors of an ancient line of fishes that dominated the world during the age of dinosaurs some 250 to 65 million years ago. Gar are air breathers: they roll frequently at the surface to gulp air and will drown if kept submerged. Gar prey on fish, frogs, and crayfish.

Bowfin are even older. They are the sole living representatives of some of the most primitive fish in the fossil record. Like the earliest known jawed fishes and like their relatives, the gars, they have bony, armorlike scales as well as bony plates that cover a semicartilaginous skull. And like gar, they, too, gulp air.

The fabled sturgeon, whose eggs are prized as caviar, is another relic from the time of the dinosaurs, a 250-million-year-old species. Two subspecies of sturgeon occur in separate ranges that include Florida: one along the Gulf coast east to the Suwannee River or a little farther; the other along the Atlantic coast from Canada to the St. Johns River or a little farther.

Florida native: Spotted gar (*Lepisosteus oculatus*) in Silver Run, Marion County. Gar swim up and down Florida's streams and often congregate in springs.

Native to Florida waters: Gulf sturgeon (*Acipenser oxyrinchus desotoi*). This mighty fish was netted at the mouth of the Suwannee River, carried up to Fanning Springs to be photographed, and then released. Gulf sturgeon can attain a length of about 15 feet, a weight of 800 pounds or more, and an age of 60 years. This animal was about five feet long.

Florida native: Greater flamingo (*Phoenicopterus ruber*). Flamingos occur naturally only in extreme south Florida, but have been made welcome in this wildlife park at Homosassa Springs in Citrus County.

Reptiles are also abundant in spring-run streams. Turtles are everywhere. The loggerhead musk turtle (not to be confused with the loggerhead sea turtle, a completely different species) secretes a vile-smelling material if disturbed. It tends to sink rather than float and can absorb oxygen from the water through its skin, so it can remain submerged indefinitely. It is often found deep in caves where other turtles could not survive. Other common spring-run turtles include the Florida cooter, common musk turtle, and the Florida softshell. They eat insects, plants, and the abundant snails of spring runs. Alligators are the dominant top predators, taking fish, turtles, and even the occasional unwary raccoon or deer at night.

Big, slow-moving manatees also visit Florida's springs. Manatees are marine mammals that range from the Carolinas to Louisiana grazing on coastal and river-mouth seagrasses. Manatees are tropical animals. They will die of pneumonia if exposed to water colder than 65 degrees Fahrenheit, and they could not survive in Florida were it not for the freshwater springs. During warm months, manatees feed, calve, and rest in the rivers; then when winter chills coastal and surface waters, they take refuge in the springs. They congregrate in large numbers in Crystal River on the Gulf coast and in Blue Springs in the St. Johns River, where they feast in the underwater pastures of eelgrass there. Each manatee can eat more than 100 pounds of vegetation per day. Several hundred manatees winter in these two spring systems, representing about 15 percent of the species' total population.

Florida's springs have been called the jewels in her crown. The state's Florida Springs Initiative of 2000 embodies the effort to preserve and protect them for future generations to enjoy.[5]

Florida native: Anhinga (*Anhinga anhinga*). This fishing bird's feathers do not repel water and as a result, the bird cannot float like other water birds. It swims with only its long head and neck above water, a practice that gives it the name *snake bird*. To catch fish, it swims altogether under water. Periodically it rests, spreading its wings in the sun to dry them.

Native to Florida waters: West Indian manatee (*Trichechus manatus*). This big, gentle mammal was photographed in Homosassa Spring, Citrus County.

Springs are the visible evidence of the ground water's quality. A spring whose water is pure and clear reassures monitors that the underlying aquifer is, too. A spring whose water is polluted is a warning sign that ground water quality is impaired.

The ground water is vital to all living things in Florida. It supplies most of our drinking water, and, together with newly-fallen rain, it replenishes streams, rivers, ponds, and lakes as they evaporate or flow away. Trees sink their taproots into it to obtain a continuous supply. The ground water even holds up the earth's surface by filling underground spaces. When the water table declines, the overlying earth may collapse. Cities, farm fields, forests, marshes, swamps, coastal bays, and estuaries all depend on ground water. It behooves us, then, to protect and maintain our springs. Not only are they beautiful and important habitats for living things; they also are the guardians of our safety and health.

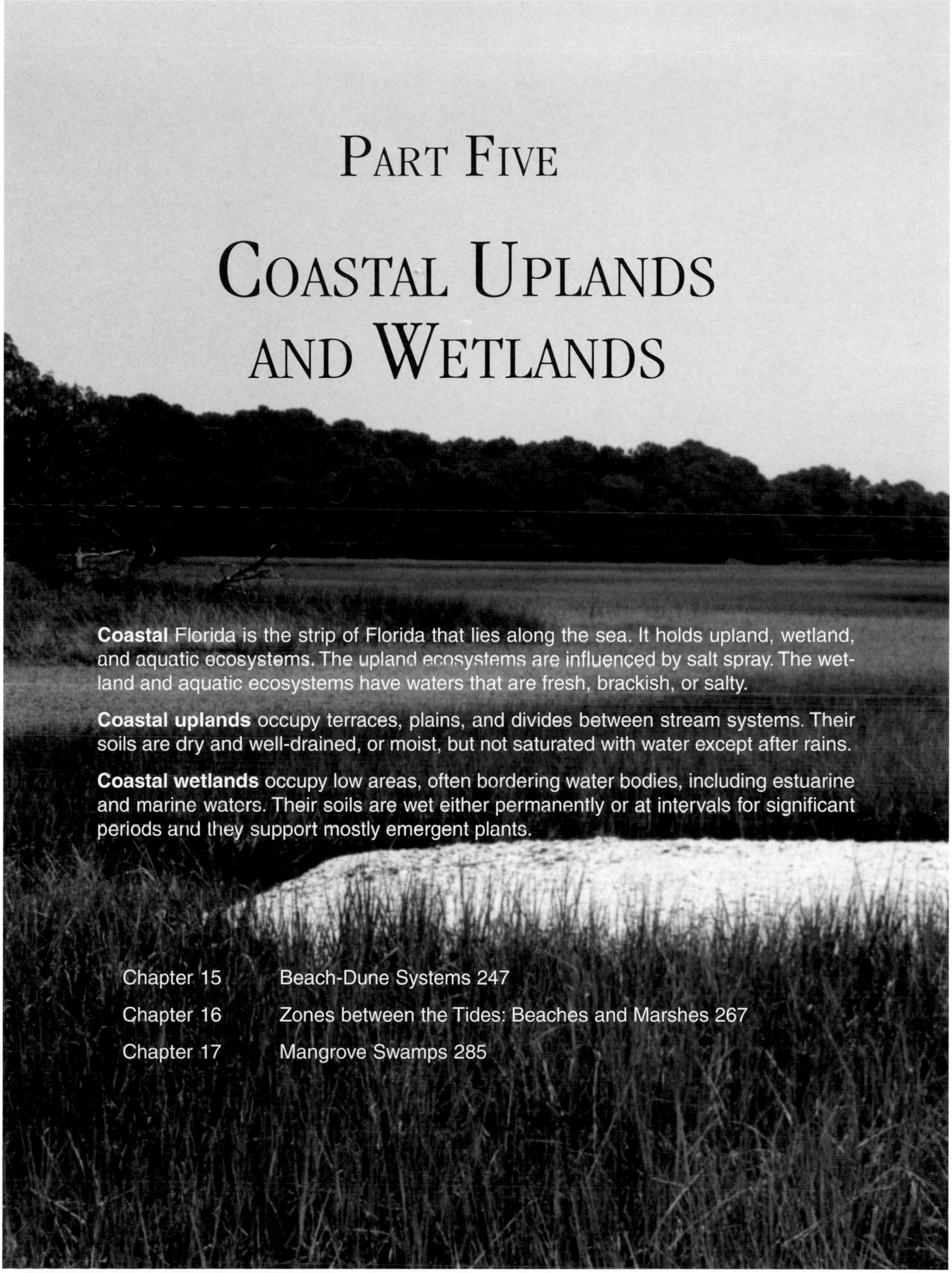

PART FIVE

COASTAL UPLANDS AND WETLANDS

Coastal Florida is the strip of Florida that lies along the sea. It holds upland, wetland, and aquatic ecosystems. The upland ecosystems are influenced by salt spray. The wetland and aquatic ecosystems have waters that are fresh, brackish, or salty.

Coastal uplands occupy terraces, plains, and divides between stream systems. Their soils are dry and well-drained, or moist, but not saturated with water except after rains.

Coastal wetlands occupy low areas, often bordering water bodies, including estuarine and marine waters. Their soils are wet either permanently or at intervals for significant periods and they support mostly emergent plants.

Tidal marsh, Guana River State Park, St. Johns County.

FPS

245

CHAPTER FIFTEEN

BEACH-DUNE SYSTEMS

Walk along an unspoiled beach anywhere in the world and you will enjoy the same familiar features. The waves roll in from offshore, the dunes are clad in their characteristic gnarled vegetation, and behind the dunes lie interdune fields and maritime hammocks dotted with ponds and marshes and rich with plant life and birds. All natural high-energy coastlines have these features, for all have been formed by the same forces—wind and waves, working on the same materials—sand and shells. Chapter 2 described how, along these shores, restless winds, waves, tides, currents, and storms constantly move sand from place to place. This chapter enlarges on that theme of constant change, because change places demands on every plant and animal in beach dune systems.

Florida native: Gulf fritillary (*Agraulis vanillae*). This butterfly occurs all along the eastern seaboard. Individuals from farther north fly south to the Gulf coast in winter.

SHIFTING COASTAL SANDS

Coastal landforms are very dynamic, compared with others. Most of the beaches, dunes, and barrier islands that fringe the world's continents have been formed just recently, in geological terms—they are no more than about 6,000 years old. All of Florida's barrier islands, large and small, have come into being in just these past six millennia. The early Indians were here long before the present-day beaches were; the ocean was 200 to 300 feet lower when they first came. Florida's beaches at that time were anywhere from a few to 200 miles farther out than the present coast, and barrier islands long since drowned lay offshore beyond them.

Today, those long-ago beaches and barrier islands lie deep under water off our present shores. Divers report that they are still decked with ridges, dunes, and even the drowned remains of ancient inland forests. Still, the shoreline looks the same, because new beaches and barrier islands have formed where the coast lies now—not everywhere, but wherever winds and waves carry enough energy to create these formations. Between the stretches of high-energy shores are coastal marshes and swamps, the subjects of the next chapters.

Why are there sandy beaches along some stretches and coastal wetlands along others? The waves coming onshore make the difference: they

> **Beaches** are long, usually narrow, strips of sand and shells heaved up by wave action along the shore.
>
> **Dunes** are mounds of sand piled up by wind.

> **Barrier beaches** face open water along seashores.
>
> **Barrier islands** are long, narrow islands parallel to the coast. Each barrier island has a barrier beach on its ocean side.

> **Reminder:** **High-energy shores** have beaches and fields of dunes.
>
> **Low-energy shores** have wetlands, notably salt marshes and mangrove swamps.

OPPOSITE: Dunes on Navarre Beach in Okaloosa County.

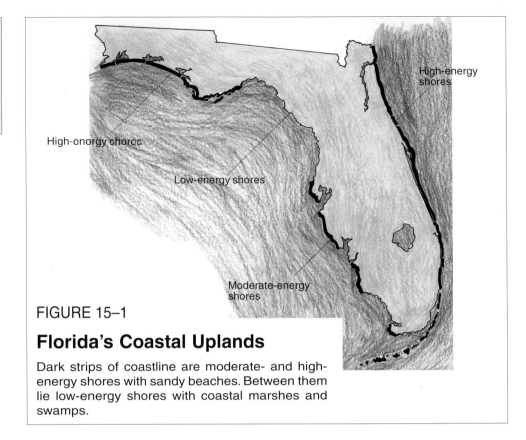

FIGURE 15–1

Florida's Coastal Uplands

Dark strips of coastline are moderate- and high-energy shores with sandy beaches. Between them lie low-energy shores with coastal marshes and swamps.

are highest and most energetic where the land slopes steeply into the ocean, and they are barely perceptible where the land and offshore seafloor are flat. Figure 15–1 shows the locations of Florida's coastal uplands, where vigorous wind and wave action rearrange coastal sands a little every day, and massively over years and during storms. Beaches may grow outward by the addition of sand, they may be eroded away, or their sand may migrate along the shore, depending on which ways the winds, waves, and currents are moving.

Broadening of Beaches. A beach grows outward in a giant step when a major storm piles up a ridge of sand just offshore and sand fills in the space between. A major storm has unusually large waves that scour huge quantities of sand from the ocean floor and then, on reaching shallow water, slow down and drop the sand in a great heap parallel to the shore. As the storm retreats, it leaves the ridge of sand above water, adding substantial area to the beach.

Once laid down, a beach ridge may remain in place for decades or even centuries until another major storm moves its sand again. For as long as it is present, the ridge protects the land behind it from erosion. Other ridges may form seaward of the first, and as the swales between beach ridges fill in with sand, the ridges seem to be moving inland. In reality, though, the shoreline is moving farther out to sea.

It is beach ridges that, over the past 6,000 years, have served as the foundation for Florida's present barrier islands and coastal beaches. Ridges that now are part of barrier islands show up in aerial photographs as parallel stripes, easy to see because the ridgetop vegetation differs from that in the swales. Figure 15–2 provides an example.

Marsh and Bay

Secondary dunes

Foredunes

Beach

Ocean

Ocean

DBM

FIGURE 15–2

Island Building

Saint Vincent Island, Franklin County. The island is now being eroded as sea level rises and storms intensify, but its beach ridges still reflect the way its beaches and dune fields were built in the past. In past centuries, each beach ridge was first shoved up in front of the island by a major storm. Between storms, wind-blown sand filled the space behind the ridge, in effect adding it to the island.

Once the water has laid down a beach ridge, the wind piles sand on top of it, forming dunes. The process of dune building occurs in small bits daily, mostly as the tide is falling (see Figure 15–3). Grains of sand dry in the wind and are blown up the beach until they strike an obstacle such as a rooted plant, most often seaoats. The sand grains fall and pile up around the oat, the oat responds by growing upward, and an oat-decorated dune forms. Sand dunes are a feature of all of Florida's high-energy shores.

FLAUSA

First, sand is left behind by retreating waves.

FLAUSA

Next, the wind blows the grains of sand up the beach and they pile up around obstacles such as seaoats.

FIGURE 15–3

Dune Building and Maintenance

FPS

Then the seaoats grow taller, the newly buried parts of their stems become roots, the roots grow longer, and more sand piles up around them, building a dune. Fort Clinch State Park, Nassau County.

JV

In all seasons, seaoats hold dune sand in place. These were photographed in fall in Caladesi Island State Park, Pinellas County.

ANIMALS OF COASTAL UPLANDS

Many animals that people never see inhabit beach-dune systems. Few visitors would guess that hundreds of species of animals are members of coastal upland communities.

On the Beach. Beach life is based on a food chain that begins with a scraggly line of seawrack along the high-tide line, where uprooted seagrasses and the remains of jellyfish, shellfish, fish, and others lie baking in the sun. The piled-up weeds are alive with a swarm of tiny animals: jumping beach hoppers, sandhoppers, other amphipods, insects, and their larvae, all feeding on the weeds and on stranded animal remains. Picking off these frantic creatures in turn are crabs and other animals, which scuttle to safety in their burrows when approached. Shorebirds forage among the weeds, catching insects and other prey. That line of weedy, smelly debris, which sunbathers avoid, is a prime feeding place for animals. Without seawrack, the beach would be much less lively.

Birds use the beach in many ways. Shorebirds seek food there, and migratory birds stop to rest before continuing their travels. Local birds settle on the sand between fishing forays, and bird couples meet and court each other. Some thirteen bird species even nest on beaches, if human disturbance is far enough away (see List 15–3). The snowy plover places her eggs in a shallow scrape of sand at the base of a dune, but abandons them if intruders approach. Terns form large colonies and defend their nests against marauding gulls, people, and dogs by flying up in a huge, shrieking mob to scare them away. American oystercatchers nest among backshore grasses and herbs and forage on mollusk reefs offshore.

At night, long after sunbathers have packed up and left the beach, many more animals appear. Tiny sand hoppers emerge from the seawrack and scour the surrounding sand for specks of food. Rove beetles come out to prey on them, and ghost crabs (aptly named) creep warily out of their burrows to prey on the hoppers and the beetles. On spring and summer

Seawrack. Hundreds of miles of Florida beaches are littered with this material, an accumulation of seagrasses, algae, and animal remains delivered by rivers and tides from interior and coastal marshes and swamps and seagrass beds.

Florida native: Sanderling (*Calidris alba*). This bird spends summers in the Arctic and winters along Florida's coasts. It takes its nourishment from seawrack, which provides important food for many shorebirds and other animals.

Florida native: Black skimmer (*Ryn-chops niger*). The bird's nest consists of nothing but a scrape in the bare sand.

Florida native: American oyster-catcher (*Haematopus palliatus*). This big bird once nested all along Florida's coast, but today is seen in only a few restricted locales.

Florida native: Ghost crab (*Ocypode quadrata*). This furtive and fast-moving animal is almost transparent against the sand, with only its eyes betraying its presence. Silhouetted against the sky, however, its features become apparent.

BIRDS THAT NEST ON BEACHES

nights, giant sea turtles may drag their bodies, heavy with eggs, out of the waves and up the beach to dig their nests above the high-tide line. From behind the dunes come other visitors: mice, rabbits, raccoons, skunks, foxes, bobcats, and white-tailed deer. Next morning, their tracks remain as clues to their diverse activities.

Among the Dunes. Plant communities among the secondary dunes offer food and shelter in abundance. The plants' seeds and berries are nourishing foods, and the trees and shrubs offer shade, a boon to animals who need to escape the punishing sun, and shelter from predatory seabirds, roving bobcats, raccoons, and others. Dune animals seldom come out by day. They are mostly nocturnal, but a search readily reveals their tracks and their burrows.

Many reptiles live in coastal uplands, notably several kinds of snakes, skinks, other lizards, and the gopher tortoise. Some exhibit remarkable adaptations to intense, desiccating heat and sun such as "swimming" in sand. The six-lined racerunner, which can tolerate daytime conditions, stands on hot sand with two legs elevated: one hind leg, and the opposite front leg. Some toads and frogs live in interdune areas and breed in temporary ponds there. Cottontail or marsh rabbits may also live among dunes, as well as small, burrowing animals such as beach mice, moles, shrews, rice rats, and cotton rats. These eat seeds and berries and use seaoats as their main food source. List 15—4 names some of the animals that live in coastal uplands.

On barrier islands live some of Florida's most interesting and threatened animals. When sea level was lower, the present barrier islands didn't exist. The land was a flatwoods. As seas reached their present levels, beach ridges were thrown up, dunes formed, and spits and barrier islands first appeared. Probably not long after that, as each island was colonized by small animals, the separated populations began diverging into separate species. (The speciation process was described in Chapter 1, Background Box 1—1.)

Each of three Atlantic coast barrier islands and each of five major barrier islands along the western Panhandle has its own subspecies of beach

**LIST 15–4
Animals of coastal uplands (examples)**

<u>Beach-Dune, Coastal Berm, and Coastal Grassland</u>
American kestrel
Beach mouse
Eastern hognose snake
Ghost crab
Hispid cotton rat
Raccoon
Red-winged blackbird
Savannah sparrow
Six-lined racerunner
Southern toad
Spotted skunk

<u>Coastal Scrub</u>
Beach mouse
Coachwhip
Eastern diamondback
   rattlesnake
Gopher tortoise
Six-lined racerunner
Southern hognose snake
Spotted skunk

<u>Maritime Hammock</u>
Gray rat snake
Gray squirrel
Ringneck snake
Squirrel treefrog

Source: Adapted from *Guide* 1990, 9–11.

The migrations of the monarch butterfly are prodigious. Monarchs breed and spend summers in the northern United States and Canada. Then in the fall, the monarchs fly south around the Gulf, along the coastlines of Florida, Alabama, Mississippi, Louisiana, and Texas and on into Mexico. There, they pile up by the millions in the forests of a high mountain retreat, they feed, and then they go dormant for the winter. When spring comes to Mexico, they awaken and return north.

It takes several generations for the monarch butterflies to make their whole north-south migration. No one butterfly can trace the whole route, because its lifespan is only six weeks long. If one adult monarch starts out from Canada, it will mate and produce eggs in, perhaps, Massachusetts. The eggs hatch there and within the next four weeks the caterpillars eat and grow, wrap themselves in cocoons, and metamorphose into butterflies that make another leg of the trip—say, to Virginia. The next generation may make it to Florida, and the *fourth* generation flies to Mexico and overwinters there. Only when they go dormant do the butterflies live longer than six weeks. In spring, millions leave Mexico to make the return journey, and it is the *seventh* generation that arrives back in Canada to start the cycle again. Each butterfly travels its segment of the route entirely by instinct. It remembers nothing of its parents or any of its previous journey.[8]

The journey is hazardous. Of the monarchs departing Mexico, only one in a thousand makes it to the Florida coast; the others all fall victim to predators, harsh winds, and other fates. Those that make it fly in low to the ground, searching for milkweeds on which to lay their eggs; no other plant will support their development (see Figure 15-11). The monarch-milkweed relationship benefits both the plants and the insects and is an instance of coevolution, comparable in antiquity to that between seaoats and fungi. It is featured in Background Box 15–1.

Left: Green antelopehorn (*Asclepias viridis*). Right: Fewflower milkweed (*Asclepias lanceolata*).

FIGURE 15-11

Two Species of Milkweed Native to Florida

Milkweeds are members of the genus *Asclepias*. More than 100 species of milkweed are native to North America.

Monarchs, Milkweeds, and Mimics

The world's flowering plants flourished rapidly after about 65 million years ago, and today they are represented by some 235,000 species. Along with them, many times more species of insects evolved, many of importance as plant pollinators.

Once an insect becomes important to a plant, evolution of both plant and insect amplifies the characteristics that make them work. Plants become increasingly attractive to insects by providing food and developing bright colors and scents that the insects can home in on unerringly from miles away. Attractive plants gain the advantage of more and more successful cross-pollination. One such partnership is that between the monarch and milkweeds. The insect is an important pollinator of milkweeds, and helps to spread their pollen all along their migratory route, favoring wide dispersal and much mixing of genes among plant populations.

Milkweeds also make a poison that, when the caterpillars eat it, is incorporated into their tissues. By this means, the caterpillars gain two advantages. First, they have a food source other insects cannot eat; and second, they develop a toxicity of their own that serves as a defense against predators. Moreover, when the caterpillars metamorphose into butterflies, the toxin persists, so that the butterflies, too, are bitter-tasting and toxic to would-be predators such as mockingbirds. A bird need taste a monarch only once to learn to avoid ever afterward the bright colors that go with that noxious experience.

Butterflies that assimilate plant toxins develop distinctive, bright color patterns—patterns that predators learn to avoid. This leads to mimicry, another example of coevolution. The monarch's nasty taste and bright colors have proved so successful in promoting its survival that birds, mice, and others avoid not only this species but any others that resemble it. As a consequence, many butterfly species have evolved almost perfect replicas of the bold yellow-and-black pattern sported by monarchs. The vice-

(continued on next page)

Monarch butterfly (*Danaus plexippus*). These butterflies have stopped on Florida's coast to rest and feed before continuing on their way to Mexico. They are feeding on the nectar of a goldenrod species (*Solidago*) to obtain the fuel they need for their flight. When they return to Canada next spring, they must find milkweed if they are to reproduce.

Viceroy (*Limenitis archippus*). This butterfly is of a completely different species and is smaller than the monarch, but mimics the monarch's color pattern almost perfectly.

Mimickry in caterpillars. At left is the monarch caterpillar. The stripes on this noxious insect are memorable and the taste is terrible. No bird that ever tasted it would try it for a second time. Thanks to the protection this color pattern affords, other species have evolved similar patterns. At right is the eastern black swallowtail caterpillar (*Papilio polyxenes*). No experienced bird would try this caterpillar either, because it resembles the monarch so closely.

roy butterfly and others, although not toxic, escape being eaten by virtue of this successful mimicry.

Not only butterflies but also their larvae, caterpillars, develop warning coloration. As with the adult form, so with the larvae: those that are successful at warding off predation are mimicked by others.

We benefit indirectly from bright coloration that has evolved in butterflies and flowers. We can surround ourselves with pleasing flowers and then win visitations by butterflies that also delight the eye.

The timing of the monarch's flight, arrival, and reproduction is synchronized with the milkweed's growing seasons. When mature, each butterfly lays its eggs on a milkweed plant that is just beginning to grow, whose leaves will provide abundant food when the eggs hatch into caterpillars. The first spring generation of monarchs is born in Florida and other southern states. Succeeding generations are born at later stopping places along the way north, and in August, as they head south again, they eat from flowers that are rich with nectar to stoke up on fuel.

The monarchs' Mexican wintering grounds are shrinking rapidly as development encroaches on their borders. The whole species may well die out, but may adapt to use other winter homes. Some monarchs are, today, wintering over on barrier islands near Tampa and St. Petersburg. Given more food plants on the barrier islands and in the remaining bits of central Florida scrub, these extraordinary creatures may survive the loss of their ancient winter home.[9]

Millions of other spectacular butterflies move through Florida as the monarchs do (Figure 15–12 shows a few examples). Bright yellow cloudless sulphurs and tawny-brown buckeyes swarm across the north Florida mainland every fall. Great southern white butterflies breed and feed on islands off the east coast where their favorite food plant, saltwort, blooms profusely in late fall. Iridescent long-tailed skippers move down the peninsula by the tens of millions, feeding on goldenrod and many different host

(A) Tiger swallowtail (*Pterourus glaucus*).

(B) Common buckeye (*Junonia coenia*).

(C) Palamedes swallowtail (*Papilio palamedes*).

FIGURE 15–12

Florida Migrants, Natives, and Host Plants

Flowers that have many blossoms in a single flowerhead ease the insects' task of gaining nourishment. From one perch, a butterfly can take nectar from blossom after blossom without having to expend energy flying from one to the next. Many other butterfly-attracting flowers also have closely grouped, multiple blossoms.

(D) Long-tailed skipper (*Urbanus proteus*).

(E) Cloudless sulphur (*Phoebis sennae*).

PLANTS: (A) Lantana (an introduced *Lantana* species). (B) Water oak (*Quercus nigra*), a convenient perch, but not a host plant. (C) Red milkweed (*Asclepias rubra*), a native milkweed and a host plant, one on which monarch caterpillars can successfully develop into butterflies. (D) Lantana (an introduced *Lantana* species). (E) Godfrey's gayfeather (*Liatris provincialis*), a host plant for several butterflies. Godfrey's gayfeather is endemic to coastal habitats in Franklin and Wakulla counties and has become very rare.

plants. Bright red-orange Gulf fritillaries follow the shoreline around the Gulf of Mexico in the early fall, often flying far out over the water.

Of all the migrants that traverse the oceans of air and water along the eastern seaboard, these fragile butterflies seem to be taking on the most impossible task. Yet they must make the trip to perpetuate their kind. They carry their itineraries in their genes, together with the instincts to seek out the food plants they need at each stopping place. Their natural habitats along Florida's coasts are crucial for their survival.

* * *

Beaches are wonderful places for people to seek relief from the stresses of life, to walk and dream, to build sand castles, to picnic and romp and play with children. They provide habitat for coastal plants and animals that are adapted to live in shifting sands and with salt spray. Beaches and dunes protect the land behind them from the assaults of storms: their sands absorb the impact of the waves. And they they inspire painters, photographers, and poets with their beauty. Perhaps less well known to most people, the wetlands that lie along Florida's other, *low*-energy shores are important and beautiful, too, as the next chapter demonstrates.

CHAPTER SIXTEEN

ZONES BETWEEN THE TIDES: BEACHES AND MARSHES

A profound and fascinating force affects the inhabitants of coastal zones: the tides. Coastal zone plants and animals all live by tidal rhythms, which have shaped their evolution for hundreds of millions of years. To survive in tidal zones, every plant and animal must have ways to withstand the constantly changing water level, and in most tidal wetlands, daily oscillations between salty and fresh water, as well.

Most of this chapter deals with tidal marshes, with the lion's share of attention given to salt marshes. However, the beach, too, has a zone between the tides, the foreshore, and this zone deserves a moment's attention. The foreshore is familiar to all beach visitors, and the life in that zone illustrates well the forces with which tidal creatures have to contend. This chapter, then, begins on the beach.

THE BEACH FORESHORE

The beach zone above the high-tide line, the backshore, was described in Chapter 15. It is a windy, desertlike environment in which only a few species of plants and animals can survive. The zone between the tides, the foreshore, is more supportive of life. It is familiar to beachgoers as the strip of beach where the sand is firm and walking is easy. High tides cover the foreshore with waves and surf, and low tides leave it above water. Daily flooding with salt water prohibits the growth of plants except for some inconspicuous algae, but animals do live along the foreshore, and they are highly specialized to exploit its dynamic high-energy environment. They occur nowhere else.

Many animals of the foreshore have bodies specially adapted for rapid digging, the only strategy that can keep them from being swept away by the waves. Beachgoers may be familiar with coquina clams, tiny colorful bivalves that are exposed just as each wave recedes, but then quickly up-

Florida native: Virginia salt-marsh mallow (*Kosteletzkya virginica*). This plant grows in salt and brackish tidal marshes as well as in freshwater wetlands.

Reminder: The **foreshore** is the zone of the beach between the high and low tides. Between tides, the foreshore is repeatedly wet by the ocean's waves; at low tide it is exposed.

OPPOSITE: Tidal marshes along the Panhandle's Big Bend. Along this coast, waves are so mild that fine particles of silt and sand remain as beds of mud that support salt marshes.

267

The very smallest members of this complex world are as sensitive to the tides as the larger members are. Even the algae migrate up and down in the sand—not with each wave, to be sure, but with each tide. They advance to the top of the sand to catch the sun's light at each low tide and then, just before the next high tide, they migrate down again and thus keep from being washed away. These algae actually anticipate the tides. They possess an internal clock that simply ''knows'' when the tides will be, and it continues to operate even when they are taken indoors, miles away.[1]

If the beach foreshore, where no plants grow, is rich in life, tidal wetlands along low-energy shores are infinitely more so. The rest of this chapter and all of the next one deal with tidal wetlands: tidal marshes in this chapter, tidal swamps in the next. First, though, a few introductory remarks that apply to both categories.

Florida visitor, winters: Short-billed dowitcher (*Limnodromus scolopaceus*). Shore-birds probe the sand to feel for the numerous, healthy creatures that live among the grains.

TIDAL WETLANDS

Tidal wetlands occupy stretches of the shoreline where the waves are not energetic enough to support beaches—essentially, those areas not highlighted in the last chapter's map of high-energy shores (Figure 15–1). They line lagoons, protected bays, and inlets; the mouths of rivers and streams; and other low-energy shores, including the lee shores of barrier islands. They are best developed where the land is most nearly flat and tidewaters travel long distances onto the land and back out onto the continental shelf. Along Florida's west coast, the terrain is so level that even a slightly rising tide crawls for miles across the land and the waves are gentle ripples, typically no more than two inches high. Florida's coastal marshes and swamps vary from about 1/2 to 4 1/2 miles in width; along coastal Citrus County, they are 7 1/2 miles across. Tidal marshes predominate in the northern half of Florida, and tidal swamps (mangrove swamps) in the southern half.

Among the stresses faced by tidal wetland inhabitants are variations in salinity, oxygen levels, temperature, and wave energy. Although tidal wetlands are classed as salt, brackish, or freshwater systems, their salinity may vary from hour to hour, day to day, and season to season. Water flowing in from the ocean may be salty, but rains and runoff repeatedly dilute the salt with fresh water. Oxygen levels in the water and soil also vary both from place to place and from time to time. Because the waters are shallow, their temperatures can vary from below freezing on a winter night to subtropical the next day at noon. Moreover, some days are utterly calm, whereas on others, fearsome winds blow in off the water. Any plant or animal living in a tidal wetland has to be able to contend with all these extremes.

Each tidal wetland has its own tidal pattern, and this distinguishes one from another. Some shores experience tides every twelve hours, others every six, and still others have a mixed pattern. These patterns are both stressful and reliable for living things: stressful, because they present such extremes; reliable, because they present them predictably. List 16–1 gives examples of some of the characteristics of the tides along different parts of Florida's shore and the plant responses to them.

Many animals and plants are unable to live in zones that are alter-

---

Tidal wetlands lie along the coast between the low and high-tide lines, and their water rises and falls with the tides.

Tidal wetlands include both marshes and swamps. Within each category, some are saltwater wetlands, some are brackish, and some are exposed only to fresh water.

Reminder: A **marsh** is a wetland dominated by grasses and grasslike plants. A **swamp** is a wetland dominated by shrubs or trees.

The highest high tides of each month are **spring tides**. They occur when the sun, moon, and earth are aligned in space, that is, at the full and new moons.

The lowest high tides of the month are **neap tides**. They occur at half moons, when the sun and moon are at right angles to each other.

LIST 16–1
Tide patterns and plant life

<u>Northeast marshes around Jacksonville</u>

Tide Pattern: Twice a day, the highest tides in Florida

Vegetation: A tall form of saltmarsh cordgrass

<u>Marshes in the Indian River Lagoon (Volusia to Martin counties)</u>

Tide Pattern: Marshes are confined behind a shore-side berm and so are inundated only during extreme high tides and when wind pushes water into them.

Vegetation: Mostly high-marsh plants mixed with black mangrove

<u>Panhandle Marshes</u>

Tide Pattern: Tides higher or lower than the moon alone would dictate. In winter, winds commonly blow from the north and northeast, pushing water out of most marshes. In summer, winds more often blow from the south and west, sometimes holding water in the marshes for a day or two.

Vegetation: Predominantly needle rush.

Sources: Coultas and Hsieh 1997, 16; Montague and Wiegert 1990, 487-490.

nately in the air and under water, but some thrive there and depend on the tides to repeat without fail. Many animal species of tidal wetlands have, in fact, evolved to use the tides to advantage, particularly the very highest "spring" tides that occur twice a month, at the full and new moons. Currents are strongest at spring tides, when large volumes of water are moving. Marine animals commonly release their eggs at spring tides because they will achieve wide dispersal this way. Transport of planktonic offspring out and away from the parents allows attached species to colonize distant habitats and later reproduce with mates from other populations. As a result, the offspring are more genetically diverse and more fit than inbred offspring would be.

ZONES IN TIDAL MARSHES

For many Florida residents, a tidal marsh is a familiar sight: a ribbon of grassland between the forest and the sea, a place of severe simplicity. Figure 16–1 shows a photo of such a marsh, in this case a salt marsh, showing its zones. Closest to open water is the low marsh, behind it, the high marsh. The distinction between "low" and "high" marsh is determined by a difference of just an inch or less in elevation. An inch is all it takes to make a big difference in plant types.

Within the high and low marshes are special habitats. Sinuous tidal creeks penetrate the marshes like arteries and veins, carrying tidal waters, the life blood of the system, deep into the marsh every few hours. There are also salt flats, sometimes called salt barrens; and there are rafts of tidal wrack—dead marsh plants heaped wherever high tides have dropped them.

The low marsh is usually a monoculture—that is, one species of grass grows there. This grass grows luxuriantly, so the canopy tends to be closed, shading the soil. Tidal currents are strong, and they sweep litter out from around the plants' roots; no litter accumulates in the low marsh. Tidal currents also bring nutrients into the low marsh, and remove wastes. The soil is often under water and evaporation occurs slowly, so the soil tends to remain waterlogged and oxygenless.

FIGURE 16–1

Zones in a Salt Marsh

Only a few species of plants dominate a salt marsh, as shown. Green ribbons of saltmarsh cordgrass twist along the banks of the salt creeks. This is the low marsh. Behind it, an expanse of needle rush sweeps across the higher marsh. Other small inconspicuous plant species, including Carolina sea lavender, grow scattered beneath the rushes. Farther back on the land is a coastal hammock with slash pine, cabbage palm, and other plants.

In the high marsh, additional plant species grow. Plant leaves and soil are warmer and dryer than in the low marsh, and some shrubs may take hold temporarily, but high storm tides kill them off from time to time. Because tides penetrate the high marsh less often, litter does accumulate, fewer nutrients flow in, and more wastes build up. Higher soil temperatures and less frequent inundation speed evaporation, so salts tend to accumulate, raising the salinity.

Many more species of algae than of vascular plants are present in a salt marsh. A thick, gelatinous coating of microscopic algae grows on soil surfaces and on all the bits of plant litter suspended in the water. These algae are only one-tenth as massive as the vascular plants in the marsh, but they are ten times more energy-rich and nutritious for plant eaters such as fiddler crabs and snails, because they are eaten fresh rather than after being shed as litter.

Although their numbers vary, bacteria and algae are always growing vigorously within the marsh, where sunlight, water, and salts are always freely available. Incoming tides lift some of these organisms off the surfaces of litter and soil into the water. There, they become food for tiny, floating plankton that live out their entire lives in the marsh. Some Florida marshes host about 20 species of plankton, numbering 100,000 individuals per cubic meter of water although weighing just over an ounce.

Like the big plants, the microscopic plants are unevenly distributed over both space and time. Some species occur in saline habitats, others in brackish; some at the periphery of the marsh, and others nearer the interior. Huge surges of different populations occur as the seasons change. Patches of blue-green cyanobacteria are especially abundant after rains (see "Tidal Marsh Food Webs," pages 273–275).

Salt marshes present the most strenuous challenges to plants, brackish-water marshes are less stressful, and freshwater marshes are relatively benign environments in which many more plant species can grow. The next sections describe the plants in the various types of marshes.

> *Reminder:* **Algae** include both single-celled and many-celled forms.
>
> Unicellular and small multicellular algae are **microalgae**.
>
> Large multicellular algae are **macroalgae**, or **seaweeds**.
>
> *Reminder:* The category of microorganisms that used to be called blue-green algae are now known as **cyanobacteria**.

PLANT LIFE IN TIDAL MARSHES

A tidal marsh looks monotonous to the untrained eye. Only a few species of grasslike plants (actually grasses and rushes) may spread for miles across a land that looks as flat as a mirror. The only breaks in the vast, grassy plain are the tidal creeks that snake through the marsh, and the empty salt barrens.

Plants in Salt Marshes. In salt marshes, conditions vary between extremes of salt and fresh, wet and dry, hot and cold. Life in an environment of such extremes is so hard that only a few plants have mastered it. In fact, just one grass and one rush visually dominate some salt marshes so completely that they appear to be the only two present. Along creek banks and levees and in the lower marsh is saltmarsh cordgrass, which is green. In the higher marsh grows needle rush, which is brown. Just these two plants may cover many square miles of marsh. A few other, less conspicuous species grow among them (see List 16–2 and Figure 16–2).

Along tidal creeks, borders of tall green vegetation thrive, thanks to the rising and falling tides that repeatedly bring in nutrients and carry away wastes. Minor creeks branch off the main creeks and different plant populations are found at each level of the hierarchy. Marsh plants along creeks also benefit from the abundance of fiddler crabs found there, which help to aerate and fertilize the soil.[2]

On relatively high ground within marshes, ground that is flushed by salt water only a few times a month, are salt flats. Between inundations, the sun bakes and simmers the evaporating water on the flats until the salts grow so concentrated that they crystallize, suppressing nearly all plant life. What little water there is evaporates even faster thanks to the presence of fiddler crab burrows all over the flats, which keep the soil loose.[3] List 16–3 itemizes plants that grow on salt flats.

Tidal wrack, that is, heaps of dead plant material, can occupy large areas of the high marsh. Wrack consists mostly of dead shoots of needle rush caught within the upper marsh. (Saltmarsh cordgrass, because it borders the creeks, is more likely to drop its litter into ebbing and flowing water that promptly transports it away.) When storm tides enter the high marsh, they lift and swirl the tidal wrack in eddies, then leave them heaped up again, elsewhere. A pile of tidal wrack may be many yards across and three or more feet deep. After it settles down, it may take a year or more to decompose, meanwhile killing all of the living vegetation beneath it. An-

Saltwort (*Batis maritima*).

Annual glasswort (*Salicornia bigelovii*).

Bushy seaside oxeye (*Borrichia frutescens*).

Sand cordgrass (*Spartina bakeri*).

FIGURE 16–2

Native Salt Marsh Plants

Saltwort tolerates ordinary salinity and grows well in salt marshes. Annual glasswort and sea oxeye are so salt tolerant that they can even grow on salt flats. Sand cordgrass grows not only in salt marshes but in brackish and freshwater marshes and along lake margins.

The first three are succulents. They have pulpy tissues that store water and enable the plants to resist dehydration, which occurs readily in salty environments.

other storm tide may lift the pile again and drop it farther up the marsh, killing more vegetation.

Where tidal wrack has lain, salt barrens may later develop and remain for long times. In one marsh, wrack killed more than a tenth of the vegetation before being moved again by tides. But wrack is not just dead material; as on the beach, it is a breeding ground for hosts of insects, arthropods, and others that become food for a myriad other animals.[4]

Brackish-Marsh Plants. Brackish marshes develop along river deltas and wherever freshwater discharges are regular and substantial. The farther inland these marshes are, the less saline they are; and the lower the salt stress, the greater the variety of plants that can grow. List 16—4 itemizes some of the plants in brackish marshes along the St. Marks and Wakulla rivers; there are several dozen species in all.

Plants in a brackish marsh respond so sensitively to salinity that investigators can easily distinguish between two kinds of brackish marsh. The saltier type has three dominant plants: needle rush (as in salt marshes), big cordgrass, and bulrush, with little else. The fresher type of brackish marsh has several dozen additional species, including Jamaica swamp sawgrass.

Freshwater Tidal Marsh Plants. Freshwater tidal marshes occur still farther inland along tidal rivers. They are never saline except during tropical storms when extreme high tides penetrate far inland, but their waters rise and fall just as the tides do at the coast, although they may lag the tides by several hours.

Marsh plants have many advantages in a freshwater tidal setting. Free of salinity, they still reap the benefits conferred by the rising and falling waters: fresh nutrients come in with every flow of the tides, and wastes wash away with every ebb. The marsh soils are permanently waterlogged and therefore lack oxygen, so trees can seldom root there; those that do, soon fall over. Also, fires occasionally burn into the marsh from neighboring forests and the fires, too, preclude tree growth. Accordingly, grasses and forbs in freshwater tidal marshes are more diverse than in salt or brackish marshes, with each plant tending to thrive in a particular part of the terrain (see List 16—5). Forests begin at the landward edges of tidal marshes, sometimes abruptly, sometimes gradually across broad zones of blending.[5]

As these descriptions show, tidal wetland plants are much more diverse than they look. Taken all together, plants both large and small and unseen organisms in the marsh water and soil support a vast food web on which all of the wetland's animals depend. The next section explores the food webs in a tidal marsh, using the salt marsh as an example.

TIDAL MARSH FOOD WEBS

A salt marsh is thick with living things engaged in intense activity. All of the green plants, algae, and cyanobacteria constantly gather up water from the mud, oxygen from the air, and energy from light, making huge quan-

LIST 16–3
Plants on salt flats (examples)

Annual glasswort
Bushy seaside oxeye
Carolina sealavender
Keygrass
Perennial glasswort
Perennial saltmarsh aster
Saltgrass
Saltwort
Spikerush

Source: Clewell 1997, 82.

LIST 16–4
Plants in brackish marshes (examples)

Awlleaf arrowhead
Carolina sealavender
Climbing hempvine
Eastern grasswort
Gulf coast spikerush
Jamaica swamp sawgrass
Marsh fimbry
Needle rush
Saltgrass
String-lily
Wand loosestrife

Source: Clewell 1997, 98. The list names 74 species in all.

LIST 16–5
Plants in freshwater tidal marshes (a sampling)

These are 15 miles upriver:
Awlleaf arrowhead
Dotted smartweed
Eastern false dragonhead
Indian rice
Manyhead rush
Pickerelweed

At the border with the river floodplain forest:
Jamaica swamp sawgrass

In patches of marsh and on river banks:
Arrowhead (2 species)
Big cordgrass
Common reed
Southern cattail

Source: Clewell 1997, 102. These plants were noted along the St. Marks River.

Tidal wrack lodged in a salt marsh. The base of the food web consists largely of dead needle rushes washed up in the marsh.

tities of their own kind and pumping energy into the system. From them, and from the detritus they generate, hundreds of species take nourishment. Complicated webs of life tangle together in ways that change from morning to night, from season to season, and from one part of the marsh to the next.

Of all the masses of plant tissue in a coastal wetland, only a small fraction is eaten by insects and other herbivores while it is still alive. This fraction supports one system of consumers known as the grazing food chain. The larger fraction of the plant tissue dies, is shed, and ends up in the water and on the marsh floor as detritus—the main food for the major system of consumers, the detrital food chain. The two chains interweave in complex patterns as energy moves from the wetland's primary producers, the plants, through the other organisms both small and large. Background Box 16–1 delves a little more deeply into the routes taken by water-borne detritus.

BACKGROUND   **16–1**

Nourishing Soup: Water-Borne Debris

Tidal marsh systems receive huge quantities of minerals (washed down from eroded mountain rock) and of plant and animal remains (washed out of interior forests, swamps, and marshes). The living organisms in the marshes combine these materials to make enormous quantities of living things.

Among the members of water-borne detritus are microbes so tiny, and present in such astronomical numbers, as to stagger the imagination. Microbes work on every particle of organic material that is carried into the water. Scientists have studied these organisms in Apalachicola Bay and have found them to be present in such numbers that, if it were possible to collect them in one place, their mass would equal that of all the bay's other creatures put together—shrimp, crabs, oysters, fish, and all the rest.

Consider what takes place on a single leaf as it floats down a river from some interior swamp. Microscopic organisms attach to the surface of the leaf in an ordered sequence. Bacteria first colonize the leaf, several species in succession, and work on it for several weeks. Then fungi take hold and spread their filaments across the leaf, digesting it further. Later, diatoms and algae take their turn. Then grazers move in to eat from the leaf surfaces, and by then they find there a highly nutritious diet. The grazers are tiny animals—mostly microscopic copepods, a highly varied group of crustaceans. As they grow, these leaf digesters also release waste materials that sink to the bottom sediments and serve as fertilizer for nearby plants.

Meanwhile, equally intense activity occurs in the bottom sediments. Myriad decomposer organisms live within the top few millimeters, coating every particle of sediment: bacteria, fungi, and other single-celled creatures. There are billions of these microfauna in every cubic inch of mud and their populations are layered vertically. Those in the top few millimeters use up all of the available oxygen, while those beneath use sulfur instead of oxygen to ferment the warm organic soup that filters down to them. (These latter organisms are responsible for the foul smell of hydrogen sulfide that sometimes arises from salt marshes when the soil is disturbed.) Together, these organisms gobble up detritus that has sunk to the bottom. They break it down, transform it, and return it to the detrital chain in new forms.

The bacteria, fungi, and others in the soil feed still other tiny animals in the sediment, the meiofauna. Worms, mussels, clams, shrimps, minnows, oysters, and the young of hosts of other animals reside in the mud or roam its surface, chewing and scraping their tiny prey from the grains, sucking up water from which to strain more of the same.

Clearly, the reason why marsh plants grow so abundantly is that they receive abundant sunlight, water, and a rich fertilizer in the marsh mud, composed of detritus and the organisms that process it. The plants feed energy into hosts of tiny animals, and the larger animals of the marsh, those people can see, such as snails, crabs, and fish, are there because they can partake of that feast of tiny life. The relationships build from the bottom of the food web as shown in List 16–6.

It is in these ways that the detritus in salt marsh water and soil becomes huge quantities of animals of all descriptions. And since the detritus comes from living forests, swamps, and marshes in Florida's interior, it follows that the health of those communities supports the health and abundance of the offshore sea life.

Source: Succession of microscopic organisms from Livingston (R. J.) 1983, 32.

Animals are known as **fauna** (plants are **flora**); and the animals that live in burrows within the bottom sediments, straining their food from the sediments and water, are **infauna** (IN-faw-na).

The tiniest burrowing animals are **microfauna**.

Those that are barely visible are **meiofauna** (MY-o-faw-na).

Those that are easy to see are **macrofauna**.

**LIST 16–6
Marine communities built on detritus**

Nutrients and detritus are lowest in the food web. Large fish are highest.

Nutrients; organic detritus
Plant plankton (phytoplankton)
Emergent vegetation
Submergent vegetation
Microbes dependent on these nutrients and plants
Animal plankton (zooplankton) and tiny larval fishes
Invertebrates in the bottom sediments
Oysters, blue crabs, shrimp
Anchovies
Larger fish (croaker, spot, sand seatrout)

Source: Livingston 1983, 24–51. Each of these levels of the food web occupies a chapter in Livingston's book.

Although dead plant tissue is by far the larger contributor to salt marsh life, living plant tissue also feeds a host of insects and other herbivores. Plant hoppers, which feed on plant juice, are everywhere. So are grasshoppers, which eat plant tissue. Numerous spiders of many species spread their nets to capture these—what better place could there be for a spider to set up housekeeping than where so many small insects are on the move? Several species of wasps are parasites on the spiders. Mites, flies, a predatory beetle, butterflies, dragonflies, all abound in the marsh. List 16–7 mentions only a few of the more than 400 species of arthropods observed in one part of one marsh at one time.

These animals attract still others into the marsh to feed upon them. Birds fly in to catch insects and spiders, and fish swim in when the marshes are flooded. Fish catch the insects that fall into the water and even leap from the water to catch insects perched or flying above it. In each of Florida's tidal marshes, several dozen species of fish prey on insects, which they harvest with great efficiency. One observer saw a single killifish make nine successful strikes in just five minutes.[6]

***

The many small animals that live in tidal marshes support thriving populations of larger ones. The next sections of this chapter describe some of the large-animal residents of coastal marshes, as well as animals that visit from elsewhere.

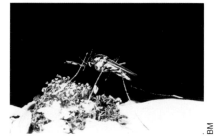

Florida native: Salt marsh mosquito (an *Aedes* species). Among the innumerable tiny animals that live on the "soup" in salt marshes are the larvae of this mosquito.

Florida native: Robber fly (*Efferia pogonias*). Dozens of species of two-winged insects are present in buzzing hordes in salt marshes.

RESIDENT SALT MARSH ANIMALS

At low tide, only a few types of animals are visible on the salt marsh floor, but they occur in large populations. A mob of fiddler crabs bustles among the fat red saltworts on an open sand flat. Thousands of periwinkle snails graze microalgae from the mud, five or six at the base of every blade of grass. Clicking and splashing noises suggest the presence of other animals, not seen.

Despite the bounteous feasts offered by a salt marsh, not many animals can stay there throughout their lives, because conditions alternate among such extremes. A few animals, however, can withstand the rigors of the marsh and therefore can take full advantage of its abundant resources. These year-round residents exhibit remarkable behavioral and physiological adaptations. Within an environment that swings from fresh to salty, airy to oxygenless, hot to cold, and high to low water and back again, they find food, attract mates, breed, and produce young that flourish and grow to adulthood, all in the marsh. One such animal is the fiddler crab, ubiquitous along Florida's coasts.

The Fiddler Crab. Millions of fiddler crabs inhabit every tidal marsh, digging burrows on salt flats, along creeks, and along marsh edges (see Figure 16–3). As the tide comes in, they scurry up the marsh, duck into their burrows, and plug them closed, remaining under water until the tide goes out again. Then they re-emerge to forage. Being aquatic animals, they breathe through gills as other crabs do, so when on land, they carry water with them in their gill chambers. They use this water also for evaporative cooling, and for sorting edible from inedible particles.

Different species of fiddler crabs live on muds and sands of different grain sizes. Each species has hairs in its mouth, whose spacing permits it to scrape microscopic diatoms off individual particles of sediment—more widely spaced for sand grains, more closely for grains of mud. The mouth parts are so specific to the type of soil in a given marsh, that they even permit the crab to sort its food, selecting the digestible parts to eat, and rolling up the rest into a ball which it replaces on the mud. (The large pellets that are everywhere in fiddler-crab colonies are mud balls, not fecal pellets.) The next high tide breaks these pellets up and gives the local microbes another go at digesting the detritus in them.

Fiddler crabs are keystone species: they not only prosper in a marsh ecosystem that is healthy, but they provide many services that keep it that way. They dig some 25 to 150 burrows per square yard of marsh, some only a fraction of an inch wide and deep, some an inch and a half wide and up to three feet deep. Thanks to these burrows, which aerate the soil, marsh plants can grow and decomposers can break down litter efficiently, freeing nutrients that the plants can take up (and grow still faster). The crabs' wastes also serve as fertilizer while the plants' detritus becomes food for the crabs, a relationship that benefits both. Each year the crabs completely turn over the top half inch of marsh mud, returning buried nitrogen to the surface.[7]

Fiddler crabs also break up the algae that grow on the surface of the mud. If undisturbed, algae would soon form a carpet so thick that it would

Male displaying
before female

FIGURE 16–3

Salt Marsh Fiddler Crab (*Uca pugilator*)

This is one of many fiddler crab species that occur in salt marshes. This particular species is a resident of Florida Gulf coast marshes, especially higher-elevation zones in needle-rush marshes.

The male crab has one oversized claw, sometimes the right and sometimes the left, which it waves to assert its claim to the territory around its burrow and to attract the female of the species. Crabs compete for sites near the upper border of the intertidal zone, where a burrow is most likely to stay wet at low tide yet not collapse at high tide. The best burrows attract the most females, which inspect many of them before deciding which to occupy. Crabs with the largest claws and best burrows are most likely to produce many offspring, so genes for the large-claw trait continue to be reproduced in each new crab population.

Fiddler crab species in general are famous for their internal biological clocks, which accurately calculate the times of the tides even when the crabs have been transported far from the ocean. They become active at low-tide times and somnolent at high-tide times, just as they did at the shore. Research on fiddler crabs has helped biologists to understand how biological clocks work.

Another fiddler crab, the Atlantic marsh fiddler occurs along Florida's east coast, where it conducts its life in a similar fashion. At the edges of its habitat, still another crab, the purple marsh crab shares its territory but is active only at high tides when the Atlantic marsh fiddler crab is in its burrow. The two eat different foods and so do not compete. Meanwhile, dozens of other crabs occupy other habitats nearby.

Note: The Atlantic marsh fiddler is *Uca pugnax*; the purple marsh crab is *Sesarma reticulatum*.

Sources: Grimes 1989; Salmon 1988.

halt other growth, but the fiddlers mix the surface at every low tide. In the process, they bury some algae, which smother, die, and become food for recyclers and fertilizer for marsh plants. The algae that remain on top are given space to keep on multiplying.

The crabs are not the kings of the marsh; rather, they are the prey of fish, birds, raccoons, and other eaters. From the hungry human point of view, this becomes of special interest when fishing boats using crab as bait return to shore loaded with red drum, sheepshead, and other prized catches. The white ibis, clapper rail, and many egrets and herons also thrive on fiddler crabs. So do other fiddler crabs, blue crabs, and a dozen or so small hangers-on, including both harmful parasites and harmless freeloaders.

The Periwinkle Snail. While the fiddler crab copes with life in an alternately submerged and exposed environment by ducking into its burrow at high tide, the periwinkle snail stays just above the ebbing and flowing water by using plants as escape towers. The snail slowly climbs a plant, keeping just ahead of the incoming tide, which carries in hungry, shell-crushing blue crabs and other marine predators. At high tide, the snails all crowd at the tips of the plants, looking like a crop of white berries. They can thrive on the plants because a lush lawn of algae grows there.

Empty fiddler holes at low tide. The fiddlers will retreat into these holes when the tide comes in.

Florida native: Periwinkle snail (*Littorina irrorata*). This animal finds a fresh crop of nourishing algae every time it reclimbs the stem of a marsh plant.

Small Fish. An incredible abundance of small fish, specially adapted to salt marsh extremes, grow in the murky tidal creeks. More than 90 percent of the fish in tidal creeks and pools are tiny killifishes, mummichogs, and other minnows. Scoop up a net full of marsh water and you will find it jumping with little silver fish.

Killifishes can handle salt, brackish, and even nearly fresh water equally well. Storm surges inundate their habitat with pure salt water and they survive. Rainstorms pour in fresh water and they still survive. During hard freezes, they burrow into the mud, lie still, and await the next thaw to resume their busy lives. In Florida salt marshes a dozen or more species of killifish may occupy different aquatic habitats and live on different prey. They make excellent bait fish, because they are the normal food of the larger fish, and they are reliably there for belted kingfishers, great blue herons, and other fishing birds.

Amphibians and Reptiles. No amphibian can tolerate salt marsh environments the year around, but two species of frogs can conduct their life cycles in brackish and freshwater tidal marshes: the green treefrog and the southern leopard frog. Among reptiles, two snake species are especially well adapted for life in a tidal marsh environment. The rough green snake feeds and lives in brackish and freshwater tidal marshes along both coasts and eats soft-bodied invertebrates. The salt marsh snake lives in tidal marshes and mangrove swamps around the Gulf and along Florida's Atlantic coast. It is almost never seen: it is active only at night. It drinks only when fresh water is available as rain or dew. When there is no fresh water, the snake resists water loss and survives anyway. It feeds primarily on small fish, occasional crabs, and other small invertebrates.[8]

Other snakes, such as the banded water snake and the cottonmouth, cannot tolerate salt water but do live in brackish and freshwater tidal marshes. Three species of turtles also occur in Florida tidal marshes: the Alabama redbelly turtle, the diamondback terrapin, and the Florida cooter.

Birds. Three bird species also feed and breed in Florida salt marshes: the clapper rail, the marsh wren, and the seaside sparrow. All three also nest in the marsh, but do not compete for space. They require salt marshes, and in the case of the seaside sparrow, salt marshes with a certain hydroperiod and frequent burns. The seaside sparrow cannot live in any other habitat.

VISITORS TO THE MARSH

Visitors to the salt marsh far outnumber the permanent residents just described. They come in when the marsh can meet their needs and depart when conditions become inhospitable.

Visitors from the Sea. Gulf coast marshes support immense numbers of young ocean fish. In Florida's needle-rush marshes alone, a total of 90 species are reported. Fish find diverse food sources in the marsh's varied aquatic habitats: algae along plant stems, plankton and insect larvae in the water column, detritus on the bottom, and small animal prey both swimming and in burrows. Shrimp use the marsh as a nursery ground.

Juvenile white shrimp enter tidal creeks, remain until they are about 2 inches long, then move offshore to spawn. Then the larvae reappear in the marsh.[9]

Figure 16—4 shows illustrations of a few of the fish that depend on salt marshes for parts of their lives. For red drum, striped mullet, and many other fish, tidal creeks offer much more food than do neighboring coastal waters. After swimming up the creeks, red drum push their way over the berms and through the grasses to feed in the interior of the marsh, probing the bottom to extract fiddler crabs from their burrows, their tails thrashing the surface as they go.

Striped (black) mullet swim far up tidal creeks and feed on algae, detritus, and other tiny organisms. They remain in the marsh for most of their lives, going offshore only to spawn. They are major converters of plant detritus and algae to fish protein within the salt marsh and near-coastal system, and they occur in such large numbers that their collective impact is enormous. They can live in very low-oxygen systems, they stir up the

Florida native: Mantis shrimp (*Squilla empusa*). Named for its praying mantis–like appearance, this member of the infauna burrows in mud bottoms below the low-tide line and slashes at passing prey with its razor-sharp claws. At night it emerges to swim and in turn becomes food for fish, small sharks, sea turtles, and other large predators.

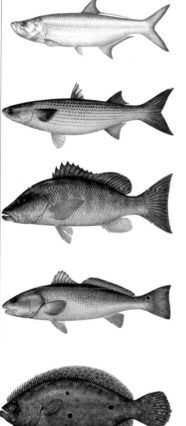

Tarpon (*Megalops atlanticus*). A striking fish with huge scales, the tarpon spawns offshore but lives primarily inshore. Often some 50 pounds when caught, it can grow to nearly 250 pounds.

Striped mullet (*Mugil cephalus*). Striped mullet spawn offshore but the juveniles soon swim into salt marshes in enormous numbers when only an inch long. They feed on detritus and ultimately attain a size of about 3 to 6 pounds.

Gray snapper (*Lutjanus griseus*). Juveniles stay inshore in tidal creeks, mangroves, and grass beds. Adults move offshore over reefs and rock bottoms. Commonly 8 to 10 pounds, these fish can grow to about 15 pounds.

Red drum (*Sciaenops ocellatus*). Red drum stay inshore in estuaries until they are about 30 inches long, then move out to spawn offshore. At 30 inches they weigh about 8 pounds; they can grow to more than 50 pounds.

Gulf flounder (*Paralichthys albigutta*). Usually found inshore partly buried in sandy or mud bottoms, these fish probably spawn offshore. After the fish hatches, the right eye migrates over to the left side early in life. Gulf flounder typically weigh up to 2 pounds.

FIGURE 16–4

Fish that Depend on Salt Marshes (Examples)

These are only a few of the 100 or so species of ocean fish that depend on Florida salt marshes for parts of their lives.

Sources: Kruczynski and Ruth 1997; Anon. c. 1994 (*Fishing Lines*); drawings by Diane Peebles.

bottom, assisting other fish in finding food, and they are eaten by most of the other fish in the system. They are also a major food source for dolphins and wading birds.

Like all others, fish populations in the salt marsh vary over both time and space. Fish diversity peaks in late summer and early fall. Populations in winter are different from those in summer. Migrants come through at some times of year and not at others. For example, in the marshes adjacent to one barrier island, Live Oak Island, five species were year-round residents, three were winter-spring residents, and three others were summer-fall residents. Day-night differences also occur, and so do high- and low-tide differences. Every fish species has its own behavior pattern and timing.

Wet-dry cycles also affect populations of animals such as fish and shrimp. Rainy times can turn the waters of a salt marsh completely fresh. Small fish and freshwater shrimp increase when marshes are flooded by rains. Then as water levels fall, the animals become highly concentrated in the remaining water. This makes them easy to catch, an advantage to birds that are rearing their nestlings at these times. But, thanks to their abundance, these dry times are not a great disadvantage to the fish and shrimp. Many young survive to swim out of the marsh and grow to maturity.

Chapter 21 describes other, larger ocean visitors that occasionally appear in tidal marshes. The West Indian manatee sometimes swims up tidal creeks to forage there; so do the bottlenose dolphin and the Atlantic ridley sea turtle. Two sharks, the sandbar shark and the bull shark, make thousand-mile migrations along the east coast, using salt marshes all along the way.

Visitors from the Air. Although the salt marsh is a required habitat for only three bird species, others can nest and feed there, and numerous visitors drop in. From nests in nearby forests, long-legged wood storks, ibises, herons, egrets, and in south Florida roseate spoonbills fly in to fish in the creeks and probe the bottom mud. Gulls, terns, and other shorebirds flock to the salt barrens to rest and feed. Hawks soar over the marsh and dive to catch small birds and rodents. Fish crows prey on the eggs of nesting birds such as seaside sparrows. Migratory bird populations peak in spring and fall when insects and spiders are most numerous. List 16—8 identifies a few of the species of birds that frequent Florida salt marshes and Figure 16—5 displays three of them.

Florida native: Black skimmer (*Rynchops niger*). Skimmers and their close relatives are the only birds with the lower mandible longer than the upper. They fly low, skimming the surface, and when they contact small fish, snap their bills shut. That they flourish around tidal marshes bespeaks the abundance of fish there.

Snowy egret (*Egretta thula*). This fish eater is a frequent visitor to salt marshes.

FIGURE 16–5

Black-necked stilt (*Himantopus mexicanus*). The stilt nests just above the high-water line in both interior and coastal marshes. When the waters rise, it frantically adds nesting material to keep its eggs above water.

Clapper rail (*Rallus longirostris*). This bird feeds on fiddler crabs, worms, snails, insects, and other tidbits from the mud and grass. Its constant, loud, nonmusical *ticket, ticket, ticket* is plainly audible all over the marsh. It nests in medium-height vegetation at the high-tide line and when threatened it runs silently through the marsh grass, seldom rising to take flight.

Bird Visitors to Florida Salt Marshes

Visitors from the Uplands. While fish and shellfish visit the marsh from the ocean, and birds visit from the air, terrestrial vertebrates visit from the landward edge of the marsh. The raccoon, river otter, southern mink, long-tailed weasel, marsh rabbit, marsh rice rat, cotton rat, and cotton mouse come in from coastal fields and forests. The eastern diamondback rattlesnake crosses tidal marshes and even open marine water to hunt for prey off the mainland and can swim as far as to offshore islands. The alligator swims from one water body to the next, stops to feed in tidal channels, and suns itself along the banks. The white-tailed deer crosses on its way from here to there, and nibbles on salty vegetation. The bobcat prowls into the marsh on occasion to surprise unsuspecting rats, mice, and birds.

TIDAL MARSH VALUES

Of all the ecosystems so far described, tidal marshes are among the richest in numbers and kinds of living things. Tremendous amounts of resources concentrate there. Rivers discharge fresh water and sediments into them, and together with ocean water, contribute many nutrients. Wind, waves, tides, and sunlight inject energy. The result is production of living matter that is among the highest per unit area for any known grassland.[10]

Salt marshes are as productive naturally as the agricultural fields that people cultivate most intensively, fields such as rice paddies and sugarcane fields. The ultimate harvest of marshes, though, is in the form of

LIST 16–8
Bird visitors to salt marshes: a sampling

More than 60 species of birds visit salt marshes.

Black-crowned night-heron*
Clapper rail*
Fish crow
Green heron*
Laughing gull*
Least bittern*
Marsh wren*
Northern harrier
Red-winged blackbird
Saltmarsh sharp-tailed
   sparrow
Seaside sparrow*
Snowy egret
Sora
Tricolored heron*
Virginia rail*

Note: *These birds occasionally nest in salt marshes. Others usually nest elsewhere and visit salt marshes to forage.

Source: Hubbard and Gidden 1997, 334–336.

animals—not large grazers as in a terrestrial grassland, but small marine species that mostly stay hidden.

Salt marshes are critical to the survival of marine animals and most are legally protected. Fisheries scientists vouch for their importance to sports and commercial fishermen. Between two thirds and nine tenths of all fish caught, as well as important shellfish, such as shrimp and crab, are species that live in salt marshes for part of their lives, mostly as juveniles. Florida's marshes are especially significant to sea life because they are so vast (see Figure 16–6).

Besides their high rate of productivity and their importance as nurseries, tidal wetlands also serve as a natural defense system, protecting coastal land from high tides, winds, and waves. A hurricane may smash seawalls, rip out dikes, and knock down whole forests, but a natural salt marsh may barely be disturbed. It quietly goes on creating and recycling detritus and, in the process, feeding and sheltering a myriad of marine organisms.

Tidal marshes are also natural purifiers and recyclers. They filter water that runs off the land, releasing it clean to seagrass beds and estuaries. On a global scale, they help return sulfur and nitrogen to the atmosphere in the mighty cycles that keep these elements available to the planet's living systems. And they export materials to adjacent marine habitats. Although some materials settle into sediments and some are taken up by the marsh's own plants, animals, and microscopic organisms, other materials depart with outgoing tides and become important resources in seagrass beds and open marine waters (next chapters). The value of seawrack on beaches has already been described—and seawrack, of course, comes mostly from tidal marshes. Tides lift from the marsh floor fine particles of detritus from needle rush and cordgrass decomposition and carry them away on the surface as an oily slick to nourish estuarine and marine life farther offshore.[11]

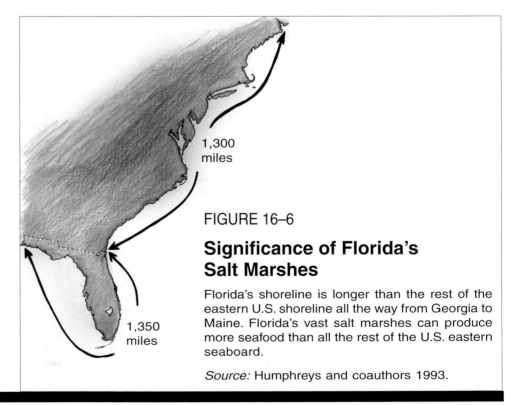

1,300 miles

1,350 miles

FIGURE 16–6

Significance of Florida's Salt Marshes

Florida's shoreline is longer than the rest of the eastern U.S. shoreline all the way from Georgia to Maine. Florida's vast salt marshes can produce more seafood than all the rest of the U.S. eastern seaboard.

Source: Humphreys and coauthors 1993.

Salt marsh, St. Marks National Wildlife Refuge, Wakulla County. The moon, high in the sky, looks tiny, but together with the sun it governs the tides, which in turn govern all life in the marsh.

To contemplate a tidal marsh with understanding is to be awed at its antiquity, at all the interdependent relationships within it, and at its connections with other ecosystems. To preserve a tidal marsh, it is necessary to maintain the health of the whole landscape mosaic of which it is a part, both terrestrial and marine. The rewards are many, not least the bounteous harvest of seafood that results.

CHAPTER SEVENTEEN

MANGROVE SWAMPS

Something green is bobbing up and down in gentle, shallow swells off a peaceful south Florida shore. It looks like a long, fat cigar floating one end up, with the other end ten inches below the surface. What is it? Where did it come from? Is it alive?

In another month, the answers will be clear: it is a mangrove embryo, dropped from a tree along the shore. As it drifts in shallow water on the outgoing tide, its tip punctures the mud and sticks there, perhaps stabilized by a bed of marsh or seagrass already present. In the next days it sprouts arching roots and fat, succulent leaves, and begins to become a tree. Floating debris snags among its roots and an island begins to form.

After several hundred years, the original tree will be surrounded by a big mangrove island on a bed of silt and peat some twenty feet deep. The island may be several hundred acres in size and hold many brackish lakes in its interior. Hundreds of species of bacteria, fungi, algae, and tiny invertebrates will busily process detritus from the leaves, stems, and roots of the mangroves and from the droppings of crabs, snails, snakes, and birds that live in the trees. Numerous, diverse fish will dart and dodge among the underwater roots, and hordes of worms and other unseen creatures will sift their livelihoods from the mud below. Besides supporting myriad residents, the swamp will also serve as nursery grounds for a host of fish and birds who emigrate to the outside world to live out most of their lives.

Mangroves grow along tropical shores the world over. They thrive where the land is level or gently sloping, the waves are insignificant, and the tidewaters flow far in and far out every half day. Gentle waves permit mangrove seedlings to root, and leave fine sediments in place, whereas rough waves snatch up seedlings before they can take hold and wash away fine sediments. Figure 17–1 shows the Florida distribution of mangroves and Figure 17–2 shows the types of communities they form.

MANGROVES: MASTERS OF ADAPTATION

Mangroves are especially notable among trees for their adaptations to saline soil, but they contend with many other challenges in their environ-

OPPOSITE: The Ten Thousand Islands along the southwest coast of Florida. Exploring these mangrove islands can entail a several-day canoe trip.

Reminder: A **succulent** plant is one that has thick, fleshy leaves and/or stems that are adapted for storing water. Succulents typically grow in deserts (cactuses are examples) and in salty environments (saltwort and glasswort are examples).

RPP

Florida native: Black mangrove (*Avicennia germinans*). The seedling has just broken through its case and begun to sprout. The "cigar" lying in front of it is a red mangrove seedling that did not survive.

285

Red mangroves in the Everglades. Where fresh water first begins to encounter salt water, sawgrass begins to give way to mangroves.

FIGURE 17–1

Florida's Mangrove Swamps

Mangroves grow not only along *low*-energy coastlines, but in inlets and along rivers inshore from *high*-energy coastlines, hence the apparent overlap with Figure 15–1 in Chapter 15. Both freezing and temperatures above 100 degrees inhibit their growth. Shrubby mangroves occur along short stretches of the Florida Panhandle, but mangroves flourish as trees south of Cedar Key on Florida's west coast, and south of Cape Canaveral on the east. Along these shores, they form a band that occupies some 470 to 675 thousand acres and in some areas is twenty miles wide.

Sources: Toops 1998, 17; Odum and coauthors 1982, 1–2; Odum and McIvor 1990.

Cedar Key

Cape Canaveral

The term **mangrove** can refer to any of some 50 species of trees and shrubs that grow in saline soil worldwide. Three are common in Florida: red, black, and white mangrove. A fourth, buttonwood, is sometimes counted as a mangrove, too. A **mangrove swamp** is a tidal wetland dominated by saline-tolerant trees and/or shrubs.

ment. Like the plants in salt marshes, they grow on shifting, sulfurous, waterlogged ground, in water that rises and falls daily and in some areas oscillates between salty and fresh. The water is often low in nutrients and oxygen, and often dark, especially following storms.

Two mangrove species dominate Florida's tidal swamps: red mangrove and black mangrove. They hold on well in shifting substrates and thrive despite the oxygen-poor soil. Because lack of oxygen limits root growth to within a few feet of the surface, they have no deep taproots to help them withstand hurricanes, but they produce extensive, interlaced, horizontal root nets that keep them anchored during all but the most powerful winds and waves. Each species also grows special organs that enhance its oxygen-gathering ability (see Figures 17–3 and 17–4).

Survival in salt water is a mangrove's specialty. No mangrove can exclude salt perfectly, and salt does accumulate in the trees' tissues, but the plants concentrate salt in their leaves and then discard the leaves one by one, all year long, returning salt to the environment. Continuous evaporation of water from the leaves draws fresh water up from below. Each species also has additional strategies to manage salt. For instance, the red mangrove excludes the most salt at its root surfaces by means of a powerful filtration system. The salt in red mangrove sap is at 1/70th the concentration in the surrounding water, but ten times that in land plants. In

PRICELESS FLORIDA

Overwash forest. Some mangroves grow on bars and islands, and water washes over their roots at every high tide.

Fringe forest. Some mangroves grow on the fringes of the land in the intertidal zone.

Basin forest. Some mangroves grow along coastlines in basins that are frequently washed out by tides.

Scrub forest. Some mangroves grow in interior basins whose water is seldom washed out by tides. These become sparse, dwarf mangroves.

Riverine (rhymes with *fine*) forest. The most extensive mangrove swamps grow along river estuaries on level land where tides travel far in and far out every day.

FIGURE 17–2

Mangrove Community Types

Mangrove communities vary in appearance. Along the Shark River in the southeastern Everglades, red, black, and white mangroves together with buttonwood trees grow to more than 60 feet in height and form a dense, dark forest. On marl prairies in the same part of Florida, dwarf mangroves are less than 5 feet tall and sparse, even though they may be more than 50 years old.

Sources: Drawing adapted from Odum and coauthors 1982, 8, fig. 4; information from Odum and McIvor 1990; Kangas 1994.

contrast, black mangroves reduce their salinity by secreting a white bloom of salt crystals from the undersides of their leaves. You can rub your finger across a leaf and taste the salt. The salt concentration in their sap is 1/7 that in sea water. Still, the salt burden becomes too heavy in some places, even for mangroves to tolerate. There are salt barrens among mangroves as there are in salt marshes.

Red mangrove prop roots.

Red mangrove drop roots.

FIGURE 17–3

Red Mangrove (*Rhizophora mangle*)

The large tree (in the mid-left) is a black mangrove; the others are red mangroves. Red mangrove grows best in frequently flooded, "low marsh" zones, although it can tolerate long dry spells during the seasonally low waters that occur in south Florida in spring. It grows to some 80 feet in height and its trunk may be 7 feet around.

Red mangrove roots form a tangle that holds the tree fast to the ground. Its prolific "prop roots" arch from its base to three feet above the mud and extend out around the trunk well beyond the branches. Roots from its branches, "drop roots," penetrate just an inch or two into the soil. These aerial roots have spongelike internal tissues that allow oxygen to diffuse downward to the lower roots. Red mangrove also has breathing pores (lenticels) in its lower trunk like those in freshwater wetland plants. The lenticels let oxygen in at low tide, and exclude water at high tide.

Sources: Toops 1998, 72; Odum and coauthors 1982, 1–2.

Mangrove seeds become seedlings before they leave the parent tree. They are called **propagules** (PROP-a-gyules).

Just as adaptations of the roots and leaves enable mangroves to flourish, the design of their offspring enables them to begin to grow in the withering brine where they must take root. Mangroves don't produce seeds as such, they produce propagules—whole embryos, buoyant and succulent, that are nearly ready to root even before they leave the parent tree.

Thanks to these adaptations, mangroves not only thrive in their present habitat but take over new territory wherever it becomes available. Like seaoats on beaches and dunes, mangroves help to form and stabilize the land they grow on. New seedlings start up most easily on silt or clay that already contains organic matter, but they can do without it at first.

SFWMD

BM

Black mangrove leaves. The leaves constantly pump salt out of their tissues, reducing the salt concentration of the tree's internal fluids.

FIGURE 17–4

Black Mangrove (*Avicennia germinans*)

Black mangrove grows best in infrequently flooded, "high marsh" habitats, although it can tolerate season-long flooding during fall's high waters in south Florida. It grows to some 60 feet in height. The tree has long "cable roots," that extend outward from the trunk some two to three feet down into the mud. The cable roots put up a bristly miniature forest of many vertical, pencil-thin poles which not only conduct air to the lower roots but help to aerate the surrounding substrate. The breathing poles are pneumatophores (new-MAT-o-fors).

Pneumatophores greatly inhibit pedestrian explorations, especially when the explorer is trying to wade through the mud at high tide.

Sources: Douglas 1947; Toops 1998, 72; Odum and coauthors 1982, 1–2.

After taking hold, mangroves demonstrate still more competitiveness by altering the ground they have claimed, making it less hospitable to other species. They shed abundant litter, especially from their fine rootlets, which break off in the shifting substrate and partially decompose in the wet sediment, turning to peat. The acidic peat then dissolves the underlying limestone rock, so the peat layer grows both downwards and upwards. As a mangrove island enlarges over time, its peaty substrate becomes a layer several meters deep.

Once mangroves have begun to take over a site, their elaborate, widespread root systems trap additional sediments. The litter they drop continuously releases nutrients as it decomposes. The branches attract birds and other animals to roost and nest, and their droppings add fertilizer to the soil.

The substrate is further altered by algae, bacteria, fungi, and small invertebrates. These organisms devour oxygen, nitrogen, and phosphate, making the sediment low in oxygen, acidic, sour with hydrogen sulfide, and hot: essentially, a place only a mangrove would love.

Once they have altered the environment, mangroves live easily between the high and low tides—and the tides do a lot for them. As in tidal marshes, alternate wetting and drying helps to eliminate less well-adapted species. Tides bring in fresh nutrients, wash away wastes, refresh water that has grown salty from evaporation, and carry off detritus that would otherwise accumulate in excess. Incoming tides push floating mangrove propagules up river estuaries where they can lodge and enlarge riverine forests. Out-

OAR/NURP

Mangrove ecosystem. The whole system, visible here, is part under water and part above, with rising and falling water connecting the two parts.

going tides disperse them along estuarine shores, where they form fringing forests. Ocean currents may carry them far from Florida. They may float for as long as a year and travel for thousands of miles, finally taking hold on tropical shores in Africa or South or Central America.

ENERGY FLOW THROUGH THE MANGROVE COMMUNITY

The physical structure of a mangrove swamp creates a rich environment in which many other living things can find niches. Leaves, bark, branches, prop roots, underground roots, and the mud surrounding them all offer different kinds of spaces. Thanks to their complex geometry, mangrove communities are among the most densely populated ecosystems on earth. An ecologist says, "It's all about structure." And where many creatures can reside, a complex food web can develop—not surprisingly, built upon detritus. This section first examines the role of detritus in the mangrove ecosystem, then moves on to explore the mangrove swamp's living creatures, both small and large.

As in salt marshes, detritus drives the ecosystem—in this case, from tree litter: leaves, bark, and wood. Litter decays fast in brackish and salt water. The process is quick-started by shredder organisms such as crabs and amphipods, and is accelerated by alternate wetting and drying conferred by the tides. The result is tons of plant detritus mixed with the droppings of insects and other animals. Not all swamps have been studied, but red mangrove swamps along rivers produce two pounds of detritus per square meter every year, or about 4 tons per acre. This detritus is the main material in a food web that, as in salt marshes to the north, is believed to nourish up to 90 percent of the area's sports and commercial fisheries.[1] Background Box 17–1 provides more detail on the routes taken by detritus through the system and illustrates that, as in the tidal marshes described in the last chapter, the microscopic life is highly specific and organized.

Microscopic Life among Mangroves

Detritus fluxes through the unseen life of mangrove systems very much as it does through tidal marshes. Each time a mangrove leaf drops into the water, fungi attach to it and grow a network of fine filaments all through its tissues, where they digest its stored carbohydrates. Multitudes of algae, bacteria, and other microscopic organisms soon assemble on the leaf and form a community in which each finds the nourishment it needs and each produces wastes that another can use. A crowd of small animals joins in, and the leaf becomes a slime-coated raft carrying tiny worms and crustaceans—minute animals that an observer describes as "little more than huge stomach pouches attached to eyespots and mouths," an efficient microscopic disassembly line. The names of the workers read like the cast of characters in a Greek opera, but with a cast of thousands. To provide just a glimpse of the system's complexity and order, List 17–1 presents just the marine fungi that digest shed mangrove leaves. Other fungi with equally exotic names coat living leaves, stems, branches, and submerged roots at each stage of a mangrove tree's life.

Communities of algae in the mangrove system are equally orderly and complex. Coats of algae on prop roots and mud are so numerous and thick that the Florida Natural Areas Inventory considers them communities, equal in status to tidal marshes and swamps. They have their own predictable structures. In sunny locations on the swamp's ocean-side edges, each prop root may support nearly a pound of algae arranged in strata. Starting at the top of a tree, four genera (which may include dozens of species of algae) grow on the exposed, above-water trunk and branches. Algae that are members of another genus make up that bright green band (which everyone sees, but no one notices) that grows around the roots at the high-water line. Below that line, between the high- and low-tide lines, are algae of three other genera. Last, on the roots that are in permanent water below the low-tide line, still other algae grow. Each algal species, then, has its own preferred habitat and plays particular roles in the complex living system that surrounds mangrove roots.

And that is by no means the end of the cast of algal characters. Red mangroves have one team of algae; black mangroves have another. The prop roots (as distinct from the tree trunk and roots) have their own team—in fact, several teams of algae: some on the permanently submerged portions, some on the moist parts; some on the mud around the roots. These algal teams include diatoms, dinoflagellates, filamentous green algae, and blue-green cyanobacteria. Black mangroves have other algae on their roots. Away from the roots, on shoals, floors of shallow bays, and creek bottoms near mangrove swamps, are still other algae. Their proportions differ regionally from Tampa Bay to south Florida. In sum, the distribution of these plants, although not simple, is very orderly.

Floating algae (phytoplankton), too, are numerous and diverse. Waves and currents mix marine, estuarine, and freshwater phytoplankton all together, permitting energy to flow efficiently through the food web.

Sources: Toops 1998; Odum, McIvor, and Smith 1982, 41.

Detritus. The base of the food web in a mangrove system is the shed parts of trees. Here, red mangrove leaves are decaying; and a smooth bubble alga (in the exact center of the photo) is growing on some of the products.

**LIST 17–1
Leaf-Digesting Fungi in
Mangrove Systems**

First, as leaves age:
Nigrospora
Pestalotica
Phyllostica

Then, after the leaves fall:
Drechslera
Gloeosporium
Phytophthora

Then, to complete the decay process:
Calso
Gliocidium
Lulworthia

Source: Odum and coauthors 1982, 40.

Native to Florida waters: Brown shrimp (*Penaeus aztecus*). The young shrimp can make a good living amidst mangrove debris.

Invertebrates. Many kinds of invertebrates exploit mangrove communities: sea worms, sea snails, shrimp, crabs, and others. Some live in the muddy substrate; some on or in the mangroves' prop roots. Some live above the water's surface among the branches and leaves; and some high in the treetops. Some come out at high tide; others, at low tide. Some emerge only by night; others, by day; and still others, not at all. All contribute uniquely to the energy economy of the mangrove system.

Hordes of insects populate the mangrove swamp. The red mangrove is the larval host plant of the mangrove skipper butterfly, one of many instances of coevolution. These include moth and butterfly larvae, grasshoppers, and crickets. On some overwash mangrove islands in the Keys, more than 200 species of insects have been counted. These insects busily cut leaves, eat leaves, eat each other, and export frass (insect droppings). One kind of moth larva develops entirely within the mangrove leaf (at the expense of the leaf). One kind of beetle attacks propagules while they are attached to the tree.

Some invertebrates dwell within the living prop roots of mangroves. The wood-boring isopod, a species of crustacean, inhabits red mangrove roots. It lays its eggs and its juveniles develop, all within the roots. Fungi and bacteria follow this invasion, and the roots may ultimately be severed. On occasion, this will kill a whole tree, but sometimes it stimulates root branching and makes the tree stronger. Either way, the community benefits, because both dead and live mangroves are useful.[2]

On and around the prop roots are other invertebrates. Coffee bean snails may number 500 per square yard among frequently inundated red mangroves; they average about 150. Other snails number at least 400 per square yard among black mangroves. The prop root complex also serves as a spiny lobster nursery. The young lobsters settle in among the roots and grow there until they are about two inches long.

The substrate around the prop roots supports other thriving communities. Fiddler crabs and several other crab species dwell in the mud and creep out to forage on exposed mud flats when the tide is out (see Figure 17–5). Others emerge while the tide is *in*, particularly at night.

Some invertebrates climb to, or live in, the canopy to feed or escape high tides. The mangrove tree crab is a tree climber among red mangroves—a surprise to unwary explorers who don't expect to see crabs at eye level. About a dozen mangrove tree crabs occupy each square yard at the edges of mangrove swamps; about half that many in the interior. The crabs are omnivorous, eating insects such as caterpillars and beetles, as well as living mangrove leaves. Mangrove tree crabs are an important food source for birds, fish, and other predators.

Finally, of course, many invertebrates swim and drift in the water column: crabs, shrimp, and their larvae; worms and worm larvae; and infant fish. Some of these swarm to the top at night to feed.

In sum, the community includes a tremendous mixture of invertebrates—even more than in salt marshes, thanks to the more complex structure of the vegetation which provides more microhabitats and niches to occupy. Some of these animals graze on the mangroves, some filter the

Native to Florida waters: Medusa worm (*Eteone heteropoda*). Like its numerous brethren, the polychaetes, this worm obtains its nourishment by filtering detritus from the water.

Mud crab (*Scylla serrata*).

Great land crab (*Cardisoma guanhumi*).

Purple marsh crab (*Sesarma reticulatum*).

FIGURE 17–5

Crabs among Mangroves

These are only three of many crab species that use the mangrove habitat.

water, and some sneak up on others and snatch them up as snacks. With all the life milling about beneath the mangroves, it should come as no surprise that fish are numerous, too.

Fish among Mangroves. An astonishing 220 species of fish use mangrove communities; Figure 17–6 shows just six of them. The mangrove swamp serves fish in two main ways. Prop roots provide protected habitat for juveniles; and the leaves, together with phytoplankton, algae, seagrass litter, and litter running off the land, support a giant food web for them. Silvery schools of striped mullet swim through the detritus-laden waters and suck mouthfuls in, converting the detritus to masses of mullet. Larger fish, such as snook, snapper, and spotted seatrout, eat the mullet. Fish populations among mangroves vary with substrate (mud versus sand), salinity, temperature, and frequency of flushing.

In basin mangrove forests on the southeast coast, where black mangroves predominate and the substrate is mud, several species of small fish live out their entire lives in waters so acidic and so seldom refreshed by tides that most fish would perish there (see List 17–2). Occasionally, high spring tides or seasonally heavy rains wash tons of these little fish out into coastal ecosystems, where they become food for snook, ladyfish, tarpon, gar, and mangrove snapper. At the opposite extreme, when the fishes' pools shrink, herons, ibis, wood storks, raccoons, mink, and others prey on them.

In contrast, riverine mangrove forests on the southeast coast, where red mangroves predominate and renewing tides flow far in and out twice daily, support ladyfish, striped mullet, and others. These forests lie between south Florida's freshwater marshes and the shallow brackish and saltwater bays

LIST 17–2
Fish among Mangroves (Examples)

IN BASIN MANGROVES

Killifish and Relatives
Goldspotted killifish
Mangrove killifish
Marsh killifish
Sheepshead minnow

Livebearers
Mosquitofish
Sailfin molly

IN RIVERINE MANGROVE SWAMPS

Year-Round
Ladyfish (a tarpon)
Striped mullet
Yellowfin mojarra

In Wet Seasons
Florida gar
Freshwater catfishes
Freshwater killifishes
Largemouth bass
Sunfishes

In Dry Seasons
Goliath grouper
Great barracuda
Jacks
Needlefishes
Stingrays

Source: Adapted from Odum and coauthors 1982, 51.

Cobia (*Rachycentron canadum*). Cobia swim inshore and near the shore in inlets, bays, and mangrove swamps, feeding on crabs, squid, and small fish. They commonly grow to 40 pounds or more.

Fat snook (*Centropomus parallelus*). These fish stay inshore among mangroves, which are their nursery grounds, seeking weakly brackish to fresh water. They seldom grow to more than 20 inches in length.

Mutton snapper (*Lutjanus analis*). Young snappers stay inshore, in mangroves and grassbeds, feeding on fish, crustaceans, and snails. Adults move to offshore reefs and commonly grow to about 15 pounds.

Common snook (*Centropomus undecimalis*). Snook swim inshore in fresh, brackish, and saline coastal waters along mangrove shorelines from central Florida south. They feed on fish and large crustaceans and weigh 5 to 8 pounds when mature. They can continue to grow to more than 50 pounds.

Tarpon snook (*Centropomus pectinatus*). These snook swim inshore in south Florida, frequently in fresh water, feeding on small fish and larger crustaceans. The young fishes' nursery grounds are along mangrove shorelines. They grow to about a pound or less.

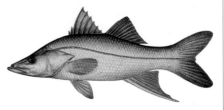

Swordspine snook (*Centropomus ensiferus*). These snook inhabit Florida's east-coast estuaries as far north as the St Lucie inlet, seeking weakly brackish and fresh water. Mangrove shorelines are their nursery grounds. They weigh less than 1 pound when mature.

FIGURE 17–6

Marine Fish among Mangroves

These fish spend their young lives in mangrove communities.

Source: Anon. c. 1994 (*Fishing Lines*); drawings by Diane Peebles.

and lagoons along the coast. They vary greatly in salinity and in fish populations. In wet seasons come rains that make them altogether fresh. Fish wash in from interior sawgrass and needle-rush marshes and swim among the mangroves: Florida gar, sunfishes, and others. Dry seasons reverse the water from fresh to salty, and then saltwater species come in from offshore: grouper, stingrays, needlefishes, jacks, and barracuda.

In the mangrove forests on the western shore of Biscayne Bay, seagrasses grow densely, the water is more saline, and there are coral reefs nearby. Fishermen catch grunt, bonefish, snook, and snappers. (Many whole families, whole sets of genera, of fish live in riverine mangrove communities and associated tidal streams and rivers.)

Fish populations in mangrove communities change from season to season with the water's temperature. Cold water drives some fish offshore to deep waters: lined sole, hogchoker, big-head sea robin, and striped mullet. Warming waters of spring and summer bring a rise in juveniles following offshore spawning: striped mullet, gray snapper, sheepshead, spotted sea trout, red drum, and silver perch. Riverine mangrove forests are the fishiest of mangrove habitats, supporting a large standing crop of fish the year around. Any single tidal stream in a mangrove swamp supports about 50 different species over the year, and taken together, all the tidal streams host 111 species. (And all of this begins with mangrove detritus.)

American white pelican (*Pelecanus erythrorhynchos*).

FIGURE 17–7

Birds among Mangroves

Pied-billed grebe (*Podilymbus podiceps*).

Tricolored heron (*Egretta tricolor*), an adult in breeding plumage.

Fringing forests along estuarine bays and lagoons, fringes along ocean bays and lagoons, and overwash islands, all are slightly different. All produce many fish of importance to local food webs as well as to sports and commercial fisheries.

Bird Life in the Canopy. When birds eat fish, detritus energy becomes airborne—many times transformed and no longer recognizable as rotten leaves from tidal swamps. It becomes pelicans, herons, ospreys, eagles, and more. The fruits and insects in the mangrove swamp, as well as the trees' tangled, strong, and sturdy branches, attract numerous birds to feed, to breed, and to rear their young. What better place could there be for a family of nestlings than a thick clump of sturdy, strong trees under lain by water, where predators find hard going? List 17–3 hints at their diversity, but no single page can do them justice. Figure 17–7 shows only three of them.

Earlier chapters showed what great rookeries freshwater swamps can support. The same is true of mangrove swamps. The many roosting and nesting places they provide, together with the gentle, shallow waters beneath them, invite all manner of birds to probe, wade, surface-feed, dive, perch, and fly about. On Lee County's Sanibel island alone, 92 bird species have been recorded among the mangroves, and altogether, some 181 species use Florida mangroves in one way or another. Clearly, mangrove communities can support a great mob of winged creatures.

Mangrove-clad ocean islands make bird sanctuaries that are far superior to the tree islands in the Everglades. The Everglades tree islands lie in a freshwater marsh, which is accessible to snakes, raccoons, alligators, and other predators, but the mangrove islands are surrounded by ocean water. As a result, they are largely predator free, a great advantage in a bird nursery. Wading birds maintain giant nesting grounds on some mangrove islands. There may be thousands of roseate spoonbills, herons, and egrets all clustered together in a single swamp.[3]

LIST 17–3
Birds among Mangroves

<u>Wading Birds</u>
Cattle egret
Great blue heron
Great egret
Snowy egret
. . . and 13 more

<u>Probing Shorebirds</u>
Clapper rail
King rail
Semipalmated plover
Sora
Virginia rail
. . . and 20 more

<u>Surface and Diving Birds</u>
American white pelican
Brown pelican
Common loon
Horned grebe
Pied-billed grebe
. . . and 24 more

<u>Birds that Search from High in the Air</u>
Bonaparte's gull
Gull-billed tern
Herring gull
Laughing gull
Ring-billed gull
. . . and 9 more

(continued on next page)

Snowy egret (*Egretta thula*). Thick mangrove greenery provides excellent shelter for breeding birds as well as fruits and insects for food.

Wading bird rookeries are messy, noisy places. Huge nests are stacked all over the treetops where a confusion of comings, goings, and awkward landings prevails. Masses of guano carpet the ground. The stink of regurgitated food and of the decomposing bodies of fallen baby birds is everywhere. The air is loud with the din of squawking, hungry youngsters and the buzz of busy flies. But they are safe nurseries, and parent birds fight aggressively to secure nesting places in them.

Mangrove islands offer another major advantage to parent birds: they can feed nearby in the shallow bays and lagoons. Low tides expose abundant fish that make easy prey. Fiddler crabs swarm over the mud flats. The birds share the resources. Brown pelicans eat estuarine fish; great blue herons eat fiddler crabs, sardines, herrings, and mullet. White-crowned pigeons nest on small mangrove islands in Florida Bay and nearby, and breed and feed in the Florida Keys and on the extreme southern tip of Florida. They seek out large hardwood hammocks where they can feed unmolested on the seeds and fruits of tropical trees and shrubs.

The integrity of the landscape—or, in this case, the land-and-seascape—proves a crucial factor to breeding birds. To succeed at growing up, breeding, and raising the next generation of their kind, birds need protected nesting places and nearby feeding grounds just as people need crime-free neighborhoods near grocery stores and shopping. Try to establish the birds where one of these parts of the landscape mosaic is missing, and their survival is no longer assured.

Migratory raptors also use mangrove swamps, and for the same reasons. The peregrine falcon, red-shouldered hawk, red-tailed hawk, American kestrel, and others all find indispensable food, shelter, and rest in stopover places in the Keys.

Other Animals. To complete the array of animals using mangrove swamps, 24 species of amphibians and reptiles and 18 species of mammals are common (see List 17–4, List 17–5, and Figure 17–8). Among

Egret rookery, Cuthbert Island, Monroe County. Nesting birds are so numerous, they look like a dusting of snow.

Ornate diamondback terrapin (*Malaclemys terrapin macrospilota*). The terrapin nests on sandy berms in tidal wetlands and buries its eggs at the high-tide line. When the tide is in, it scouts watery channels, feeding on carrion, green shoots, fiddler crabs, and other crustaceans. At night, it completely buries itself in the sand.

Mangrove salt marsh snake (*Nerodia fasciata compressicauda*).

Squirrel treefrog (*Hyla squirella*).

FIGURE 17–8

Amphibians and Reptiles in Mangroves

amphibians, at least three species use mangrove swamps. Among turtles, one, the ornate diamondback terrapin, lives almost exclusively in mangrove swamps. Young loggerhead and Atlantic ridley sea turtles hide among mangrove roots during their early lives. Green turtles feed on roots and leaves. Hawksbill turtles eat all parts of red mangroves: fruits, leaves, bark, and roots.

Among reptiles, three species of anoles live in the treetops and six species of snakes glide among the roots. One, the mangrove salt marsh snake, is completely dependent on mangrove habitats. In addition, both the alligator and the crocodile use mangrove swamps. While the alligator is common throughout Florida, the crocodile is increasingly rare. Undisturbed red and black mangrove thickets may be critical breeding places for the crocodile.

As for mammals, you might not think that foxes, bears, and skunks would find it easy to maneuver among tangled mangrove roots, but the eating is so good as to make it worth the trouble. Some panther sightings occur in mangrove swamps, too. Even the white-tailed deer often visits, picking its way daintily among the roots and feasting on the leaves.

VALUE OF MANGROVE COMMUNITIES

Imagine what Florida's southern shores would be like without mangroves. Without mangrove trees to nest in, the diverse arrays of birds that raise their young in the swamps would be gone. Without the rich detrital soup made from mangrove roots and leaves, uncountable tiny recyclers would die out, depriving hordes of shrimp, crabs, snails, and mullet of their nourishment, so that they too would disappear. When storms swept the coast, the great peat beds on which the mangroves grow would be eroded away.

Florida native: American crocodile (*Crocodylus acutus*). Crocodiles inhabit brackish, mangrove-lined waterways at Florida's southern tip. After laying her eggs, the female stays nearby, and both male and female guard the hatchlings. Young crocodiles die in salt water, but adults swim long distances in the ocean.

Crocodiles and alligators are both reptiles, but differ in many ways. Most obviously, the alligator's snout is broad and flat, whereas the crocodile's is longer and narrows to a point at the tip. The crocodile is endangered: fewer than a thousand remain in the wild in Florida. The wild alligator population is estimated at about 1.5 million.

Winds and waves would scour the coast and eat into the interior. Faraway ecosystems, too, would be disrupted. The Everglades marshes in the interior would suffer saltwater intrusion and the offshore waters of Florida Bay would be less well protected from pollution. The once-great schools of mullet, snook, snapper, and red drum would dwindle and life around south Florida's coral reefs would lose diversity.

Ecologists say that to manage mangroves wisely means simply to let them grow: "Bottom line: don't trim mangroves at all."[4] Mangroves themselves, undisturbed, are better managers of south Florida's coastal and offshore environments than any human-contrived system could be.

Passage into Florida Bay through overarching mangroves.

PRICELESS FLORIDA

PART SIX

COASTAL WATERS

Florida's **coastal waters** are brackish or salty and have floors of soft or hard sediments with or without obvious, visible living communities on them.

PG/NOAA

A seagrass meadow beneath the blue waters of the Gulf of Mexico.

CHAPTER EIGHTEEN

ESTUARINE WATERS AND SEAFLOORS

This is the first of a three-chapter series that explores the waters and sea-floors of the continental shelf that surrounds Florida. The continental shelf extends at least five, and up to 200, miles offshore before it slopes down to the deep seafloor. Underwater habitats on that broad, productive terrain are varied and full of life. Even where the seafloor appears to be nothing but bare sand, mud, or marl, tens of thousands of animals and millions of microorganisms may dwell on and in each square yard of the seafloor. Still more living things are present where there are limestone outcrops on the shelf. And when, in turn, the living things form structures themselves, such as oyster or coral reefs or seagrass meadows, they become immense communities of immense complexity.

Most of the communities described in these chapters occur in both estuarine (brackish) and marine (salt) water, although they may do better in one environment than the other. The Florida Natural Areas Inventory (FNAI) classifies only estuarine communities, and its scheme is followed here; it covers some representative groups of each type. FNAI sorts estuarine communities into three categories: mineral-based, plant-based, and animal-based. In mineral-based communities the foundation material is rock, sand, mud, or marl, and the community members dwell on or in this material. In plant-based communities, seagrasses and algae provide most habitats. In animal-based communities, mollusks, sponges, or corals build the structures among which the community members live.

List 18–1 shows FNAI's categories. This chapter deals with communities on mineral seafloors and with mollusk reefs and worm reefs. Chapter 19 continues with seagrass and algal beds, and Chapter 20 is devoted to sponge beds, limestone-based communities, and coral reefs. Figure 18–1 presents the coastal landmarks to which these three chapters refer, and Figure 18–2 highlights significant communities on the continental shelf.

DN/GSML

Native to Florida waters: Margined sea star (*Astropecten articulatus*). This starfish lives on sand bottoms and in seagrasses in the northern Gulf of Mexico.

LIST 18–1
Communities in Florida's Estuarine Waters

<u>Mineral Based</u>
   Loose sediments
   Solid bottom (limestone)
<u>Plant Based</u>
   Seagrass beds
   Algal beds
<u>Animal Based</u>
   Mollusk reefs
   Worm reefs
   Sponge beds
   Coral reefs

Source: *Guide* 1990, pp. 60–73.

OPPOSITE: Florida Bay. Shallow estuarine waters gently bathe mangrove swamps and seagrass beds from the coast of south and southwest Florida to the farthest Keys.

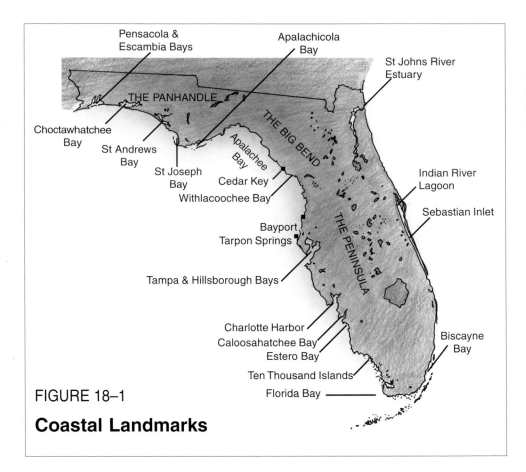

FIGURE 18–1

Coastal Landmarks

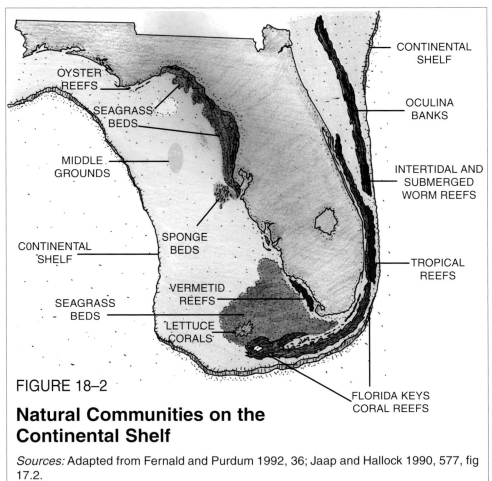

FIGURE 18–2

Natural Communities on the Continental Shelf

Sources: Adapted from Fernald and Purdum 1992, 36; Jaap and Hallock 1990, 577, fig 17.2.

CHALLENGES OF LIFE IN ESTUARIES

Much of the water along Florida's shores is estuarine (brackish), consisting of salt water from the ocean that is constantly mixing with fresh water off the land. Typically, brackish water is found only in bays and inlets near the mouths of rivers. This is true along Florida's Atlantic coast; the estuaries are clearly bounded within Miami's Biscayne Bay, Sebastian Inlet, the Indian River Lagoon, and the mouth of the St. Johns River. Along Florida's Gulf coast, however, estuaries are much more extensive. The land's slope is gentle, the continental shelf is broad, fresh water runs out all along the shoreline, and the water is brackish even far from rivers. More than two million acres off the coasts of south and west Florida and the Panhandle are estuarine.[1]

No number can be assigned to the salinity of estuarine water; in fact, the distinguishing characteristic of an estuary is *variation* in salinity, both seasonally and daily. Plants and animals in some parts of estuaries may find themselves sometimes in wholly marine water and sometimes in nearly fresh water. Take, for example, an oyster reef near the shore. In seasons of drought, when only a little fresh water is running off the land, the ocean's salt water advances over the reef. Then in seasons of flood, rain water and river water pour off the land and nearly fresh water surrounds the reef. And every day, the water is most saline at high tide, when salt water is pushing into estuaries, and freshest at low tide, when salt water withdraws and river water flowing off the land becomes dominant. To survive, the oysters and other reef dwellers must be able to tolerate not only the mixture of salt and fresh water, but wide swings between the two.

Another distinction marks estuaries: turbidity. Estuarine water is especially murky near the mouths of alluvial rivers that are delivering heavy loads of suspended fine clay particles to the coast. Where fresh water forces its way out of a river, and marine water pushes in against it, the fresh water floats, forming a top layer, and the salt water sinks, forming a bottom layer. Such layering may not occur in shallow areas where winds keep the fresh and salt water well mixed, but in deep basins, it may be very pronounced. Where the water is layered, fine clay particles delivered by the river sink to the bottom of the fresh water, but when they arrive at the salt water layer, they can sink no further. The particles accumulate at the plane between the layers and clump into frothy aggregates, forming a dark ceiling below which neither light nor oxygen penetrates. Growth of both plants and animals is limited beneath that ceiling because plants can't grow in the dark, and animals can't live without oxygen. If winds stir the water enough to help return some oxygen to the depths, they also make it murkier.

A marine biologist encountered first-hand the daily alternation between fresh and saline, light and dark water that creatures in estuaries have to cope with. She took her class snorkeling in Saint Andrew Bay at Panama City after one January's hard rains :

> *The normally clear salt water in the bay had been flooded out by fresh water off the land. The boat basin was filled with opaque, black water and looked like a cypress swamp. What is lovely in a cypress forest was a problem here, though. Visibility was zero.*

DN/GSML

Native to Florida waters: Red-footed sea cucumber (*Pentacta pygmea*). Sea cucumbers are members of the same phylum as starfishes, the echinoderms.

I didn't think we'd find anything today, but having come so far, we launched the boat anyway and ran out to a line of jetties offshore to have a look. And as we left the lagoon, the black swamp water abruptly gave way to the usual green and blue sparkle of salt water. A wavy boundary separated the two different worlds of swamp and ocean. The water over the jetties was clear. We could dive after all. . . .

As we swam along the base of the jetties, it was obvious that the swamp water had been here earlier. The sea urchins, which fare best in salt water, had been hit hard by the freshwater flood. Normally they cover the rocks by the thousands where they graze on seaweed. Now, the spines of the victims littered the sandy bottom.

Then the tide began to fall. As it did, a wall of black swamp water advanced down the channel, replacing the retreating salt water. There was almost no mixing. A line of foam marked the boundary—one side was black, the other was clear. It passed our anchored boat and enveloped the rocks. . . . Every day, it seemed, the sea urchins we'd seen got a few hours of relief from the fresh water at high tide and then got nailed again when the tide turned.[2]

Oscillations in salinity and light pose a great challenge to organisms in estuaries. There are wide swings in temperature, too. In summer, shallow coastal water may be very hot at low tide; then as the tide rises, the flux of incoming sea water cools it. In winter, the opposite is true: the water may be extremely cold at low tide; then the rising tide warms it. Some plants and animals can withstand all these stresses of life in estuaries; many cannot; and some, such as oysters, can thrive for long times *only* in estuaries, as explained later.

Despite the stresses in estuaries, they produce a tremendous mass of life. They are biological factories that support huge crops of plankton, thanks to the massive flows of detritus and mineral nutrients from inland rivers and tidal wetlands. And plankton are the raw material for estuarine productivity.

ESTUARINE PRODUCTIVITY: PLANKTON

Plankton occupy every cubic inch of ocean water. They are a fascinating mixture of peculiar-looking, tiny plants and animals that float or weakly swim in the water. Some are single-celled and remain microscopic all their lives. Others are fertilized eggs at first and develop into larger plants and animals in time. By the time they take up their mature lives, they have become an incredibly diverse assortment of living things and they comprise most of the life in the sea. Some become sponges and form large colonies on the seafloor. Some turn out to be worms that burrow in bottom sediments. Some become jellyfish, squid, shellfish, or fish. Most never grow up at all, because they become food for other creatures. Figure 18–3 displays some plant members of this enormous community, and Figure 18–4 shows some of the animal plankton, the largest ones, at their actual size.

The phytoplankton, that is, the plant members of the plankton, are the staple commodity in the food web that supports all of the life in the ocean. These tiny plants, by the millions of tons, derive their life from the sun

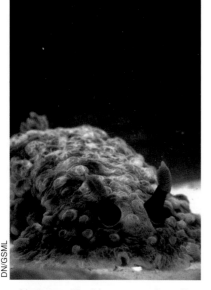

DN/GSML

Native to Florida waters: A nudibranch. Although it doesn't look like a snail or clam, this animal is a bottom-dwelling mollusk.

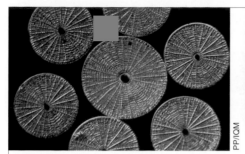

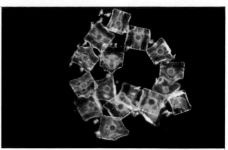

Diatoms (an *Actinocyclus* species).

Diatoms (a *Pleurosigma* species).

Chain diatom. This is a chain of single-celled algae. The nuclei within the cells are clearly visible.

PP/IQM

FIGURE 18–3

Ocean Plants: Phytoplankton

All three of these specimens are diatoms—single-celled algae with cell walls that are made of a silicon compound. These plants and their relatives are at the base of the food chain on which virtually every animal in the ocean depends.

FIGURE 18–4

Ocean Animals: Zooplankton

The majority of plankton are microscopic; the ones shown here are giants relative to most others. The true sizes of these animals are shown in the center of the figure.

A An arrow worm, 9 mm long, species unknown. A voracious eater, this animal can seize others with its grasping spines and devour them. It does not hesitate to eat its own kind.

B A crustacean, 10 mm, species unknown. This is a crab-to-be, although at this stage it resembles a shrimp.

C A nudibranch, 10 mm, *Phylliroe bucephala*. This planktonic swimming mollusk, when alive, is brightly bioluminescent.

D A rock lobster larva, 6.5 mm, *Pallinurus*.

E An octopus, 4.5 mm, species unknown. Visible are two small fins at the back of the head, and the large eyes, positioned at the base of the tentacles.

F A polychaete, 5 mm, species unknown, with its proboscis out. (The proboscis is a muscular organ that can be turned inside out to grasp prey, and then pulled back inside.)

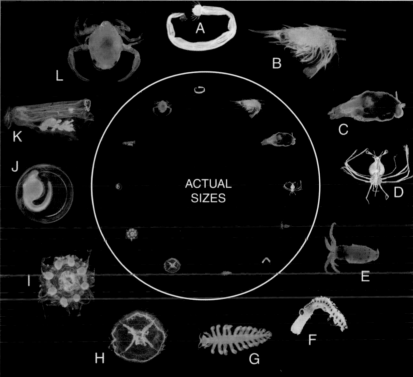

ALL PHOTOS BY MAB

G A polychaete, 3.6 mm

H A medusa, 7 mm, species unknown. Its four oral arms are loaded with stinging cells with which it can paralyze its prey. Then it pulls the victim into its mouth (in the center of the bell).

I A medusa, 3 mm, a *Nausithoe* species. The round, yellow structures on its underside are its gonads.

J A fish egg, 3.1 mm. The developing fish's head, tail, and yolk sac are clearly visible.

K A tunicate, 4 mm, species unknown. The tunicates are chordates, members of the phylum to which the vertebrates (fish, amphibians, reptiles, birds, and mammals) also belong.

L A crab, 5.7 mm across, species unknown. All the appendages including well-developed claws are clearly visible.

that shines down on them in surface waters, and from nutrients that run down to them off the land. Then, when consumed, they pass on their energy and substance to support the growth and development of millions of tons of animals, from tiny zooplankton to great whales. The phytoplankton known as diatoms are among the most numerous organisms in the oceans, and it is through their activities that the oceans trap carbon from the atmosphere and offset excessive warming of the planet.

Diatoms and other plankton are hardly a household conversation topic, but their work is vital to all life on earth. As phrased by Gulf of Mexico scholar Robert H. Gore, "The small, the nondescript, and the numerous will always support . . . the few and the large."[3] Again and again, the chapters to come refer to "filter feeding"—the ways aquatic animals nourish themselves by straining plankton from the water.

Plankton are more numerous in estuaries than in any other part of the sea, and it is thanks to their presence that the estuaries can produce immense quantities of shrimp, crabs, oysters, and other seafood. Nearly all of the animals that swim over the continental shelf start their lives in estuaries; it is not an exaggeration to say that Florida's estuaries are the birthplace of most of the sports and commercial fish and shellfish that fishermen catch around Florida. Other animals, too, depend on estuaries. Manatees, sea turtles, and wading birds forage in the grassbeds and marshes; ospreys and bald eagles dive there for their prey. These large and well-known creatures thrive thanks to the thousands of small, obscure marine species that flourish in the estuaries. But before going on to look at them, an overview is needed. Each Florida estuary is different from all the others, and areas within estuaries are diverse, too. Background Box 18–1 offers brief glimpses of Florida's major estuarine systems.

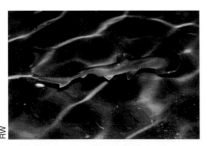

RW

Native to Florida waters: Bonnethead shark (*Sphyrna tiburo*). This small shark dwells inshore in bays and estuaries, feeding on fish. It grows to a final size of about 3 to 4 feet.

The next major estuary to the east, Apalachicola Bay, presents a contrast. Like Pensacola Bay, it is confined behind barrier islands, but it is shallow. It receives a larger volume of alluvial river water than any other Florida bay, and its inhabitants enjoy long periods of low salinity that keep marine predators away. Its waters are often turbid and darkly colored, so bottom-rooted plants are uncommon. Phytoplankton, which have access to the sun by floating, are the major photosynthesizers in Apa-

(continued on next page)

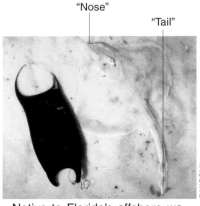

Native to Florida's offshore waters: Clearnose skate (*Raja eglanteria*). The black "purse" is an empty egg case. The newly hatched skate is almost perfectly concealed in the sand to its right (see pointers).

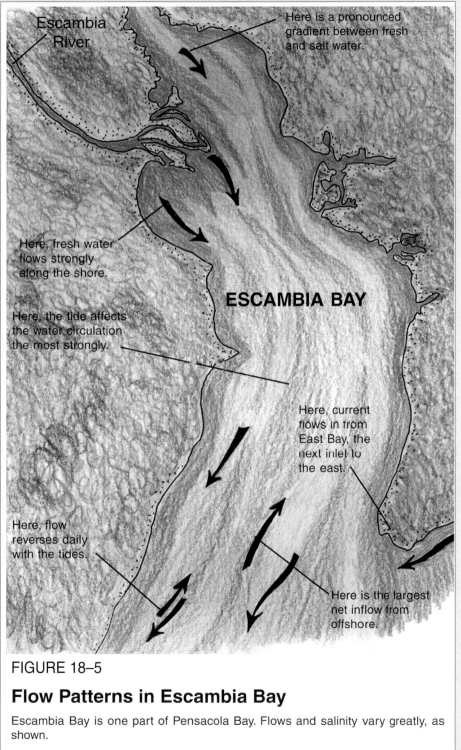

Escambia River

Here is a pronounced gradient between fresh and salt water.

Here, fresh water flows strongly along the shore.

ESCAMBIA BAY

Here, the tide affects the water circulation the most strongly.

Here, current flows in from East Bay, the next inlet to the east.

Here, flow reverses daily with the tides.

Here is the largest net inflow from offshore.

FIGURE 18–5

Flow Patterns in Escambia Bay

Escambia Bay is one part of Pensacola Bay. Flows and salinity vary greatly, as shown.

Source: Jones and coauthors 1990, 19, fig 7.

Note: Figure 13-1 (page 217) shows the locations of most of these rivers.

Source: Livingston 1990.

lachicola Bay. Nearly half of the bay supports oyster bars on a mud floor, and these provide habitat for many other organisms. Of all of Florida's estuaries, Apalachicola Bay is the richest in both detritus and minerals.

The next estuary, Apalachee Bay, presents another contrast. It is not bounded by barrier islands; in fact, no physical features visible on the map reveal where its boundaries are. It consists of all of the shallow water offshore that lies on the broadest part of the continental shelf. Its calm, brackish waters reach far out into the Gulf, and even 8 to 10 miles offshore the water is only about 60 feet deep. Apalachee Bay extends all the way from the Ochlockonee River almost to Tampa Bay and receives freshwater flows from sixteen rivers (see List 18–2). Still, Apalachee Bay receives less total river water than does Apalachicola Bay, and with lighter loads of minerals and detritus, so it is both more saline and less turbid than Apalachicola Bay and produces fewer phytoplankton. Sunlight penetrates deeper into the water, and the shores and seafloors hold vast marshes and grassbeds. Marsh and sea grasses are both the main habitat and the main source of nutrients for the bay's fish and other animals.

South of the Big Bend are many more estuaries. Major rivers feed into the Gulf in Tampa Bay, Sarasota Bay, Charlotte Harbor, Caloosahatchee Bay, and Estero Bay. Each has its own character. They vary in salinity and nutrient load, and each is complex. As an example, Tampa Bay is a subtropical estuary with a bottom consisting largely of soft sediments dominated by calcium carbonate and oyster shells. Streams carrying little or no sediment deliver clear, alkaline water. The bay is less than ten feet deep and is protected from the ocean's influence by the surrounding land. Frequent rains cleanse the bay at intervals, and where the shorelines have been left natural, fringing marshes and forests filter runoff flowing in off the land. Ground water also feeds into the bay from springs and underlying aquifers, helping to keep its water composition constant and suitable for estuarine organisms. Oysters are numerous in the bay, although less so than in past times, and oyster reefs support attached macroalgae and marine animals.

The tremendous diversity of organisms produced by Tampa Bay is apparent from the numbers in List 18–3. Note how many kinds of phytoplankton there are. Note that the zooplankton have not even been counted.

On the southwest coast of Florida, sheet flow of Everglades water supports yet another great estuary in Florida Bay, which lies inside the Florida Keys and extends far westward into the Gulf. Swirling eddies off Florida's southwest coast keep estuarine water circulating throughout the whole area and the estuarine floor holds vast seagrass beds.

Continuing around the Florida peninsula, the continental shelf narrows to a mile in width off Palm Beach. Because the east coast has higher-energy shores than the south and west, most of the shelf is bathed in marine water. Farther north, it is broader, but major estuaries lie only within enclosed bays and inlets, notably the Indian River Lagoon, Sebastian Inlet, and the mouth of the St. Johns River. Each estuary holds a large, complex, unique ecosystem. The Indian River lagoon, for example, has a dozen species of seagrasses and other plants, to which 41 different species of algae are attached like epiphytic plants on trees. There are numerous species of small seagrass-associated animals, several hundred species of mollusks, and many other groups. Just that one lagoon has *700 species of fish*, and those in its low-salinity portion are unique for North American waters. At least five of its fish species may have no other North American breeding populations.

Clearly, Florida's estuaries are wonderfully rich communities. However, most of those on the Atlantic coast have experienced profound impacts from human activity and industry, and only a few natural areas of any significance remain. The Gulf coast estuaries, in contrast, especially along the Big Bend, are among the most nearly pristine coastal waters along the lower 48 states, and they receive most of the focus in the rest of this chapter.

Sources: Livingston (R. J.) 1990; seagrasses from Dawes, Hanisak, and Kenworth 1995; algae from Hall and Eiseman 1981; associated animals from Virnstein 1995; bacteria from Garland 1995; mollusks from Winston 1995 and Mikkelsen, Mikkelsen, and Karlen 1995; statistics on Indian River lagoon from Florida Biodiversity Task Force 1993, 3.

ANIMALS ADAPTED TO ESTUARIES

Given such wide swings between extremes of salinity, light, and temperature, one might expect that no animals could survive in estuarine environments. Indeed, only a few species live permanently there: diversity is much lower than in pure salt water. However, for those that can cope, the rewards are generous. Floods reliably deliver heavy loads of nutritive material from the land every year. Ground water flows out constantly, delivering minerals from underground springs. Tides repeat predictably, bringing fresh nutrients from the ocean. Plankton are everywhere in the water. With relatively few species competing, and with this endless superabundance of foodstuffs, those animal species that can reside in estuaries can thrive by the billions.

Not even these populations thrive at all times, however. When they are stressed, individuals may die by the billions, but the species populations are able to bounce back quickly. They are opportunists: they are as adapted to natural disturbances as are weeds on land. Oysters exemplify their resilience. In the fall of 1985, two hurricanes destroyed more than 90 percent of Apalachicola Bay's major oyster bars—yet by late spring of 1987, they were back in greater abundance than before. Two factors permitted their dramatic recovery. The surviving oysters produced great clouds of offspring; and other species, unable to cope with the stresses of estuarine life, failed to compete for their attachment sites.[4]

For most species, however, estuaries pose too great a challenge to overcome. If any freshwater animals blunder into estuaries, they die. As for marine species, the overwhelming majority of those present at any given time are just passing through or visiting for a spell. A few of these transients and visitors are described in the next paragraphs; then the rest of the chapter is devoted to permanent estuarine residents and communities.

Native to Florida's near-shore waters: Gulf toadfish (*Opsanus beta*). Florida's productive, shallow, near-shore waters are ideal for this fish, whose prey include crustaceans, annelid worms, mollusks, and fishes. The toadfish rests on the bottom, often hiding in seagrass or near rocks or reefs, then snaps up its prey and chomps them with its powerful jaws.

Native to Florida's offshore waters: American shad (*Alosa sapidissima*). This fish lives in salt water offshore, but crosses estuaries to make a late winter spawning run up Florida's east coast rivers, notably the St. Johns. The young remain in fresh water until they are 2 to 4 inches long, then move out to sea, feeding on plankton. Shad reach a mature weight of 2 to 3 pounds, sometimes 5 pounds.

Source: Anon. c.1994 (*Fishing Lines*); drawing by Diane Peebles.

Native to Florida's near-shore waters: Gulf flounder (*Paralichthys albigutta*). Flounder swim in towards shore on the rising tide, following small fish, and conceal themselves in the sand to ambush prey. Only the outline of this fish can be seen (its mouth is near the magenta bryozoan at left).

Transients Crossing Estuaries. Imagine being a fish or a crab attempting to swim or crawl from the open ocean across a river estuary to get into the river. Look again at the map of Escambia Bay in Figure 18–5. The turbulence, the low visibility, and the changing currents, salinity, and temperatures present formidable barriers. Still, many animals cross estuaries repeatedly throughout their lives. Sturgeon, gar, eels, and others pass through estuaries during breeding migrations. Flounder and needlefish make their way from the ocean across the Suwannee River estuary and up the river to springs miles from the coast. Mullet, flounder, blue crab, and others make prodigious migrations 100 miles up the St. Johns River to Silver Spring. Most people are familiar with the feats of the salmon that leap up rapids and waterfalls to go to their spawning grounds at the heads of mighty rivers on both coasts of North America, but Florida's freshwater animals make equally perilous transits across Florida's estuaries, and they are equally well adapted by long periods of evolution to manage their journeys.

Visitors to Estuaries. Currents and waves bring some visitors into estuaries daily. Many fish and invertebrates use the reversing tidal currents to ride in and out of the shallows, moving inshore to feed on rising tides and drifting back offshore on falling tides. Other visitors enter estuaries monthly, timing their spawning efforts to coincide with the highest high tides as described earlier, so that extra-strong currents will transport their planktonic larvae over wide areas.

Many animals visit estuaries at longer intervals—for certain seasons of the year, or for parts of their life cycles. Seasonal visitors stay when the habitat is right for them and move out when conditions change. For example, many fish feed and grow in north Florida's shallow coastal estuaries all summer; then, as the first strong cold front of fall chills the water, they pour out into the Gulf's warmer waters in waves. They spawn there, producing their tiny larvae out at sea, where salinity and temperature are more constant than in estuaries. The young start their lives as plankton, and the next year, as young fish, they ride ocean currents to the shore and enter the estuaries to forage and grow. Similarly, along south Florida's shores, many shrimp, crabs, and lobsters flourish and feed in estuaries in summer and remain until the water turns cold in midwinter. Then they depart, to spawn over lime rocks in the reliably warm, calm waters of the continental shelf. At the depths where these rocks lie, the brief three-month winter chill doesn't penetrate. White shrimp feed in estuarine marshes in summer; then move out to barrier island beaches in fall and return the next spring. Pink shrimp, as juveniles, feed in seagrass beds on the west coast during summer; then migrate west, far beyond the Florida Keys, to spawn in the Tortugas. Stone crabs, as juveniles, grow up in Apalachee Bay and other estuaries along southwest and south Florida, either in seagrass beds or among shells or rocks. As they mature, they move to deeper water and burrow in soft sediments or hide in seagrass.[5]

Blue crabs use estuaries in a distinctly different way. They do mate there, in summer. Fertilization takes place internally, so the newly fertilized eggs are protected from fluctuating estuarine conditions. Blue crabs wander all over the continental shelf down to a depth of more than 100 feet, but they

FIGURE 18–6

Blue Crab (*Callinectes sapidus*)

The blue crab occurs along both coasts of Florida. On the Atlantic side, a different crab population occupies each major estuary. They mate in the estuary; then the female moves near the mouth of the estuary carrying the fertilized eggs attached to her body. She buries herself in the bottom and remains there until the following spring. Then, when the fertilized eggs are ready to hatch, the female releases her larvae to float as plankton in near-shore waters on the continental shelf.

On the Gulf side, mating also takes place in every estuary, and the females bury themselves over the winter, but a migration takes place in the spring. When it is time to release their young, the females migrate north, looking for shallow muddy and sandy bottoms and seagrass beds. Some swim several hundred miles to Apalachee and Apalachicola bays.

The blue crab is adapted to the murky water of estuaries.The female blue crab, when ready to mate, releases a chemical attractant into the water. The male, once he detects the chemical, orients into the current and follows the odor gradient upstream. He will find the female, even though he may never see her.

Source: Livingston 1990.

When fertilized eggs are released by aquatic species, they are called **spawn**. The act of release is called **spawning**. In most aquatic species, spawning and fertilization take place at the same time. The eggs and sperm are released into the water, fertilization takes place, and the spawn promptly disperse.

For a few species, such as the blue crab, fertilization is internal and the female stores the fertilized eggs for a time. Spawning occurs later, when the eggs hatch and the female releases the larvae.

are considered estuary-dependent, because they reproduce there. Figure 18–6 features the blue crab.

Among the most conspicuous visitors to estuaries are large fish, including those that fishermen like to catch. Some feed on the bottom, some forage above it over bare floors, in seagrass beds, and especially around structures such as oyster reefs. Figure 18–7 illustrates the ways in which a few fish use estuaries.

From these few sketches, it must be clear that Florida's estuaries are, for the most part, grand central stations where populations are changing constantly. Some populations pass across the estuaries into rivers and later cross back to return to the ocean. Others visit the estuaries for a spell and then turn around and go back to the ocean. Each species has its own timing. There are, however, some organisms that reside in estuaries throughout their lives, and the next sections turn to these.

RESIDENTS ON MINERAL FLOORS

Some estuarine floors are of sand, some of mud, and some of marl. Along the western Panhandle, where the waves are energetic, seafloors are largely

Goliath grouper (*Epinephelus itajara*). The largest of the groupers and mostly a south Florida fish, this fish may grow to an age of 30 to 50 years and a weight of nearly 800 pounds. It stays inshore, swimming in estuaries and especially around oyster bars, where it feeds on crustaceans and fish.

Atlantic sharpnose shark (*Rhizoprionodon terraenovae*). This shark can thrive inshore, even in surf. The young grow up in bays and estuaries, feeding on small fish and crustaceans. Mature when 2 to 4 feet long, the adults swim offshore.

Atlantic croaker (*Micropogonias undulatus*). The young of this species grow up in central and north Florida estuaries along both coasts. Mature at 2 to 3 years and weighing 1 to 2 pounds, adults move to deep offshore waters in winter, but return to bays and estuaries in warmer seasons.

Sheepshead (*Archosargus probatocephalus*). Young sheepshead up to 2 pounds swim inshore around oyster bars, near seawalls, and in tidal creeks, feeding on fiddler crabs and barnacles. They spawn near shore in late winter and early spring, and the adults swim offshore, reaching 8 pounds in weight.

Sand seatrout (*Cynoscion arenarius*). Mainly a Gulf species, this fish feeds on small fish and shrimp inshore in shallow bays and inlets and spawns inshore in spring and early summer. Mature when nearly 1 pound in weight, the adults move offshore in winter.

Florida pompano (*Trachinotus carolinus*). In warm seasons, this fish swims near shore along sandy beaches, around oyster bars, and over grassbeds. It feeds on mollusks and crustaceans, especially mole crab larvae. It moves in and out with the tides, often in turbid water. In cold seasons it moves to water as deep as 130 feet. It usually weighs less than 3 pounds when caught.

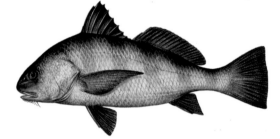

Black drum (*Pogonias cromis*). This fish swims on the bottom both in and off shore, often around oyster reefs. It spawns near shore in late winter and early spring. It feeds on oysters, mussels, crabs, shrimp, and occasionally fish. It lives to 35 years and commonly attains a weight of 30 pounds.

Ladyfish (*Elops saurus*). This fish lives in bays, estuaries, tidal pools, and canals, sometimes in wholly fresh water, but spawns offshore in fall. It feeds predominantly on fish and crustaceans and attains a mature weight of 2 to 3 pounds.

FIGURE 18–7

Fish that Use Florida's Estuaries (Examples)

Source: Anon. c.1994 (*Fishing Lines*); drawings by Diane Peebles.

of sand. Along the Big Bend, where the waves are gentle, there remain finer sediments from alluvial rivers and the bottoms are muddy. In Florida Bay, the bottom is marl laid down by microscopic algae mixed with shell hash, the broken-up remains of seashells. A different assortment of creatures lives within the bottom sediments of each of these areas, and these creatures near the bottom of the food chain influence the rest of the life that is present there.

As an example, consider the floor of Apalachicola Bay, where the water is usually turbid. Although visible plant life is severely limited by lack of light, tons of fine mineral particles and tons of detritus yield phytoplankton that become food for an extensive living world. In such an environment many a hungry animal resorts to filter-feeding, either by straining the water for food or by vacuuming food off of sediment grains. In the top inch of the sediment, crustaceans, worms, and shelled amoebas that are too small to see obtain their nourishment by ingesting whole grains of mud or sand and digesting off the coatings of still tinier living things. Then they excrete the sediment grains, now clean and ready to grow a new crop of food. The concentration of these tiny creatures, the microfauna and meiofauna, can range from 50,000 to a million individuals per square yard of seafloor. Below the top inch of sediment in which these creatures are concentrated, life is sparser, except where larger infaunal species pump oxygen down from the surface.

The larger animals in the sediments, the macrofauna, are big enough to see, but they live in burrows. Their lifestyle centers on straining their nutrients from the water. They have peculiar bodies (peculiar, that is, to people who are familiar only with land animals) and they obtain their food in ways that seem bizarre. A clam buries itself in mud and secretes strands of sticky mucus that flow across its gill tissue and then into its stomach, carrying trapped food items somewhat as sticky tape picks up lint. Another innovative clam cultivates single-celled plants in gardens that grow inside its own tissues. The throngs of invertebrates in bottom sediments have no common names, but to specialists, their scientific names evoke associations of form, growth habits, and behavior, just as to a gardener the names chrysanthemum and geranium bring pictures instantly to mind. Each of them, *Corophium*, *Ampelisca*, *Grandidierella*, and all the others, have distinct ways of eating, mating, producing young, and maturing to repeat the life cycle.

Plumed Worms. Among the small burrowers on estuarine floors are polychaete and oligochaete worms—by the millions. They are a highly varied group and many are surprisingly beautiful; a few of the polychaetes are shown in Figure 18–8. They are also versatile: they can travel at will, produce energy, obtain oxygen, and dispose of waste efficiently, giving them an enormous range of lifestyle options. Some polychaetes are stationary; some swim; and some eat their way through the seafloor sand or mud as fast as coquina clams dig in the surf. Polychaetes are among the most common animals in the sediments of all coastal environments and about 80 species are found along the Florida coast. Wash several shovelfuls of sand or mud through a screen, and you will almost always find several of these important worms as well as other burrowing species. In a healthy estuary, these busy animals are pumping water through their bodies around the clock, filtering food from it while remaining hidden from predators. The burrows they make often serve later as habitats for other animals.

Because they occur in such enormous populations, marine worms are a major food resource for thousands of species of fishes and crustaceans. In shallower parts of soft mud bottoms, birds also feast on these bottom dwellers. If the marine worms disappeared, thousands of other species that depend on them for food would also disappear.

Clams and Other Mollusks. Many species of mollusks also live in bottom sediments and filter feed. Mollusks have been mentioned before,

Reminder: The **infauna** are animals that live in burrows they create within the bottom sediments. They strain their food from the sediments and water. The tiniest are **microfauna**, those that are barely visible are **meiofauna**, and those that are easily seen are **macrofauna**.

Polychaete (POLLY-keet) and **oligochaete** (AH-lee-go-keet) worms are members of the segmented-worm phylum, whose most famous members are the earthworms.

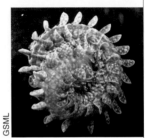

A polychaete worm on a Keys reef. At left, the fine, featherlike structures that enable the worm to filter volumes of water can be seen. At right, the tube, into which the worm can retreat, is buried, with only its top portion visible.

A group of polychaetes with their tubes buried in coral rubble on the seafloor.

A feather-duster worm mounted on a coral in the Keys reefs.

Red scaleworm (*Levensteiniella kincaidi*). This worm has left its old burrow and will make a new one elsewhere.

FIGURE 18–8

Polychaete Worms

but because they are so important in estuaries, they are given special attention here. When we think of mollusks (if we do at all), we might think of clams or oysters or the snail slugs we see in the garden: "lowly" or "primitive," shelled crawling creatures, but this is a major misconception. Mollusks are the second most successful group of animals on earth: their basic body plan is found in some 125,000 different species. Only the arthropods (insects, spiders, crabs, and other animals with jointed external skeletons) outnumber them, with 1 million or so described species. How could we be so unaware of this mind-boggling assortment of living animals? The answer is that most are in the ocean. The world's snails number 75,000 species, and two-thirds are marine. The clams amount to some 25,000 species, and nine-tenths are marine. And of the other mollusks, all are marine, so most people never see them. Background Box 18–2 offers some additional information on these remarkable creatures.

BACKGROUND 18–2

Mollusks: Soft Bodies with Hard Shells

There are known to be some 125,000 species of mollusks. Of those, 75,000 are species of snails; 25,000 are species of clams. To put these numbers in perspective, the group that includes *all* of the vertebrates numbers only about 50,000 species. That group includes all of the fishes, frogs, lizards, turtles, birds, and furry mammals, all of the forms that are displayed in our zoos, and humans as well. About half of the vertebrates are fishes, and even among those, the majority of fish species are (no surprise) in the ocean. And yet we arrogantly classify animals into "vertebrates" and "invertebrates"—that is, animals with bony skeletons and everybody-else-lumped-together. This makes no more sense than if a resident of tiny Sopchoppy, Florida, were to divide the world into two equal parts: Sopchoppy versus the rest, with "the rest" including not only Miami but India as well.

The mollusks include snails, clams, octopuses, squids, and several other groups. Why are they classed together—what does a hard-shelled clam buried in the mud have in common with a jet-propelled rainbow-colored squid? Most of them have a soft body that, on the inside, contains the internal organs, and on the outside secretes a shell of calcium carbonate. The soft parts are folded into a variety of body shapes, ranging from the spiral shapes of snails and the flattened shapes of clams to the streamlined, tentacle-bearing shapes of octopus and squid. (Octopus and squid also have the shells, but only remnants that they conceal inside their bodies to facilitate rapid movement.) The history of the phylum Mollusca is one of endless efforts to balance the mass of the heavy protective shell with mobility.

Among the most thickly armored mollusks are the snails, or gastropods, animals with a single shell (such as snails and conchs) that can move by virtue of a muscular "foot" that they can protrude from the shell and use to creep along the seafloor. Thanks to the heavy shell, they can survive being pounded by sand and surf and thanks to the foot, they can scour the seafloor for food and move to new areas when it gets scarce. The success of the combination is reflected in the huge number of species that are alive today. Figure 18–9 shows photos of some gastropod shells commonly found on Florida beaches.

Another big group of mollusks is the bivalves, those with a pair of

Native to Florida's offshore waters: Flamingo tongue snail (*Cyphoma gibbosum*). This inch-long marine mollusk eats corals: this one was photographed on a sea fan in the Florida Keys coral reefs. Its shell is actually opaque and white, but the snail extrudes a thin skin of tissue that both covers it and serves as gills, obtaining oxygen from the water. When threatened, it pulls this tissue into its shell and "turns" white.

The **gastropods** are a class within the phylum Mollusca (see the box, "Categories of Living Things," on pages 13–14). Many gastropods have heavy, single shells with a single opening. Through that opening they can extend both the muscular "foot" that they use to pull themselves along, and the body parts that they use to eat with. A child might say "they eat with their feet" (**gastro** means "stomach" and **pod** means "foot").

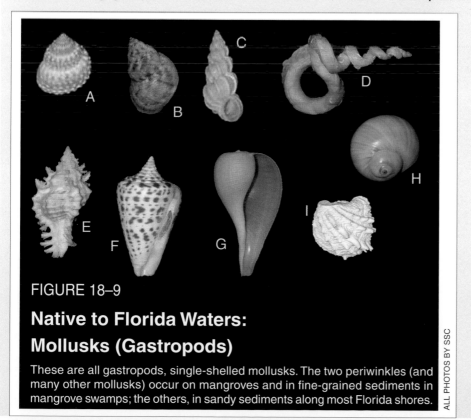

FIGURE 18–9

Native to Florida Waters: Mollusks (Gastropods)

These are all gastropods, single-shelled mollusks. The two periwinkles (and many other mollusks) occur on mangroves and in fine-grained sediments in mangrove swamps; the others, in sandy sediments along most Florida shores.

ALL PHOTOS BY SSC

KEY:

A. Beaded periwinkle (*Tectarius muricatus*), 1" long.

B. Angulate periwinkle (*Littorina angulifera*), 1" long.

C. Wentletrap (an *Epitonium* species), 1" long.

D. Worm seashell (*Vermicularia knorri*), up to 4".

E. Lace murex (*Chicoreus dilectus*), up to 2.5".

F. Alphabet cone (*Conux spurius*), up to 3.5"

G. Paper fig seashell (*Ficus communis*), up to 4"

H. Shark eye or moon snail (*Polinices duplicatus*), up to 2"

I. Spiny jewel box (*Arcinella arcinella*), up to 1.5"

Native to Florida's offshore waters: Horse conch (*Pleuroploca gigantea*). The pink body is the mollusk's "foot," which enables it to creep across the hard bottom. Its big, beautiful shell has earned it the honor of being named "Florida's state shell."

Conchs on the move. A whole population of conchs moves across the seafloor, scouring the sediments for food.

Bivalves are mollusks with paired shells connected by a hinge.

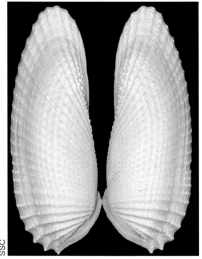

Native to Florida waters: Angel wings (*Cyrotopleura costata*). This bivalve buries itself in mud flats about one to two feet below the surface. It grows up to six inches long.

hinged shells. The bivalve shells of clams allow them to burrow smoothly through sediments, buried and out of reach of many potential predators. Scallops, which lie visible on the surface in seagrass, can, if threatened, "fly away" through the water by repeatedly clapping their shells together. Margin photographs through the rest of this chapter display some of the extraordinary variety of bivalves that inhabit Florida's coastal and offshore environments.

Shell collectors comb Florida beaches, seeking treasures to take home as reminders of the beauty of the sea. Most of the shells they find are mollusk shells—cockles and scallops, predatory conchs, olive shells, pink-striped sunray Venus clams, and many more. Those most numerous in Florida's estuaries are the mussel and, especially, the oyster. These animals are reef builders and their reefs attract many other creatures, forming communities.

Oyster Reefs and Others. Oysters have built extensive reefs around Florida. At least one large population of oysters occurs in purely marine waters of the Gulf of Mexico near Anclote Key (Pinellas County), but most are in estuaries, to which they are extraordinarily well adapted. They flourish best in water less than 30 feet deep with a firm bottom of mud or shell, notably among the mangroves of the Ten Thousand Islands, in some parts of Tampa Bay, and on the floor of Apalachicola Bay.

The oyster is the quintessential example of an animal that can thrive in an estuarine environment. Oysters filter their food from the turbid, dark water; they don't have to find it by sight. They build their hard shells from the minerals, and their soft bodies from the organic materials, in the water. They don't have to see to mate, either, because they spawn; that is, the males and females release their sperm and eggs into the water and fertilization takes place there. The oysters' tiny, planktonic larvae (known as spats) float in the plankton for a while, and then lodge on a hard or soft seafloor. On arrival, each oyster attaches to the estuary floor by the center of one of its two bowl-like shells; that way, most of the sediment particles that drift by with the current will slide beneath the lip of the shell, rather than smothering the oyster. If the water gets choppy and stirs up the sediment, the oyster claps shut.

It takes only a few spats to start a new oyster community. The reef or bar it forms keeps growing year after year, as spats settle out of the plankton and attach to the shells of older animals. An oyster can live for years, completely covered by other animals, then see the sea again when a chunk of the colony breaks off in rough waters. But over time, those that are farthest down in the aggregate are smothered and die, while living oysters are concentrated on the outer surfaces of the pile. Currents shape the oyster bars or reefs into narrow, linear structures, and the individuals themselves also grow long and narrow. Their shells open toward the food-bearing currents that pass across the bar. Each oyster "inhales" water through one siphon and filters its food, plankton and nutrients, from suspended sediments. Then it ejects the sediments through another siphon.

If exposed above water, oysters can close their shells, so they can occupy intertidal areas. Intertidal, "coon" oysters are usually stunted, because they can ingest food only when submerged at high tide, but if transplanted to deeper subtidal sites where they can feed continuously, they grow larger. When the tide goes out, coon oysters snap shut and squirt their water out; people walking along the shore as the tide goes out can see them "spitting." Figure 18–10 features a colony of coon oysters and a handful of oyster shells from a reef.

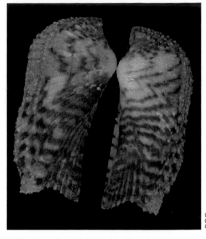

Native to Florida waters: Turkey wing, or zebra arc (*Arca zebra*). This bivalve, whose shells are commonly found on beaches, grows to a length of 4 inches.

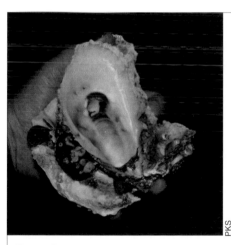

Part of an oyster reef: a cluster of oyster shells cemented together.

"Coon" oysters in the intertidal zone. They are stunted, because they are growing in a less-than-ideal site, but they are otherwise normal. They attract throngs of shorebirds.

Native to Florida waters: Sunray venus (*Macrocallista nimbosa*). This is a large clam, up to 5 inches long, that buries itself in sandy bottoms beneath shallow water.

FIGURE 18–10

American Oyster (*Crassostrea virginica*)

Native to Florida waters: Jingles (*Anomia simplex*). This animal may be up to 1 1/2 inches long. Its shells often break apart at the hinge and are commonly found singly, typically near shore in shallow water.

Oysters use the water to convey several chemical signals. Mature oysters release a chemical into the water, and this attracts spats that are searching for attachment sites. Thus the oyster bar grows and the colony stays together. When one male spawns, he releases another chemical into the water. The surrounding males and females detect it, and spawning is triggered in the whole population. This ensures that plenty of eggs and sperm are present simultaneously and makes fertilization efficient, producing large numbers of larvae. The oyster colony also adjusts its sex distribution constantly to ensure that there will always be enough individuals of both sexes to reproduce. After the males have shed their sperm, they slowly turn into females; after the females have released their eggs, they may revert to being males.

Oysters can tolerate both marine and nearly fresh water equally well, but in salt water, marine predators can attack them. Their worst enemy is the oyster drill, a mollusk that bores into their shells and digests the oysters. In Apalachicola Bay, when the river's flows are average, fresh water dilutes the salt water enough to drive out or kill the oyster drills, but with abnormally low river flows, the water becomes salty and the marine predators stay alive, killing oysters. Thus, high river flows are protective. Major rainstorms and floods, which may not be welcome to people on land, are necessary to cleanse the bay and support its ecosystems. Withdrawals of water from the river system can severely impact the oyster fishery.

Oysters filter up to eight times more sediment from the water than would settle out due to gravity alone. As a result, they efficiently turn turbid water clear as it flows over them. After using the nutrients they need to build their soft bodies and hard shells, the oysters drop clean sediment particles on the bottom, laying down a broad, flat floor and releasing clearer water to the seagrass and algal beds that lie farther out in the estuary. In clear water, seagrasses and algae can obtain enough sunlight to thrive.

Like every intricate structure in nature, an oyster reef provides hiding and attachment places for many other species. Among the several dozen species of animals that find habitats on oyster reefs are mussels, boring clams, barnacles, shrimp, polychaete worms, anemones, burrowing sponges, lightning whelks, and many kinds of crabs. Populations of one tiny mud crab on oyster bars can reach densities of 1,000 per square yard.

For completeness' sake, it should be mentioned that there are other kinds of reefs in some of Florida's coastal zones. In north Florida's Apalachicola Bay, mussels have built reefs on some hard-bottom areas. Around south Florida, worm reefs occur on the soft floors of several shallow, intertidal zones. The worms make a protein that cements sand grains together into tubes, and these build up, one upon another, as new worm larvae settle on them. Because they are in shallow water and can add material only when submerged, worm reefs grow mostly horizontally, forming broad, rocklike platforms that attract lobsters, fish, and other animals. Other reefs (called vermetid reefs, pronounced VER-me-tid) were built by a species of wormlike mollusk on sediment floors outside the Ten Thousand Islands. They are no longer active, but their rocky structure attracts juvenile and adult stone crabs and sometimes some fish. The reefs tend to catch floating debris and may form the nuclei of new islands.[6]

Native to Florida waters: Blue mussel (*Mytilus edilis*). Mussels grow on some hard bottoms in Apalachicola Bay.

Perspective on Estuaries

This relatively brief overview of Florida's estuaries only hints at the huge arrays of plant and animal species they support. There are many more: we don't even know how many. We know that among the fishes alone, more than 1,000 species occur in near-shore habitats around Florida. We know that of these, 30 percent have populations only in, or mainly in, Florida, and that these represent about one-fourth of the fish species recorded for the entire western hemisphere north of the equator. According to a state committee on biodiversity, "Indications are that the fish [diversity] of this state [is far greater] than that of any comparable area in either North or Central America."[7]

As for invertebrates, consider just the oligochaete worms. Some 350 species of oligochaetes occupy coastal environments off eastern North America. More than half of these are found off Florida's shores. Or take the crustaceans: about 900 species are known all along the Atlantic coast from Florida's tip to northern Canada. Of these, more than 700 are found in Florida waters, and 35 are probably endemic. Thus, Florida's near-shore waters are a major biological resource. They deserve respect, preservation, and where needed, restoration.[8]

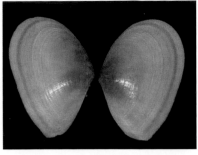

Native to Florida waters: Rose petal tellin (*Tellina lineata*). This animal, which is seldom more than 1 1/2 inches long, is common on bay shoals in sand.

CHAPTER NINETEEN

SUBMARINE MEADOWS

This chapter continues the series about natural communities on the continental shelf around Florida. Chapter 18 dealt with estuarine communities on mineral floors and with mollusk and worm reefs; this one describes grassy realms where sunlight penetrates the water far enough to reach the bottom, and where seagrasses and large algae can attach, grow, and spread. Numerous other organisms then find habitats on and among the plants and form communities of great complexity. To a small invertebrate animal, a meadow of seagrass is a wonderland of different kinds of foliage and a forest of blades on which to feed, hide, mate, and travel from place to place. To carnivorous fish and shellfish, it is a jungle in which to find abundant prey. To the sea turtles and manatees for which two of the grasses are named, a seagrass meadow is a luxuriant pasture in which to feed to the heart's content. And to tiny travelers such as shrimp and small fish, seagrass meadows provide safe passage across the ocean floor.

SEAGRASSES AND MACROALGAE

Vast meadows of seagrasses blanket some estuarine and marine seafloors. Florida's two largest seagrass beds lie in Apalachee Bay and off the peninsula's southern tip in Florida Bay (shown earlier in Figure 18–2, page 302). The beds in Apalachee Bay occupy some 750,000 acres; those in Florida Bay almost 1,250,000 acres. These underwater meadows spread in a wavy band from just beyond the freshwater outflows of rivers up to 70 miles offshore, and in south Florida, they grow luxuriantly along tidal channels in mangrove islands.[1]

Seagrasses are true flowering plants. They bloom just as land plants do. They release pollen, and it drifts through the water, cross-pollinating other plants. They produce seeds that may take root and spread, and they also spread by growing long, horizontal rhizomes that put up new shoots from every node. Thus they can colonize increasingly broad sand and mud seafloors and come to dominate large areas. Figure 19–1 shows the seagrasses that dominate Florida's estuarine and marine floors down to about 100 feet of depth.

OPPOSITE: A soft coral growing in a seagrass bed.

> **Seagrasses** are not true grasses but aquatic flowering forbs that evolved long ago from terrestrial ancestors.

> *Reminder: Algae* may occur as single cells, as filaments, as thin sheets, as calcified crusts, or as bushy, leaf-shaped forms.
>
> Tiny algae, or *microalgae*, are single-celled and small multicellular plants that float among phytoplankton and grow on sediments, rocks, and plants.
>
> Leafy-looking, large multicellular algae (**seaweeds** or **macroalgae**), hold to rocks, hard bottoms, corals, or seagrasses, or float free.

DN/GSML

Native to Florida waters: Pipefish (a *Syngnathus* species) from near-shore waters off Florida's Big Bend. Pipefishes are closely related to the fishes called seahorses. Their slender shape and mottled color enables them to hide easily in seagrass beds.

Shoalweed (*Halodule wrightii*).

Turtlegrass (*Thalassia testudinum*).

Manateegrass (*Syringodium filiforme*).

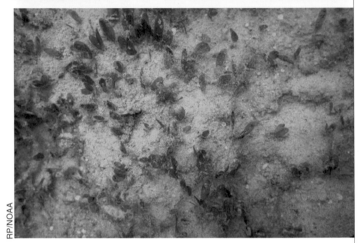

Caribbean seagrass (*Halophila decipiens*).

FIGURE 19–1

Estuarine and Marine Seagrasses Native to Florida Waters

Each seagrass has a different habitat preference. Just as each tree species in a forest grows best at a certain elevation along a slope, each seagrass dominates a different part of the seafloor. Shoalweed grows closest to the shore and around the mouths of rivers; it withstands low salinity the best, so it can tolerate the fresh water running down rivers and runoff from rain along shores. It can also tolerate exposure to the air during low tide, hence its name. It is a pioneer, the first to colonize newly available areas or to recolonize disturbed areas.

Shoalweed has brittle, cylindrical leaves that break off easily, but its root-rhizome system branches prolifically and stabilizes the sediment, so like its terrestrial relatives on dunes, seaoats, shoal grass builds up land. Once shoalweed has begun to grow, sediment accumulates around it, forming mounds and ridges and enabling other seagrasses to take hold. Thus, shoalweed paves the way for other grasses.

Turtlegrass lacks some of shoalweed's adaptations, but has others of its own. It dies if exposed to the air for more than brief periods, but it is a tough grass with broad leaves and branching roots and rhizomes that firmly stabilize the sediment in which it grows. It can form dense stands just below the low-tide mark and down to about ten feet of depth. Beyond ten feet, it can still grow, although more sparsely, all the way down to about thirty feet if the water is clear. Turtlegrass cannot tolerate low salinity, though, so shoalweed remains dominant near shores and near the outflows of rivers.

Other grasses often grow with turtlegrass in mixed beds (like biodiverse prairies on terrestrial Florida). Manateegrass may grow mixed with turtlegrass, or may assume dominance in waters more than ten feet deep. Like turtlegrass, manateegrass grows only below the low-tide line and does not tolerate exposure to the air. Because its broad blades enable it to catch light efficiently, another seagrass, Caribbean seagrass, grows as an understory plant beneath the taller blades of turtlegrass and manateegrass and also occupies the deepest, most dimly lit parts of seagrass meadows out to a depth of about 100 feet. Several other species grow mixed with, or around the fringes of, these seagrasses.

Seagrasses grow well on muds and mucks, the sorts of sediments that accumulate in relatively quiet waters. They prevent the current from stirring up the bottom, so while they are flourishing (during the growing season) the water remains free of turbidity, permitting maximal light to reach the grasses. When winter arrives along the northern Gulf coast, the grass blades may die back and slow down wave action less effectively. Then the water grows too turbid to allow much light to penetrate, but the grasses' metabolism has slowed and they no longer need as much light. In spring, they grow back from their subsurface roots and rhizomes and again the water becomes clear.[2]

Seagrasses exhibit many other adaptations. Those that grow in shallow water have flaccid leaves that lie flat at low tide, so they remain immersed even when the water is low. Those that grow in deeper water have stiffer, more erect leaves, which are less likely to bend in the current, so they can reach higher for light under prevailing conditions.

Like phytoplankton, salt marsh grasses, and mangroves, seagrasses deliver to other organisms much of the energy they capture from sunlight. Some energy, they yield up directly to herbivores, notably sea urchins, sea turtles, and manatees, but most they release as detritus, which feeds organisms both within and beyond the grass beds. Because they grow fast—up to half an inch a day in strong light and clear, warm, slowly moving water—seagrasses are prolific producers of both green tissue and detritus. During the day, while photosynthesizing, they may release oxygen so fast that it is easy to see bubbles escaping from their leaf margins. They also pump up nitrogen from bottom sediments, where bacteria are always binding ("fixing") nitrogen into compounds that are usable by other living things. The seagrasses release some of this usable nitrogen through their leaves into the water, where other organisms have access to it. Some fixed nitrogen washes out to sea and becomes available to other communities.[3]

Seagrasses add structure to the underwater environment, and as in all ecosystems, the more surface area is available, the more hangers-on can attach. A square inch of "bare" mud or marl on the seafloor may hold 5,000 organisms, but a few leaves of seagrass growing on that square inch may hold several hundreds of thousands: single-celled algae, bacteria, invertebrate grazers, filter-feeders, and predators on these. Moreover, seagrass beds have many nooks and crannies that can accommodate huge numbers of swimming plant cells (dinoflagellates); phytoplankton; long filaments of attached algae; and even large, free-floating clumps and cylinders of drift algae that roll along the bottom. Feeding on these are zooplankton, shellfish, small fish, and many others. Descriptions of food webs based on seagrass beds quickly grow long.

The *several hundred species* of single-celled algae that grow on seagrass blades are a microscopic lawn that feeds a multitude of tiny animals, providing nutrients that make fast growth possible. Some single-celled algae fix nitrogen as bacteria do in bottom sediments. Some make skeletons or coats of calcium salts from the sea water. More than 30 species of tiny snails and dozens of species of barely visible shrimp and other crustaceans roam over seagrass leaves and partake of their algal coating. These animals find such nourishment in the lawns of algae on seagrass blades that they live out their entire lives there and grow a thousandfold in size from newly hatched larvae to adults. When seagrass washes up on the beach, people step aside to avoid its slimy, gritty coating; they call it "seaweed," but it is in fact a grand pastureland for grazers and the "grit" is evidence of their presence.

Some underwater meadows are composed of *large* algae (macroalgae) in greater numbers than seagrasses. These algae, or seaweeds, are not vascular plants—that is, they do not have internal vessels to circulate their fluids, and they do not have true roots that can absorb nutrients or take firm hold in soft sediments. However, in deep water, macroalgae can grow as tall as trees, and they can attach to rocks with creeping, rootlike structures called holdfasts. Limestone seafloors, limestone outcrops, oyster reefs, and mussel reefs may be cloaked over many acres with macroalgae—either thick algal mats or tall, leafy, swaying, green filaments.

Some large algae make skeletons of calcium carbonate and when they die, add them to the sediments (limestone is still forming on the seafloor, just as it has done for millions of years past). Some algae disintegrate into sand-grain-sized plates, some into powderlike sediments such as the fine mud inside the coral reefs off south and southwest Florida. Like seagrasses, attached algae move nitrogen from sediments to the water column.[4] List 19—1 names some bottom algae and Figure 19—2 displays three of them.

Holdfasts are structures by which seaweeds attach to the rock, coral, seagrass, and other hard surfaces.

Holdfasts don't enable large algae to attach to soft substrates, but they still manage to occupy soft-bottom areas: they attach to the seagrasses that grow there: they use the grass as a sort of mooring rope. In the Indian River lagoon, fully 70 percent of all the algae are filamentous forms attached to grasses. By shading seagrass leaves, attached algae shorten the lives of the plants on which they depend, but many snails and others graze the leaves and help to keep them clean. Still, most seagrass leaves are shed along with their attached algae within sixty days of first emergence. The shed leaves of manateegrass form large floating mats; turtlegrass leaves wash in below them; and wave action breaks them up into detritus. Then the whole complex of disintegrating leaves, algae, and grazers becomes food for a host of sea and land animals.

Shed grass leaves with their algal dressing make a nourishing food for

A green alga (a *Caulerpa* species).

A red alga (a *Hymenella* species).

A brown alga (a *Sargassum* species).

FIGURE 19–2
Macroalgae Found in Florida Waters

detritus eaters. In the water, bacteria, fungi, and zooplankton disassemble and ingest the leaves. In bottom sediments, tiny detritus eaters eat whole leaf particles, or strip bacteria and fungi from them, or ingest the fecal pellets dropped by other detritus eaters. Shrimp and small fish twitch up tiny clouds of debris from the bottom, suck them in, and grow on the nutrients they contain. Pink shrimp, voracious omnivores, find a feast on grassbed floors. Blue crabs, too, are prodigious pickers-up of litter. Even the seagrasses themselves take nourishment from their own detritus.

SMALL LIFE IN SEAGRASS BEDS

To this point, except for seagrasses and macroalgae, most of the grassbed inhabitants described here have been creatures too small to see. Just above the visibility threshold are hundreds more species of invertebrates. Lists 19–2, 19–3, and 19–4, which accompany Figure 19–3, identify some of the animals found. To give an idea of the numbers involved, the grassbeds in Tampa Bay support up to 3,000 polychaete worms per square yard, and the grassbeds in Apalachee Bay hold up to 1,600 amphipods per square yard as well as dozens or hundreds of populations of other species.

Besides offering a permanent home to many residents, seagrass beds are prime nursery grounds for many more animals: bryozoans, starfishes, sea anemones, worms, mollusks, crabs, and others. These creatures make of the seafloor a magical world. Says one snorkeler, swimming over grass beds in St. Joe Bay:

> *Sunlight sweeps over the grass in blazing golden lines as lacy piles of rose-colored algae roll past. Red sea urchins sit scattered about, grazing on broken fragments of grass and using their suction-cuplike tube-feet to wave bits of shells over themselves as camouflage. Scallops hold their shells wide open, their tiny, sapphire-blue eyespots sparkling in the sunlight as they filter food from the water. A flounder, buried in the sand with only its eyes uncovered to betray its presence, waits in ambush for an unwary small crab or fish to pass by—then it will lunge up out of the sand, all teeth, to seize it. Here and there, the red crowns of tube anemones sit at the openings of their elevator-shaftlike tubes. Upon being touched, an anemone rockets to the bottom of its tube so fast that an observer wonders if anything was ever there. . . .*[5]

LIST 19–1
Algae on bottom sediments

These algae are attached to hard substrates off the west Florida coast:

Brown Algae
Dictyota dichotoma
Padina vickersiae
Sargassum filipendula
Sargassum pteropleuron
. . . and others

Red Algae
Chondria littoralis
Digenia simplex
Gracilaria species
. . . and others

Drift Algae
Laurencia species
Members of these families:
  Acanthophora
  Gracilaria
  Hypnea
  Spyridea

Source: Zieman and Zieman 1989, 44.

FIGURE 19–3

Seafloor Occupants (Tampa Bay area)

Of the three habitats shown, seagrass beds are by far the most densely occupied. Hundreds of invertebrate species may be present in a seagrass bed. Each animal has its own lifestyle. For example, an amphipod (*Cymadusa compta*) grazes algae from the tops of the grass blades at night, and eats detritus from the seafloor in the daytime.

Small differences in season or location alter the proportions in which these animal populations occur, and they also differ from south to north. An equally varied array with a somewhat different assortment of organisms occurs in more northern, temperate coastal waters.

Source: Zieman and Zieman, 1989, 48.

Since there are hundreds of species of colorful, oddly shaped invertebrate animals in Florida's seagrass beds, it is impossible to do them justice, but Figure 19–4 presents a few, to illustrate their diversity.

Not only are these animals diverse, they are also present in numbers so astronomical it is hard to imagine them. Consider, for example, just one crustacean, the pink shrimp, which makes a juicy morsel for dozens of predators that share its waters, and travels long distances across the Gulf and back, largely in the cover of seagrass beds.

The pink shrimp engages in a behavior that enables some individuals to survive in sufficient numbers to maintain the species population. Figure 19–5 shows the impressive journeys this animal makes from its spawning grounds to its nursery grounds and back.

How do these tiny marine animals manage their migrations across the open Gulf? Millions of them migrate from the cluster of little islands known as the Tortugas, nearly 70 miles west of Key West, to Florida's coast and back during their life cycle. Mature pink shrimp spawn in early spring in the open ocean off the Tortugas and produce minuscule larvae, only one one-hundredth of an inch long, barely visible. The larvae look like tiny snowflakes drifting in the water and, during the next day and a half, they go through five transformations. Then they begin to feed on planktonic plants and animals that are even tinier than they are, and go through several more stages of development. After about three weeks, they are beginning to look like shrimp, but are just over one-eighth inch long.

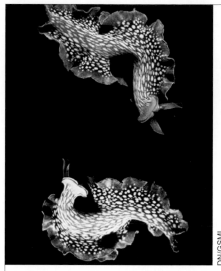

Lettuce nudibranch (*Tridachia crispata*). This soft-bodied mollusk flutters through the water, a little like a butterfly. Although named for lettuce, it may be brown, yellow, or magenta as well as green. It easily conceals itself by nestling in the algae on which it feeds. It grows to a length of three inches.

Spotted cleaner shrimp (*Pericli-menes pedersoni*) on pink-tipped anemone (*Condylactus passiflora*). Many species of cleaner shrimps dwell on other animals such as anemones and corals and serve their hosts by eating parasites and predators off their tissues.

Like the shrimp, the anemone is an animal, but a radially symmetrical one. The stinging cells at the tips of its tentacles paralyze small prey, including fish, that happen to swim by; then the tentacles feed the prey into a mouth at the center of the animal. The shrimp, though, is immune to the anemone's poison.

Reticulated brittle star (*Ophionereis reticulata*). This animal roams sea-floors, seagrass beds, and coral reefs. It captures and feeds on smaller invertebrates. When it captures one with its five arms, it turns its stomach inside-out and digests the animal while pulling it inside.

Calico scallop (*Argopecten gibbus*). Scallops abound in seagrass beds. Each animal has numerous eyes, tiny blue "pearls" that form a line along the soft tissue that is extruded from between its shells.

A sea squirt. This animal appears not at all like a close relative of ours, but its early development is similar to the development of vertebrate embryos. It is classed as a chordate, a member of the same phylum as the vertebrates.

FIGURE 19–4

Invertebrate Animals in Florida Seagrasses

Staghorn coral (*Acropora cervi-cornis*). Corals look like branched shrubs, but they are invertebrate animals (see next chapter). Corals that start out in seagrass beds may ultimately build massive reefs, such as those off the Florida Keys.

Fire nudibranch (*Hypselodoris edenticulata*). The bright stripes on this animal warn would-be predators that it is noxious.

During these transformations, the shrimp drift at sea and are largely at the mercy of ocean currents because they are capable of only a little movement. Unless favorable currents carry them to inshore nursery grounds, where the next stage of their life cycle is spent, they will perish. Also, on any given day, some 20 percent of the population is lost to predators. The next day, of the population that remains, another 20 percent is lost.

On their way across the ocean, nearly all of the shrimp are either eaten by fish, jellyfish, and larger shrimp, or are carried by currents away from their inshore nursery grounds and die. Some 9,999 individuals out of ev-

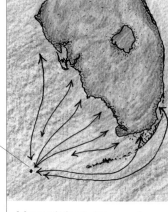

Tortugas

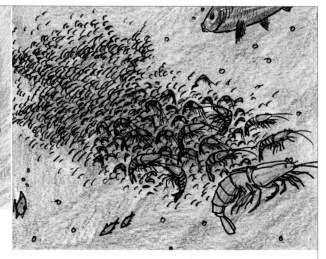

Most pink shrimp make a round-trip journey between the Tortugas and Florida's west coast, as shown. Some range along the whole of Florida's west coast and as far north on the east coast as North Carolina.

FIGURE 19–5

Travels of the Pink Shrimp (*Penaeus duorarum*)

ery 10,000 are killed. Still, some 4.3 billion shrimp make it all the way to feed and flourish in the lush seagrass beds off Florida, and northward into the rich bayous of the Gulf states. List 19–5 displays the menu of tidbits they find in these rich feeding grounds. Now about 1/4 inch long, they bury themselves every day in the soft, estuarine mud and emerge every night to continue feeding, molting, and growing.

The young post-larval shrimp stay in the nursery grounds, in areas of shallow brackish to almost fresh water for a period of four to eight weeks. As they grow, they leave the shallow estuaries, marshes, and lagoons for deeper rivers, creeks, and bays. By July they are 2 to 3 1/2 inches long and by the end of August they are 5 inches long; they have doubled their weight in three weeks.

As the water grows colder, the shrimp cease growing. They remain in the shallow bays until the cold drives them out to seek deeper waters where temperatures are more stable. (This is the winter shrimp "run" that brings hordes of people with nets and lanterns to bridges.)

Then, as winter turns to spring, the young shrimp begin to develop sexually and this is when their offshore spawning migration occurs, which takes the sexually mature shrimp back to their oceanic spawning grounds near the Tortugas. Some 21 to 35 days after they arrive back at their birthplace, fully mature, and seven inches long, they are ready to spawn and launch the next generation.

At spawning, each female shrimp can release from 300,000 to a million eggs at a time. The new population will be large enough to withstand all the pressures of predation and keep the species going.

The pink shrimp's migratory behavior is genetically programmed to fit the contours, distances, and seasons of the Gulf of Mexico. As a result, it can survive the hazards this passage presents and take advantage of the

LIST 19–5
Menu for a pink shrimp

Amphipods
Copepods
Detritus
Foraminiferans
Isopods
Juvenile shrimp
Mollusk larvae
Nematodes
Ostracods (tiny crustaceans)
Polychaetes
Sand
Shrimp larvae

Source: Zieman 1982, 58.

environment's resources. Millions of the ocean's days, nights, ebbs, and flows have molded and maintained the behaviors of this creature.[6]

LARGER ANIMALS

Because so many small animals are present in seagrass beds, many larger ones seize the opportunity to prey on them: conchs and whelks, spiny puffer fish, blue and golden cowfish, pinfish, horseshoe crabs, seahorses, pipe-fishes, tulip snails, shrimp, jacks, snapper, blue crab, stone crab, sea urchins, starfish, spider crabs, anemones, scallops, hermit crabs, and octopuses. Fish are so numerous in healthy grassbeds that they defy attempts to count them. Some are residents and some are just passing through. Some are there as juveniles and move out to sea as adults. Some come in to forage for a season. Some nibble algae off seagrass blades, a few eat the grass itself, some eat detritus, and of course large fish eat smaller fish and other animals. Nearly 50 *families* of fish are found just in west Florida's turtlegrass beds. Figure 19–6 depicts a few common species.

Sea turtles visit seagrass beds; in fact, turtlegrass is named for the green sea turtle, which eats so voraciously it has been called the seafloor buffalo. The turtle often snips the leaf at the bottom and lets the top float away, stimulating the plant to make new leaves. The behavior is reminiscent of the way deer prune shrubs and stimulate the growth of new foliage. Turtle-pruned leaves as well as turtle droppings add nitrogen to the system. Sea turtles receive attention in Chapter 21.

Manatees also visit seagrass beds. Manatees are primarily tropical animals, but in summer, Gulf coast waters and river mouths are warm enough for them. In winter, they take shelter in spring runs such as the Homosassa and Crystal rivers, whose translucent waters permit freshwater vegetation to grow luxuriantly. Manatees forage singly or as mother-calf pairs in estuaries and offshore seagrass beds. They spend six hours a day feeding and each one eats more than a hundred pounds of grass a day. They use their stiff facial bristles to clear the ground around the plants, then uproot them and shake them free of sediment.

Birds, too, dine under water over shallow seagrass flats. Herons and egrets wade, snatch up swimmers, and probe the bottom for infauna. Geese and ducks swim and dive to eat the underwater plants, invertebrates, and fish. The double-crested cormorant pursues fish under water; the osprey, bald eagle, and brown pelican fly and dive for them. Feeding in grassbeds is efficient for birds, because there are many fish in a small area. When rearing their offspring, pelicans have to deliver 120 pounds of fish to each nestling if it is to grow enough to fledge successfully. This is a tall order, but one that can be met by the forage in rapidly growing seagrass beds.[7]

PERSPECTIVE ON NEAR-SHORE ECOSYSTEMS

These two chapters have treated near-shore communities that harbor great biodiversity. Seagrass beds alone are more diverse than any other underwater ecosystems, even coral reefs, and they are more important as

Menhaden fry, growing up in estuaries and along rivers, are important in the food web. Like herrings, they are bony and oily, and wherever they are abundant, larger fish thrive. Those that survive predation may grow to 25 pounds:

Yellowfin menhaden (*Brevoortia smithi*).

Large, roaming predator fish swirl in, snatch up their prey, and streak away. Top carnivores include king mackerel, barracuda, several sharks, and many more:

Blacktip shark (*Carcharhinus limbatus*). This fish grows to 6 feet in length.

FIGURE 19–6

Marine Fish in Florida's Seagrass Beds

Source: Anon. c.1994 (*Fishing Lines*); drawings by Diane Peebles

Snappers, seatrout, and others begin their lives as juveniles in seagrass beds and grow to about 3 or 4 pounds as adults:

Spotted seatrout (*Cynoscion nebulosus*).

Larger fish in seagrass beds include stingray, permit, and others:

Stingray (*Dasyatis* species). The stingray swims over sand and mud bottoms in deep waters, and visits grassbeds to forage. Typically it grows to about three feet, but a few giants are six feet wide and ten feet long.

Permit (*Trachinotus falcatus*). This fish usually stays inshore around grass flats. It grows to 25 pounds.

Great barracuda (*Sphyraena barracuda*). This fish grows to about 5 feet in length.

King mackerel (*Scomberomorus cavalla*). Mackerel weigh about 20 pounds when mature.

nursery and feeding grounds. They are, themselves, part of a broader landscape that functions as a unit. Many organisms migrate from one seagrass bed to another. Along west and north Florida shores, fish that are moving between salt marshes and deep offshore waters can swim under cover of seagrass beds and feed as they go. In Florida Bay, juvenile and adult animals can travel back and forth between mangrove swamps and coral reefs without leaving the cover of seagrass beds.[8]

That animals migrate so extensively around the Gulf reflects the fact that the waters along the shore are all part of a single system that supports marine life around the year. Upland ecosystems far inland around stream headwaters affect the health of coastal waters. Clean forests and marshes purify the waters of rivers, creeks, and runoff from on shore. Freshwater flow determines the layering, mixing, and oxygen content of estuarine waters and is crucial to community structure. Streams must pulse as they have done historically, to deliver the seasonal highs and lows and the nutrient and sediment loads to which estuarine organisms are adapted.[9]

If one area becomes polluted or otherwise disturbed, its loss has broad impacts, not unlike the loss of a major wetland used by migratory birds. Losses of the forests alongside a river, diversion of a river's water, filling of marshes along the shore, or uprooting of seagrasses in an estuary have

PRICELESS FLORIDA

equally major repercussions. Not only communities within those environments, but neighboring and distant communities as well, decline and may even die out.

Besides supporting a profusion of life, seagrasses serve other vital roles. They help to protect shorelines from erosion. They reduce wave energy, accumulate sediment, and hold it in place. Storm tides may rise high along the shore and hurl battering waves against the land, dragging down sea walls and undercutting houses. But unless the water is so shallow that the grasses are exposed to the wind and scouring waves, the grasses hold on, and stabilize the bottom sediments. After the storm has passed, the seagrasses still remain, perhaps tattered, but able to quickly regrow. And if monster storms uproot the grasses, they still are useful. Thrown up on land as seawrack, they nourish beach life and vary salt marsh habitats. Even while floating, seagrasses are useful—small animals and fish hide and feed in floating grass.

CHAPTER TWENTY

SPONGE, ROCK, AND REEF COMMUNITIES

The major communities that occupy the shallow waters and seafloors nearest the shore, notably mineral floor communities, mollusk reefs, seagrass beds, and algal beds, were covered in the last two chapters. This chapter covers the major ones that are farther out on the continental shelf: sponge beds, limestone outcrops, coral banks, and coral reefs.

Figure 18–2 (page 302) depicted the locations of some of these communities, but structures composed of, or derived from, living organisms such as sponges, limestone outcrops, and coral formations are scattered all over the continental shelf. Many support commercially important populations of fish and shellfish (see List 20–1). A few are selected for emphasis in this chapter, primarily the sponge communities on the west-coast continental shelf and the coral reefs off south Florida. Along the way, Background boxes offer information on some of the diverse and versatile sea creatures that inhabit them.

SPONGE COMMUNITIES

On the shelf off the west coast of Florida, old shorelines from times of lower sea levels remain as ridges that run for many miles in parallel bands at increasing depths and distances from shore. Between them lie vast plains of sand, broken in places by scattered limestone outcrops up to about six feet in height. The sand plains are relatively sparsely populated, but the limestone outcrops are like rich oases that support diverse communities of marine plants and animals. Leafy macroalgae and soft corals wave in the currents among colorful sponges, bushy bryozoans, flowerlike anemones, and transparent, berrylike tunicates.

All of the organisms just named are animals, but they lack the heads, arms, and legs that earth creatures like ourselves expect animals to have. Some of these animals grow attached to seafloors or outcrops in shapes that resemble plants, and in fact most people think they are underground

Native to Florida waters: Spiny candelabrum (*Muricea muricata*), a coral in the Florida Keys reefs.

Communities of organisms such as sponges, corals, and bryozoans growing on bottom sediments are called **live-bottom communities**.

OPPOSITE: A coral reef in John Pennekamp Coral Reef State Park, Monroe County. Although they look for all the world like ferns and flowers, these formations consist of multitudes of tiny animals.

333

LIST 20–1
Fishes and shellfishes associated with Florida's banks and reefs (examples)

Keys, Tortugas Banks—
   Snapper

Keys, Tortugas Sanctuary
   —Shrimp

Southwest Florida, Florida Bay
   —Lobster and crab

Southwest Florida, 10,000
   Islands—Crab

Central Florida, Sanibel
   Grounds—Shrimp

Big Bend, Middle Grounds
   —Shrimp, fish, scallop

West Florida—No major banks

Florida-Alabama, Southeast
   Grounds—Fish

Florida-Alabama, Timberholes
   —Snapper

Note: Noncommercial fish also swim on the bottom: gobies, blennies, clinids, toadfish, skates, rays, batfish, and stargazers near shore; scorpion fish, sea robins, eels, and others farther out, and in southern waters tropical species such as damselfish, angelfish, wrasses, parrotfish, and others.

Source: Gore 1992, 115, table 3.

ferns, shrubs, and trees. Other marine animals are shaped like sheets, plates, bowls, or mounds, depending on the species. Part of the fascination of life beneath the sea is the extraordinary variety of body forms seen in animals. Background Box 20–1 explains the advantages of a body plan seen in many marine animals: the colonial form used by sponges, bryozoans, and others.

BACKGROUND   **20–1**

Animals Unlike Animals: Sponges and Bryozoans

Unlike land animal species, in which each animal is a separate independent individual, many marine animals form colonies of small, interconnected individuals. Sponges, bryozoans, corals, and sea squirts, which are among the most abundant of all aquatic animals, are all colonial forms.

The colonial lifestyle resolves a basic conflict for animals that filter microscropic prey from water. To obtain food successfully this way, it is an advantage to be relatively small, but small creatures are vulnerable to being devoured themselves. By building a body made up of repeated little units, the colonial filter feeder preserves its tiny food-gathering form, but collectively grows into a large organism whose size makes it somewhat resistant to being altogether devoured. If a colony is attacked by a predator, some of the individual units may die, but the undamaged members can give rise to new ones by budding or fission and quickly repair the damage.

Sponges are colonies composed of many individually self-sufficient animal cells. If a living sponge is put through a shredder, the cells can swim about separately, refind each other, and regroup themselves into a whole sponge again—a trick no vertebrate animal can perform.

Because of their simplicity, sponges are sometimes dismissed as a failed attempt to construct a complex animal, but in terms of species number, they are as successful as the mammals. On rocky areas of the seafloor, they often outnumber all other animals. In samples taken from hard bottoms off the Florida coast, sponges routinely make up by weight up to three-fourths of the living material found.

Sponges lack mobility and have no jaws or claws to defend themselves, but they have a subtle and sophisticated set of chemical weapons: toxic molecules that render them inedible to all but a few predators that have evolved to tolerate them. Many of these molecules, which are so complex that they remain beyond the ability of chemists to synthesize, show promise as potential cancer drugs.

More than 60 species of sponges reside in Florida's coastal waters; a few are shown in Figure 20–1. They are among the most colorful animals of the Florida coast. People may assign a "sponge color" of beige or gray to them, but some species are blue, brilliant yellow, orange, bright red, lavender, or even snow white. They come in many shapes. Chimneylike boring sponges rise above shoalweed beds near shore. Huge loggerhead sponges, taller than a man, sit farther out on the continental shelf. Sponges shaped like vases, barrels, bowls, stalks, and bunches of cigars cluster on the undersides of floating driftwood out at sea.

Some sponges are ingeniously adapted to live in special environments. The larval stage of the boring sponge bores tiny holes in the shells of other sea animals and creeps inside them, then grows large enough to deter predators. Along sheltered beaches where the wave action is gentle, a green or orange sponge the size of a fist may grow on a tiny hermit crab shell and then gradually dissolve the shell away. Finally, the crab is left shellless, forced to drag the sponge around wherever it goes. Eventually, the sponge becomes too heavy to carry and the crab either dies or moves out, but for a time, the sponge achieves greater mobility and access to more food than it could achieve on its own. (Perhaps this is what it means to sponge off someone?)

Many other sets of marine animals are unlike animals to the human eye. Bryozoans (see Figure 20–2) often resemble bushes or mosses (the prefix *bryo* means "moss"). Corals are often shaped like shrubs. That these and many other filter feeders look like plants is no coincidence, for they must solve the same mechanical problems that plants must solve. Plants must arrange their small light-gathering units (leaves) into patterns that accommodate as many units as possible without letting any one unit block access of the others to the resource, sunlight. Filter-feeding colonial animals must arrange their small food-gathering units (polyps and others) into patterns that permit access for all to the resource, plankton and detritus borne along in the water current. The optimal solu-

(continued on next page)

Bryozoans (bry-oh-ZOH-uns) are marine animals of two different phyla that form colonies shaped like trees, bushes, sheets, or coils of lace.

A yellow finger sponge. This sponge grows abundantly along the Gulf coast of Florida.

Fire sponge (*Tedania ignis*). Yellow boring sponges (*Siphonodictyon coralliphagum*) are growing on the fire sponge's surface.

Lavender tube sponge (*Spinosella vaginalis*). This sponge can grow to some three feet in height. The fringe on the tips of the sponge is a coating of colonial anemones (*Parazoanthus parasiticus*).

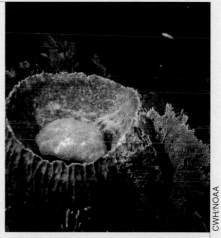

Vase sponge (*Ircinia campana*). A star coral (*Favia fragum*) is growing in the bowl of the sponge.

A basket sponge. The pores, in which myriad other animals live, are clearly visible.

FIGURE 20–1

Sponges Native to Florida Waters

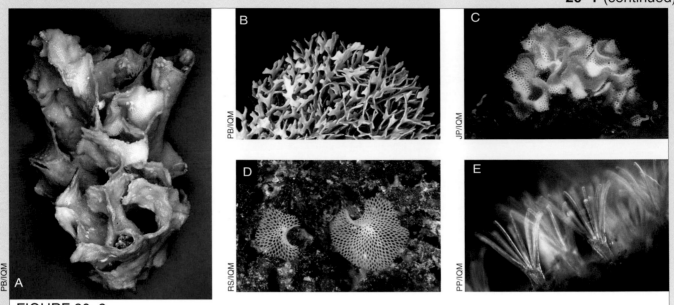

FIGURE 20–2

Ocean Animals: Bryozoans

A and B are the "skeletons" the animal colonies made before they died, showing how each tiny animal was attached to the others, forming a structure that allowed for access of every animal to the sea water. (A = *Hippomenella vellicata*; B = an *Adeonellopsis* species; C = a *Lodictyum* species). D is a live colony of lace bryozoans (a *Triphyllozoon* species). E is a colony of bryozoans in which each tiny animal is extending its tentacle-like organs to obtain oxygen and filter food from the water.

tion to this challenge, the tree or bush shape, has evolved independently in both situations.

Sponge communities occur below the low-tide line in both estuarine and marine waters of the Gulf from about 6 to 150 feet down. They are especially prolific on the west-coast continental shelf, where they are mounted by the thousands on the seafloor, on shell piles, and on limestone outcrops. Together with corals, bryozoans, and others, they form structures that support many other attached and mobile species. Some sponges specialize in boring into rocks and create hollows for themselves beneath limestone ledges. Later, fish and invertebrates use these caverns as habitats. Today, due to overharvesting, only a few sponges remain in some areas, but they have some potential for recovery.

A close look at a single sponge can give a sense of the complex, interacting community it can contain. This sponge (let us say) is three feet across and weighs thirty pounds when full of water. Although at first glance it appears to be simply an inert lump on the seafloor, it is actually host to a prodigious number of residents. It is perforated by a maze of hollows and tunnels, and an overwhelming number of other animals swarm all over it, in it, and under it, using its living tissues as holdfasts, homes, and even food. A sea whip is growing up through its middle. Plumes of hydroids and bushy bryozoans are attached to its outer surfaces. Algae, seagrasses, and corals are attached near the base. Several spiral worm shells and a marine snail have burrowed into its tissues, and several thousand snapping shrimp are hiding in all available spaces. A tiny green tunicate has spread colorful branches all over it, and masses of clear tunicates of several kinds have settled beneath it. Small anemones of two different species grow in the canals. A small, flattened crab with greatly modified claws scurries in and out of the crevices within the sponge and also travels from sponge to sponge. And there are still other residents: brittlestars, clams, and even fishes like gobies and blennies. In short, a single sponge can support a major marine community. Quiet as it looks, it is as busy as a city.

All of the residents possess special adaptations to life in their perforated host. Some have forms so modified that they cannot survive without its protection: a sponge-dwelling barnacle; a peculiar pale snapping shrimp that has lost all its pigment and spends its entire life cycle within the sponge's canals; and another tiny shrimp that is transparent. Greenish, warty nudibranchs chew out pits in their host and lay their eggs in frilly white ribbons. The hairy mud crab hunts for bristleworms and scaleworms in the cavities and crannies. Tiny pink mussels are embedded in the chambers, and there may be a host of polychaete worms—up to hundreds of thousands in one sponge.

For the most part, the sponge meets these creatures' needs exactly as they need it to. Because the sponge constantly feeds itself by pumping a stream of nourishing water through all its chambers, it also delivers oxygen, dissolved nutrients, detritus, plankton, and bacteria to its residents. Each tenant partakes in its own way. The barnacles have legs with hairlike structures that filter food particles from the water. The hairy brittlestar perches near the surface of the sponge and gently waves its furry arms, sweeping up bits of plankton and detritus and wiping them off into its mouth. Occasionally, the star takes a bite of its host to vary its diet.

Waste disposal is managed the same way. As the community members take up their oxygen and nutrients from the passing fluid, they release wastes that the constant stream of water sweeps away. The water also disperses the tenants' eggs and sperm, casting them into currents that sweep them out to sea. There, they can be fertilized, grow and mature in the plankton, and then find other sponges in which to live.

The sponge community changes constantly. Some members come, others go. Larvae from elsewhere enter and grow to maturity in the sponge, never again seeing the outside world. Others first mature outside and then crawl in: red cleaning shrimp, blennies, gobies, and tiny young toadfish among them. Fish come around, snap up whatever creatures venture forth, and move on.

For all these creatures to thrive, the sponge must thrive. It must continue pumping to nourish the colony and dispose of its wastes. But as the community grows in size, it requires a greater and greater influx of nutrients and oxygen. Like a sprawling city, it experiences waste-removal problems. Pockets of urban blight develop, and community members become stressed.

The aging sponge grows increasingly sensitive to changes in its environment. If temperature, salinity, or acidity changes too abruptly, its pumping action slows beyond the point of recovery. Food builds up, decays, and

A sponge, two feet across. This big animal serves as an apartment house for dozens of other animals. It was seen in shallow water on the sea-floor north of Cedar Key.

Native to Florida waters: Red cleaning shrimp (a *Lysmata* species) from near the shore of the Gulf of Mexico.

The **coelenterates** (se-LEN-te-rates) are a large phylum of marine animals with a radially arranged body plan.

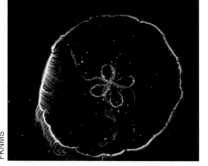

Native to Florida waters: Moon jellyfish (*Aurelia aurita*). This free-swimming medusa is famous for its painful sting. Common along Florida beaches, it can grow to more than a foot across.

produces toxic wastes. The sponge begins to turn into a mushy, rotten mess. Then its residents—worms, crabs, mussels, tiny anemones, snapping shrimp, and all—perish with their host. Or when the sponge is ripped from the bottom by a storm and hurled up on the beach by the frothing waves, then all this life comes to an end. But elsewhere, new sponges are just beginning to grow.

CORAL COMMUNITIES

Coelenterates are the animal phylum to which corals belong, and not a single member of that phylum occurs on land. Because they are the most important animals in the rest of this chapter, they are featured in Background Box 20–2.

BACKGROUND   **20–2**

More Un-Animals: Corals and Their Kin

Coelenterates are utterly strange to most people. They look more like plants than like "real" animals, because they are radially symmetrical, like many sponges. They have no legs, no eyes, no features that are recognizable as traits characteristic of land animals. They include jellyfish, hydroids, corals, and sea anemones among their numbers, and people may dismiss them as insignificant and primitive; in fact, the word jellyfish is almost synonymous with primitive. Yet the approximately 9,000 species within the coelenterate phylum represent 9,000 different ways to make a living using the simple and elegant radial body design of the group.

Coelenterates have neurons, basically the same as the brain cells in so-called "higher" animals. They also have muscle cells, which enable them to pulsate and move through the water. In addition, they possess deadly stinging cells, used for warfare against their own and other species. They are one of the few groups that can still hold the human superpredator at bay. When jellyfish move in on swimming beaches, people flee. All coelenterates have these stinging cells: microscopic harpoons, with the barb and line coiled up in a pouch. The slightest touch triggers the harpoon, sending it flying into the flesh of its prey. People can handle safely only those whose barbs fail to penetrate our skins.

Coelenterates do all this with a body composed of only two layers of living cells with a gelatinous layer between. They come in two basic models. One, the swimming jellyfish or medusa, has a mouth surrounded by stinging tentacles that hang down as the creature pulses its way through the water. The other, the sea anemone or polyp, sits attached to the bottom with the mouth and tentacles facing upward. Many species have life cycles that alternate between free-swimming jellyfish stages and attached polyps that look more like bushy plants than animals. In other species, thousands of tiny polyps interconnect to form colonies; it is these that build spectacular reefs in tropical waters. Each coral polyp grows within a tiny limestone cup, which is attached to the cups already there (see Figure 20–3).

Three types of coelenterates dominate reefs: anemones, soft corals, and stony corals. The stony corals are the ones that build reefs; the soft corals are the ones that sway gracefully in the water. Lists 20–2 and 20–3 identify some of each.

Clear water is vital to reef building because reef-building corals have symbiotic algae housed inside their cells, which conduct photosynthesis there. The algae use the carbon dioxide and other wastes given off by the corals. The corals use the oxygen freed by the algae and derive 90 percent of their energy from sugars that the algae produce. The algae also make calcium carbonate and help with sugar and protein metabolism. This back-and-forth trading recycles virtually all the resources involved and enables the duo to thrive in nutrient-poor water. In fact, nutrient-poor water is ideal for them, because other organisms can *not* thrive there.

Few other reef-building organisms can grow in such conditions. Oysters, mussels, and reef-building worms need to live in water that is enriched with nutrients and detritus, but coral reefs thrive in the most nutrient-poor waters in the world. There, they have all other conditions going for them: a tropical climate, shallow water, and rocky structures of their own making, on which they can grow. But given an excess of nutrients, nonsymbiotic algae overgrow the corals and prevent new coral larvae from settling on them.

Corals withdraw during the day and feed at night, waving their tentacles, stinging their prey with tiny poison darts, and ingesting them. They

(continued on next page)

Staghorn coral (*Acropora cervicornis*).

FIGURE 20–3

Coral Polyps

Cavernous star coral (*Montastrea cavernosa*).

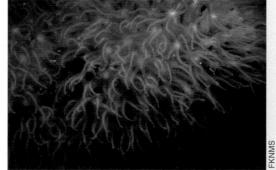

A soft coral.

Hundreds of animals are shown in these photographs. Each looks like a tiny anemone, seated in its own limestone cup, which is attached to the cups made by others.

The organisms that dwell within corals are known as the **zooxanthellae** (zoh-oh-zan-THELL-ee). They are like animals in being motile, but like plants in being able to photosynthesize. Here, they are classed as algae.

LIST 20–2
Stony corals on Florida reefs

Boulder coral
Brain corals
Branching coral
Elkhorn coral
Large-grooved brain corals
Lettuce-leaf corals
Rose coral
Staghorn coral
Star coral
Starlet corals
Thick finger coral
Thin finger coral
Tube coral
Yellow porous coral
. . . and others, 63 species and subspecies in all.

Sources: Jaap and Hallock 1990, 593–594, table 17–3; Kaplan 1982.

LIST 20–3
Soft corals and others on Florida reefs

Bushy soft corals
Deadman's fingers
Encrusting soft coral
Knobby candelabra corals
Sea blades
Sea fan
Sea feathers
Sea plume
Sea rods
Spiny candelabra corals
. . . and others, totalling 42 species and subspecies in all.

Source: Jaap and Hallock 1990, 592, table 17–1; Kaplan 1982.

defend themselves against suspended sediments, pollutants, even excess food, by secreting mucus, which the waves carry away.

Each coral grows best at a particular depth, determined by light availability and wave energy. Branching corals find optimal conditions at about 15 feet below the surface, brain and star corals at about 50 feet, and plate corals below them down to 100 feet or more. These depth preferences account for the zones seen on reefs.

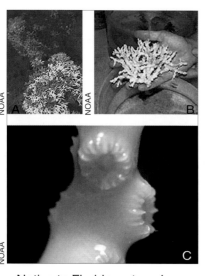

Native to Florida waters: Ivory Tree Coral (*Oculina varicosa*). A = the tree coral on site, deep on the continental shelf. B = a coral fragment. C = a closeup of the coral skeleton, showing the limestone cups in which the polyps grow.

Sponge communities and coral communities often grow together, with one or the other organism dominant. Corals (with some sponges) occupy many parts of the outer, and especially east-coast continental shelf. In some areas, ivory tree corals grow in thickets some four to five feet high. They can grow in areas as shallow as ten feet, but many stand on deeper limestone structures at the very edge of the continental shelf where they form solitary ridges and mounds in some places and extensive thickets in others. The latter, known as Oculina banks, cover many acres of the seafloor. The tiny, colonial animals that comprise the deepest corals grow slowly in their dark waters, only about a tenth of an inch a year. They depend on down-drifting phytoplankton to bring them energy from the surface.

Even in the depths at which they grow, the Oculina banks serve as hiding, feeding, breeding, and nursery grounds for crabs, mollusks, worms, anemones, small fish, shellfish, and large fish. A single coral colony may support hundreds of animals. Of decapods alone (crustaceans with ten appendages), researchers have identified 50 species living on these corals. Most of the community animals eat zooplankton and organic detritus. Some take nourishment from streams of mucus, which the corals constantly secrete to clear their crannies and crevices of suspended sediments and excess food particles.[1]

TROPICAL CORAL REEFS

Florida's famous tropical coral reefs are built of hundreds of species of animals. They vary from low, flat domes to long ridges. Multicolored, swaying sea fans and sea whips, rippling anemones, and darting schools of fish animate these structures. They are best developed outside, and parallel to, the Keys from eight to fifteen miles offshore, where the conditions they require are well met. The water is not only clear and low in nutrients but also warm, because the Keys provide a barrier against the relatively colder water of Florida Bay (Figure 20–4).

These shallow-water, tropical coral reefs are by far the most complex and extensive reef communities around Florida. More than 100 bank reefs stand outside of, and parallel to, the Keys and extend southwestward, to the Tortugas. Scattered among them are patch reefs, low, hollow domes of coral standing on the bottom.

Each bank reef is 300 feet long or more, with its top about three feet below the surface, and its seaward side about thirty feet down. Each is filled with old coral skeletons and rubble, the reef's own sediment, within

FIGURE 20–4

Currents across the Keys

The bank reefs lie parallel to the coast opposite Key Largo and from Big Pine Key to Key West. Because the Keys protect the reefs from the relatively colder water flowing out of the Everglades and Florida Bay, the reefs are best developed outside the longest Keys. Seaward of gaps between the Keys, there are fewer reefs.

Native to the Florida Keys reefs: A knobby purple sea rod.

which organisms can grow. Each has channels across it between the taller corals. These channels permit sediments carried by the current from inside the Keys to wash through the reefs without damaging them while cleansing them of metabolic wastes and delivering fresh oxygen and plankton. Figure 20–5 shows the structure of a bank reef and the corals that predominate at each level.

Scattered in the rubble around the bank reefs are smaller, patch reefs, surrounded by seagrasses. As they grow, patch reefs become roughly dome-shaped and about six feet high. Their stony corals close in, forming a canopy, the top flattens out due to wave action, and soft corals colonize the outside and undulate in the waves. Star corals and others grow on the margins.[2] Figure 20–6 depicts a typical patch reef.

Patch reefs are hollow inside, so they are a good place for fish and other animals to hide in. Around each dome, a zone of bare sand in the surrounding seagrass gives evidence that herbivores such as sea urchins and parrotfish often venture out to feed. The space inside the dome is at such a premium that it is time-shared. One set of fish and other animals lives in it by day and eases out at night to feed. The other set stays in by night and forages during the day.[3]

The diverse assemblages of corals on tropical reefs make stunning multicolored displays that resemble lush undersea gardens. Ocean swells surge back and forth across them and their leafy fronds and plumes sway to and fro.

Like natural old-growth forest communities and seagrass beds, Florida's coral reefs provide both shelter and food for hundreds of species of plants and animals. Shelter is ample because the reefs have a canopy (of branching corals), an understory (of other corals), and a substory (sediment). Food is abundant because the corals with their algae are, like trees on land or phytoplankton in water, primary producers of energy. Fish and other animals also bring energy into the community, having taken their nourishment from environments outside the reef. List 20–4 gives a sense of the diversity on the reefs.

Native to the Florida Keys reefs: Soft corals.

Native to the Florida Keys reefs: Pillar coral (*Dendrogyra cylindrus*).

At the foot of the reef on the landward side are rubble plains with coralline algae and mustard hill corals.

Partway down are isolated outcrops with lettuce coral and others.

Six to 20 feet down are elkhorn coral on tops, lettuce coral on sides.

FIGURE 20–5

Zones in a Bank Reef (continued opposite)

The clearer the water, the deeper corals can live. Under optimal conditions, reefs can grow down to about 100 feet and corals with zooxanthellae can survive down to about 150 feet.

Among plants on a reef, single-celled algae live not only in the stony corals but in sponges and other animals that are growing on sediments of dead reef material. Multicellular algae form colorful crusts that help to cement the reef together, or dwell within the sand in reef crevices, or coat inactive older coral skeletons. Large algae form tiny plates within their own tissues and then, on dying, add calcium fragments to the reef rubble. Seagrasses grow shoreward of, and marginally into, the reef base.

Among animals, sponges of all shapes and colors are numerous, each a community in itself as described earlier. Diverse starfish, shrimp, anemones, mollusks, plumed worms and bristle worms, crabs, spiny lobsters, and too many others to name inhabit the reefs.

Tropical fishes are the butterflies of the reef. They are as showy as river fish are drab: they bear brilliant, almost fluorescent stripes and spots, and

FIGURE 20–6

A Patch Reef

Patch reefs are hollow, so they offer sanctuary to resident animals. The populations within them differ from day to night and from season to season.

Source: Adapted from Voss 1988, 29, fig 7.

Shallow zone: golden sea mat, bladed fire coral, green sea mat, and false coral.

20+ feet down: star coral in big mounds.

Forereefs: Some spurs have additional growth on forereefs with corals, gorgocorals, and sponges.

Source: Adapted from Voss 1988, 29, fig 6.

Native to the Florida Keys reefs: Knobby candelabrum (*Eunicea mammosa*).

SFWMD

they swim in flashing schools among the corals. More than 150 species of tropical fish inhabit the Florida reef tract, twice as many as dazzle the eye, because, especially in the patch reefs, as many or more fish are concealed within hiding spaces in the limestone as those one can see. Figure 20–7 depicts a few of the tropical fishes on the reefs.

INTERACTIONS ON THE REEFS

With so many species of so many different groups, and with millions of years of evolution behind them, reef interactions are numerous and intense. Many organisms eat the coral mucus and in the process they clear the corals of debris. One crab induces the corals to mold themselves around its body, creating a custom-designed cave for protection. Then it uses its

Hamlet fish (a *Hypoplectrus* species).

Spotfin butterflyfish (*Chaetodon ocellatus*).

Spanish hogfish (*Bodianus rufus*).

Queen angelfish (*Holacanthus ciliaris*).

FIGURE 20–7

Tropical Fish in the Florida Keys Reefs

specialized legs to clear mucus, which it eats, from the coral. But not all relationships are beneficial to both participants; in many cases one exploits the other. For instance, a species of snail lives in pairs on sea whips, eating the polyps and leaving only enough so that the sea whips do not die. Some barnacles, too, digest corals to create spaces for themselves. Wormlike copepods live inside coral polyps. A damselfish destroys coral tissue, creating bare limestone skeletons on which algae can grow, then fiercely defends its territory like a land-starved farmer. The long-spined sea urchin grazes on reef algae, leaving cleaned limestone surfaces on which coral larvae can settle. A marine worm feeds on many coral species, as do many gastropod mollusks and fish: parrotfish, spadefish, damselfish, and butterflyfish. Bryozoans overgrow them, snails and worms prey on them. And after small animals have tunneled into the reef limestone, other animals may come to live in the spaces they have made. Crustaceans, mollusks, worms, crabs, and lobsters crawl through these spaces. Fish swim through them.[4]

Sponges have all kinds of relationships with corals. Some sponges bore into corals and weaken them, some bind living corals to the reef structure, and some protect the reef from other boring organisms. And mutual benefits are traded between corals and fish. The coral canopy offers shelter in which small fish can evade predator fishes that cruise around the reefs. Filter feeders and detritus-eating fish help to keep the water clear, enabling a maximum of sunlight to reach the coralline algae. The fishes' excreta drop to the rubble layer and feed polychaetes, bryozoans, and other detritus eaters there.

The corals themselves are predators. Some corals even attack and ingest each other. Fire corals colonize living soft coral branches. Stony corals extrude filaments that digest neighboring corals and then grow over them. Some corals defend themselves with specialized sweeper tentacles. Others employ chemical defenses and offensive weapons.[5]

The corals are susceptible to many disease-causing organisms. If winds and waves are calm and the weather is hot, bacteria may grow in the mucus and stress the corals. Black band disease is caused by several such bacterial species. Many more interactions among coral reef organisms remain to be discovered and understood.

THE REEFS IN CONTEXT

Like all Florida ecosystems, south Florida's tropical coral reefs are knit into the landscape and dependent on its natural functioning. Above all, they require protection from water that is rich in nutrients, because nutrients enable coral competitors to grow. The lay of the land naturally affords that protection, beginning at the Kissimmee River, where the rain that falls on the Florida peninsula begins its long journey southward. All the way south, the land's natural ecosystems help to keep the moving water clean and pure: the Everglades marsh, coastal mangroves, and seagrass beds all intercept and trap pollution. Ecosystem manager Maureen Eldredge did not exaggerate when she said, "All the ecosystems of south Florida are linked, from the Kissimmee River to the coral reefs."[6] They are linked in

Native to Florida waters: Yellowtail snapper (*Ocyurus chrysurus*) with seafan in the Florida Keys reefs.

FIGURE 20-8

Spiny lobster parade.

Caribbean Spiny Lobster (*Panulirus argus*)

The spiny lobster needs many intact and connected marine ecosystems to complete its life cycle. It lives on coral reefs and spawns off the reefs in deeper water. Its larvae develop in the plankton for many months, then settle among seagrasses. They grow up in live-bottom or mangrove communities, then move to patch reefs, and later as adults move to the margins of bank reefs. Adults conceal themselves in reef dens by day and move out to seagrass beds and sediment bottoms by night to feed.

The single-file formation protects every lobster (except the last one on line) from attack at the tail end, which is vulnerable. The parade takes place every fall, when newly mature spiny lobsters proceed from their homes among the corals to deeper water to mate. Then in late winter they line up and return home. They navigate using a sophisticated magnetic compass sense oriented to the earth's magnetic field.

Sources: Lyons and coauthors 1981; Jaap and Hallock 1990; Bennington 2003.

other ways, too. Mangrove swamps serve as nurseries for young fish and other animals that later will live around the reefs. Swamps and seagrass beds provide food for reef fishes and lobsters that emerge from the reefs to forage. Seagrasses offer protective cover for fish and other animals that move back and forth between the mangroves and the reefs. Some organisms use many adjacent ecosystems, including the reefs, as their habitat; the spiny lobster is an example (Figure 20–8).

As if in return for all these favors, the coral reefs protect the Keys, the surrounding seagrass beds, and the coastal mangroves from storm-caused erosion. The rubble from natural breakdown of the reefs washes up on beaches and helps replace other sand that has been washed away. Finally, all these systems protect the interior.

Given favorable environmental conditions, reef health and diversity can be outstanding. The reefs then serve not only as a sanctuary for fish and other animals, but as a self-supporting community in a challenging environment.

Beyond the mangroves, the seagrasses, and the coral reefs, beyond the edge of the continental shelf, lies the open sea, deep and mysterious. This realm, too, is part of Florida's province, as the next chapter reveals.

CHAPTER TWENTY-ONE

THE GULF AND THE OCEAN

Florida's connections with the Gulf of Mexico and the Atlantic Ocean extend beyond the state's borders, indeed, beyond the edge of the continental shelf. Many migratory species, including whales, sharks, marine turtles, and eels, spend parts of their lives in Florida or in our near-shore waters, and parts in the far ocean, thousands of miles away. But in truth, even without these connections, it would seem right to end this series of ecosystems with the oceans. If the Gulf of Mexico and the Atlantic Ocean are not within Florida's province, Florida is within theirs, and the wonders they contain draw us on to explore them before concluding this book on Florida ecosystems.

The oceans cover nearly three-quarters of the earth's surface and hold more than 100 times more space for living things than do the continents (Figure 21–1). They appear on the surface to be one huge body of water, but they have as many different habitats as the land. They have warm sunny surface waters, cold dark deep waters, and currents swirling past each other in different directions both horizontally and vertically. On the bottom, they have deep briny lakes, high mountain ridges, hot thermal vents, tall rock chimneys, and level plateaus. In deep ocean basins, cold water from both poles, which is denser than warm water, flows toward the equator along the abyssal seafloors. On the surface, major currents carry warm water from the equator towards the poles. From these, other currents spin off and move in huge, slow circles.

The oceans are up to seven miles deep, and they are inhabited from top to bottom by all manner of living things: plankton, algae, other plants, and every kind of animal—some swimming or drifting or floating, others attached to, or roaming on, the bottom. The diversity of life in the oceans may be even greater than on land—no one yet knows. Certainly the diversity of animal phyla is immense. Scientists estimate that the deep seas teem with as many as ten million species, most of them on the bottom and still undiscovered, but even explorers who never dip below the surface find great mobs of animals swimming and drifting in the waters off Florida.[1]

Of marine plant and animal populations, some resist the ocean's roaming currents and stay permanently within certain boundaries. Many species reside all their lives over the continental shelf and never stray beyond

Native to Florida waters: Bottlenose dolphin (*Tursiops truncatus*). Known for their playfulness, these animals frolic and hunt fish in near-shore waters along both coasts of Florida.

Reminder: Taxonomic diversity among plants is greatest on land, but taxonomic diversity among animals is much greater in the ocean, which is inhabited by many phyla of animals not found on land.

OPPOSITE: Sunset over the Gulf of Mexico, Grayton Beach, Walton County.

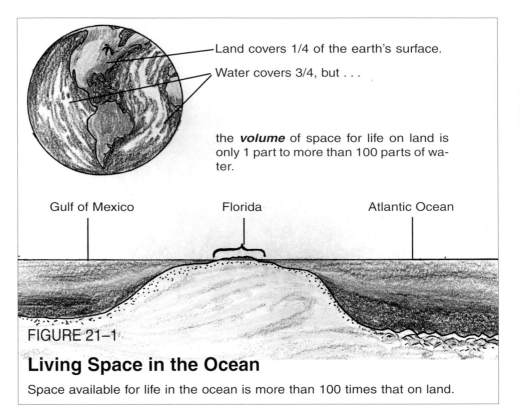

Land covers 1/4 of the earth's surface.

Water covers 3/4, but . . .

the **volume** of space for life on land is only 1 part to more than 100 parts of water.

Gulf of Mexico Florida Atlantic Ocean

FIGURE 21–1

Living Space in the Ocean

Space available for life in the ocean is more than 100 times that on land.

its edge. Many others stay in the deep ocean and never visit the continental shelf. There is even a well-defined, narrow zone directly over the continental slope, where certain populations, not found elsewhere, are concentrated. Presumably, this zone is different from other ocean habitats because cold water, rich in mineral nutrients, wells up continuously along the slope boundary.[2]

Some populations ride ocean currents, shifting locations daily, moving from inshore to offshore and back or from surface to deep water and back. Other populations shift with the seasons, just as do migratory birds and butterflies. (Chapter 18 described some of the animals that move in and out of Florida's estuaries daily, or seasonally, or during different parts of their life cycles.) Figure 21–2 provides a few examples of animals that occupy different spaces in the oceans.

Lanternfish (a *Diaphus* species). Some 200 species of these fish, which can make their own light, swim in the ocean's depths.

Yellowfin tuna (*Thunnus albacares*). Schools of yellowfin pursue their prey throughout the Gulf and Gulf Stream and in turn are preyed upon by human hunters.

Roughback batfish (*Ogcocephalus parvus*). This flattened fish "walks" on the bottom with its modified fins and attracts its prey by displaying a bright red color around its mouth.

FIGURE 21–2

Ocean Animals around Florida (continued opposite)

Because the mosaic of marine environments around Florida is so complex and knowledge about them is so lacking, this chapter samples just a few parts of the ocean with which Florida has a special relationship. Of particular interest are the Gulf of Mexico, the Gulf Stream that runs north along the Atlantic Coast, and the Sargasso Sea.

THE GULF OF MEXICO

North of the Caribbean Sea, the Gulf of Mexico lies within the arc embraced by Florida on the east and Mexico's Yucatan peninsula on the south. Cuba lies across the opening to the Atlantic Ocean and the Caribbean. The Gulf is about 1,000 miles across from east to west, and somewhat oval shaped, with an area of about 600,000 square miles. It holds a huge mass of water, 800,000 cubic miles in volume.[3] Its floor displays the complex terrain shown in Figure 21–3.

Except for game fish and conspicuous animals, living things in the Gulf of Mexico, as in most of the world's seas, are mostly unstudied and even undiscovered, but the number of invertebrate species in the Gulf probably exceeds 10,000.[4] Estimates are shown in List 21–1.

Gulf Swimmers. The Gulf's pelagic fish are much better characterized than the invertebrates. They include some of the world's largest fish species: billfish (marlin, swordfish, sailfish), flying fish, offshore game fish (cobia, dolphin, king mackerel, wahoo), jacks, and mackerels (including Atlantic bonito and tuna), as well as many inconspicuous fish that swim in the water column and inhabit floating lines of seaweed. Figure 21–4 displays three of these spectacular fish.

Many marine mammals are present in the Gulf, too, more than most people realize. Dolphins and manatees are well-known and well-loved visitors along the shores and in the mouths of rivers. However, whales around Florida are a relatively recent surprise, even though they were here long before we humans ever came. Surveys beginning in the early 1990s indicate that sperm whales, once harpooned to the edge of extinction, are now, perhaps, on their way back along both coasts of Florida. Killer whales also occur off both coasts; they roam in pods over the deep waters. List 21–2 names these and other marine mammals known to swim in the Gulf, for a surprising total of 30 native marine mammals altogether.

LIST 21–1
Invertebrates in the Gulf of Mexico (numbers of species)

Mollusks	450–500
Crustaceans	1,500
Polychaetes	600
Oligochaetes	200+
Echinoderms	400
Trematodes	200
Other worms	200
Cnidarians	600
Sponges	100
Other	5–10,000

Note: These numbers include zooplankton species. Among them are flagellates, ciliates, amoebas (forams and radiolarians), sporozoans, opalinates, and many more.

Source: Gore 1992, 139–140.

The **pelagic** ocean is the deep, open ocean, not over the continental shelf. A **pelagic fish** is one that swims in the open ocean.

Balloonfish (*Diodon holocanthus*). This fish inflates to swim and deflates to rest. It is shown here, resting on the floor of the Atlantic Ocean off the Florida Keys.

Golden crab (*Chaceon fenneri*). Golden crabs are limited to the continental slope off Florida and elsewhere and are the largest crustaceans found there.

Lobed ctenophore (*Bolinopsis infundibulum*). These animals, which roam all over the ocean, are translucent and give off a bioluminescent glow.

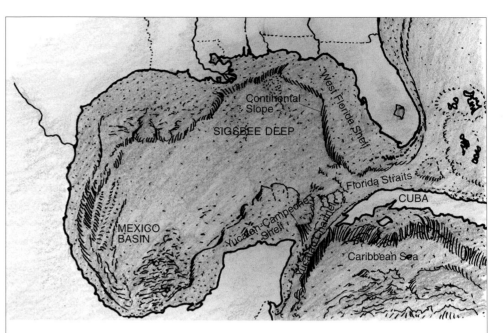

FIGURE 21–3

The Gulf Floor

The continental shelf breaks abruptly downward at the continental slope, which plunges from its edge, some 600 or more feet down, to its base, 10,200 feet down. The continental slope varies in steepness and is notched at intervals by valleys.

At the base of the slope lies a plain that holds two deep areas: the Mexico Basin and the Sigsbee Deep. The Mexico Basin holds scattered flat-topped mesas and ridges. The Sigsbee Deep supports hummocks with deep-sea corals and coral-based communities.

Elsewhere, off northwest Florida, cold water spews from major springs and seeps. Cold fresh water pours forth from the deep seafloor and cold salt water seeps from the base of the slope.

Source: Gore 1992, end pages.

Animals on the Gulf Floor. Explorers who plumb the depths of the Gulf of Mexico make strange and wonderful discoveries. In 1996, two researchers returned from a trip to the seafloor with descriptions of a salt dome–associated ecosystem never before known. Salt domes are thought to have begun forming some 250 million years ago as the land that is now the Gulf floor first stretched apart, thinned out, and sank down, so that sea water began to run into it. The Gulf Basin was shallow at first, and for about 100 million years, salty water ran in, dried up, and accumulated again, creating massive layers of salt. These salt layers compacted and sediment running off the continent covered them with a heavy overburden (which geologists call caprock). As the basin floor sank down further and the Gulf grew deeper, water dissolved the salt out from beneath some of the domes and oil seeped in. And in some domes, the salt became liquefied and pushed up through faults in the caprock, forming mushroom-shaped columns and domes. There are 500 salt domes in the Gulf, some up to two miles across.[5]

The explorers found an extensive ecosystem surrounding the dome they visited. It is many miles across and probably at least centuries, if not mil-

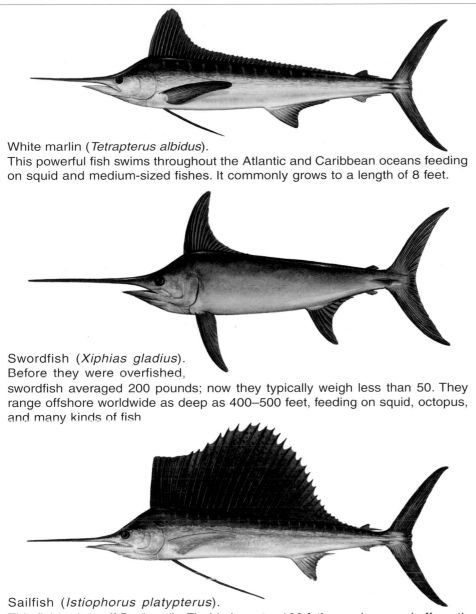

White marlin (*Tetrapterus albidus*).
This powerful fish swims throughout the Atlantic and Caribbean oceans feeding on squid and medium-sized fishes. It commonly grows to a length of 8 feet.

Swordfish (*Xiphias gladius*).
Before they were overfished,
swordfish averaged 200 pounds; now they typically weigh less than 50. They range offshore worldwide as deep as 400–500 feet, feeding on squid, octopus, and many kinds of fish

Sailfish (*Istiophorus platypterus*).
This fish swims off Panhandle Florida in water 100 fathoms deep, and off south Florida near the Gulf Stream. It commonly grows to 7 feet in length and can swim at 50 knots. It feeds on the surface or at middepths on smaller fishes and squid.

FIGURE 21–4

Big Fish in the Pelagic Ocean

Source: Anon. c.1994 (*Fishing Lines*); drawings by Diane Peebles.

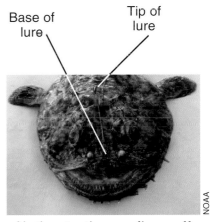

Native to the seafloor off Florida: Humpback anglerfish (*Melanocetus johnsoni*). This deep-sea fish attracts its prey with a lure that is mounted on a stalk between its eyes. Bacteria living within the bulb at the tip of the lure are biolumi-nescent, so the lure glows in the dark ocean.

The bumps below the eyes are the male fish's enlarged chemical-sensing organs. He uses his chemical sense to find the female, who releases hormones into the water to aid the search.

lennia, old. Mussels, living almost entirely without oxygen, cluster wherever icy methane gas and oil are seeping out. Encircling the mussel beds, some fifteen miles out from the center, is a ring of tube worms, apparently a new species of polychaete worm, which the explorers describe as "flat, pinkish, centipedelike creatures . . . one to two inches long."[6]

Around the dome are "pockmarks" in the seafloor holding extremely salty lakes of brine where several species of shelled invertebrates, including clams and snails, live in enormous colonies. They appear to depend on bacteria to meet their energy and nutrient needs. The bacteria trans-

IM/OAR/NURP

A hydrocarbon seep on the deep floor of the Gulf of Mexico. Shown are deep-sea mussels, worms, and a spider crab.

form sulfur compounds available in the brine into hydrogen sulfide, and this serves as an energy source for the tube worms. The worms exist in colonies hundreds of square miles in area, at depths below 1,200 feet. The bacteria, mussels, and tube worms clearly form part of a food web: other animals feed on them. Lobsters prey on the mussels, snails eat the bacteria, and eels eat everything they can catch. The rest of the food web is unknown but may be extensive and important.[7]

Amazingly, fish also live more than a mile down on the Gulf floor. Among them are anglerfishes, brotulids, and many other grotesque, rarely seen species. Eels from the land's freshwater systems pass across this terrain en route to a rendezvous site in the Atlantic Ocean. Their story is told later in this chapter.

ATLANTIC CURRENTS: FISH, SHARKS, AND WHALES

One of the currents of interest in the Atlantic is the Gulf Stream, which flows out of the Gulf of Mexico around the tip of Florida and up the eastern seaboard, turning east when it arrives at Cape Cod and crossing the Atlantic to bathe the British Isles and environs with its warm water. Many ocean animals ride north along the Atlantic seacoast on that current. Then, to return south again, they instinctively swim against that same current. Fish, whales, sharks, and others make great migrations along the U.S. east coast, visiting Florida's estuaries and salt marshes along the way. These journeys resemble the long migratory flights of birds and butterflies, but because they take place out of our view, they are less well known.

Tuna, swordfish, and marlin migrate thousands of miles in the course of their lives. They travel with the Gulf Stream to cool northern waters to feed in summer, taking advantage of huge food stocks that develop there each year during the short but intense growing season. (Cold water contains more gases, notably oxygen and carbon dioxide, than does warm water. The oxygen allows faster respiration and the carbon dioxide allows more photosynthesis, hence the great productivity despite colder temperatures.) Then they return to warmer equatorial waters as winter comes.

Bull sharks migrate more than a thousand miles to find the resources they need to complete their life cycles. In winter, they move up the St. Lucie River towards Lake Okeechobee in Florida, and in summer, they travel as far north as Chesapeake Bay (see Figure 21–5). Among their prey are young sandbar sharks.

To escape their foes, sandbar sharks travel farther, both north and south. The young are born in estuarine tidal marshes around Delaware Bay and New York harbor; then as they grow, they migrate to warm, low-salinity estuaries of the Gulf of Mexico. When sexually mature, they travel several thousand miles back to their northern tidal-marsh pupping grounds to bear their young.

Young sandbar sharks grow slowly, as slowly as an inch a year, but they may live for more than 50 years. One sandbar male, tagged and released in Delaware Bay in 1965, was recaptured off Destin, Florida, in 1989. It had grown only 24 inches in 24 years.

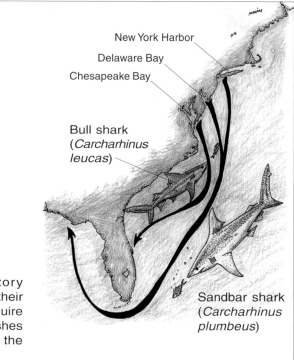

FIGURE 21–5

Migrations of Two Sharks

To complete these migratory routes, which are integral to their life cycles, these animals require naturally functioning salt marshes as stopping places all along the way.

New York Harbor
Delaware Bay
Chesapeake Bay

Bull shark (*Carcharhinus leucas*)

Sandbar shark (*Carcharhinus plumbeus*)

Whales, too, migrate, and for northern right whales, near-shore waters off the east coast of Florida are a critical habitat. Northern right whales are the most depleted species of large whale in the world, and the only one that is in danger of becoming extinct in the near future. Due to hunting pressures, right whales now number only 600 or fewer individuals, as compared with more than 10,000 for the humpback whale. Only recently have marine biologists learned that these right whales, seen in summer off the coasts of Maine and Nova Scotia, spend their winters off the coast of Florida. It is here that they give birth to their calves, specifically in a narrow strip of water very near the coast, extending from Florida's Sebastian Inlet to the Altamaha River halfway up the Georgia Coast. Off Georgia, this strip is 15 miles wide; off Jacksonville, the calving ground is only 5 miles wide. More right-whale sightings occur along the Florida coast than anywhere else in the world's oceans.[8]

Other animals traveling to and from Florida swim all the way around the Atlantic Ocean, or to and from the Sargasso Sea. As background for these travels, Box 21–1 introduces the Sargasso Sea.

Native to Florida waters: Northern right whale (*Eubalaena glacialis*). The flukes are raised high to power the huge animal's dive. The whale weighs an average of 60 tons when mature.

BACKGROUND

The Sargasso Sea

21–1

Imagine that a ship is sailing to Africa from any port on Florida's east coast. First it travels about 500 miles across open water; then it encounters a vast meadow of floating seaweed. It travels nearly 1,500 miles across this meadow before reaching open water again. Finally, after 1,000 miles across open water, the ship reaches Africa's west coast. Figure 21–6 shows that the seaweed meadow occupies a huge oval, 2 million square miles in area, as big as the Caribbean and Mediterranean seas combined. It is named for Sargassum, a group of eight species of float-

(continued on next page)

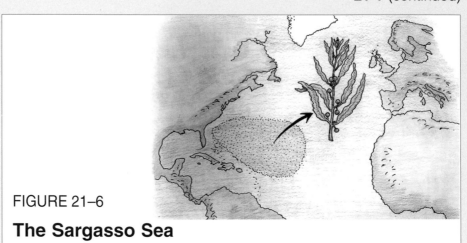

FIGURE 21–6

The Sargasso Sea

ing plants that circle endlessly in a clockwise loop around the Atlantic Ocean, drawn along on the west by the Gulf Stream, and on the north, east, and south by three other transatlantic streams.* The sargassum weeds may have originated long ago on a continental coast but now are a self-perpetuating community of floating plants that reproduce vegetatively in the ocean.

Several factors combine to make the Sargasso Sea unlike any other body of water in the ocean. It is said to be "the cleanest, purest, and biologically poorest ocean water ever studied." Within the Sargasso Sea, the water is warmer and therefore lighter—less dense—than in the surrounding ocean. This water forms a lens that floats on the salt water: its bottom is 3,000 feet deep and the top is three feet higher than sea level. This warm-water lens is bounded on all sides and below by cold, nutrient-rich water that moves around and under it, but doesn't migrate into it. As a result, the Sargasso Sea's water is nutrient poor—and because evaporation exceeds rainfall over this area of the ocean, it is also more salty than the surrounding water.

This unique environment supports some species of plants and animals that find in it the perfect habitat. Some live there permanently. Others seek it out for parts of their lives. Diatoms mass on the sargassum weeds. Sea turtles and eels from Florida's coastal sands and interior waters spend important parts of their lives there.

Note: *The other three transatlantic streams are the North Atlantic Current, the Canary Current, and the North Equatorial Drift Current.

Source: Worthington 1980.

LAND TO SEA MIGRATIONS: SEA TURTLES

A heavy female loggerhead turtle, laden with eggs, lumbers out of the ocean and up the beach. She scoops out a nest in the sand, drops her eggs into it, and covers them up, then swims away again. Some 60 days later, 100 little hatchlings emerge and scramble for the open ocean.

When ready, the hatchlings all emerge, practically simultaneously. It may be that they hear each other stirring about when hatching time ap-

Native to Florida waters: Almaco jack (*Seriola rivoliana*). The jack ranges widely in offshore waters and spawns offshore as well. A relatively small fish for the open ocean, it reaches a weight of about 20 pounds at maturity. Its young spend their early lives in sargassum.

Native to Florida waters: Dolphin (*Coryphaena hippurus*). Not to be confused with the ocean mammal shown earlier, this fish swims in warm ocean currents offshore and feeds on flying fish and squid. It reaches an adult weight of 30 pounds, and can attain a speed of 50 knots. Dolphin juveniles spend their young lives floating among drifting sargassum weeds.

Source: Anon. c. 1994 (Fishing Lines); drawings by Diane Peebles.

proaches, and the commotion within each egg arouses nearby babies to get busy, too. They benefit from each other's activity: when the eggs start thrashing about, the sand around them slips down as the turtles dig their way up, and the hatchlings easily escape the nest. Then they all scurry frantically toward the sea in a mob.

Another behavior favors their survival. The hatchlings nearly always emerge at night, rather than by day. Perhaps the heat of the sun keeps them quiet until nightfall. When they finally make their mad dash to the sea, their sudden appearance and hasty run for the water in the dark maximizes the chance that at least some will escape predation.

Once out at sea, the young turtles pick up ocean currents and swim for several years around the Sargasso Sea as shown in Figure 21–7, feeding and growing to maturity. To rest between feedings, they sleep, floating on the waters with their flippers tucked over their backs.

How the turtles find their way to and from their nesting grounds is a riddle that remains unsolved. "As soon as they hatch," a biologist muses, "sea turtles swim across hundreds of miles of featureless ocean. As adults, they navigate home to nest. How do newly hatched sea turtles find their way?"[9] He has sought to find out by watching and tracking turtle hatchlings on Florida's east coast with these questions in mind: Do they use the positions of the sun or stars? polarized light? odors? wind direction? the sound of waves breaking on the beach? the earth's magnetic field? Other migratory animals are known to do some or all of these things.

It is known that when newly hatched sea turtles first break through the surface of the sand, they are guided by light from the sky that is reflecting off the surface of the ocean. Under natural conditions, this reflected light

Visitor to Florida's shores: Loggerhead turtle (*Caretta caretta*). This loggerhead was photographed digging her nest and laying her eggs in Bill Baggs Cape Florida State Park, Miami-Dade County.

Sea turtle eggs.

Born on a Florida beach: Baby loggerhead turtle (*Caretta caretta*). New hatchlings head straight for the surf and swim out into the open ocean.

FIGURE 21–7

Loggerhead Turtles' Migratory Routes

Source: Adapted from Lohmann 1992.

is brighter on the water than on land, and the hatchlings hasten toward it in a mob and plunge into the waves. Then the waves guide them: they dive for the undertow and swim for the open ocean. (This is known because, during rare weather events such as hurricanes when the waves flow away from shore, the hatchlings swim the wrong way, into the waves and towards the land.) Lacking the guidance of light, wind, and waves, though, they still can orient using a magnetic sense that enables them to perceive the earth's magnetic field. (Under testing conditions, when the magnetic field is reversed, they reverse direction.)

When they are ready to mate, the turtles return to the beaches where they were born, or to the near vicinity. There, they mate offshore. Again there is a puzzle: How do they find their way back? Part of the answer may be that they use a chemical sense to orient to the chemistry of river waters that flavor the beaches they were born on. Probably, too, they follow magnetic "stripes" on the ocean floor. However they do it, these ancient creatures are highly sophisticated navigators.

Five species of sea turtles occur in U.S. coastal waters today and all five visit Florida's shores. They are shown and described in Figure 21–8.

FRESHWATER TO SALTWATER MIGRATIONS: EELS

Eels are not snakes, as many people believe; they are fish. Snakes are air-breathing reptiles related to lizards and turtles, and they use lungs to breathe. Eels, like all fish, have gills for obtaining oxygen under water. American eels are freshwater fish; other eels, such as the moray eel, are marine. In direct contrast to the sea turtles just described, American eels spawn at sea and return to coastal and inland habitats to mature. They are the only catadromous fish in the Americas.

Few people have any idea how many eels live in Florida's rivers, lakes, and underwater caves. They may outnumber all other fish. Elsewhere on the U.S. east coast, they populate some streams at 1,500 eels per acre.[10]

Eels remain segregated by gender throughout their adult lives. The females live in inland ponds, lakes, streams, and caves; the males in estuaries and coastal marshes. Both sexes mature at some ten to thirty years of age, and then undergo changes that enable them to navigate in dark water and to withstand marine salinity. They develop enlarged eyes that enable them to see deep in the ocean and special cells that rid their bodies of excess salts. Their digestive tracts shrink, their swim bladders toughen, and their skulls change shape. Their slime coating thickens. Previously mud-colored, they turn bronzy along their undersides and purple-black on top, camouflaged for ocean life. Now the females, ready for marine life, come down the rivers and join the males in coastal marshes.

Then, for reasons no one can explain, these fish from eastern North America, as well as their cousins (a different species) from Europe and the Mediterranean Sea, swim across the continental shelf, drop off its edge, and head for the open ocean. They congregate in the Sargasso Sea to mate, spawn and die, leaving behind hundreds of millions of tiny eel larvae.

Reminder: Of the western-hemisphere fish that travel back and forth between fresh and marine waters, all except the eel are **anadromous**: they mature in salt water and return to brackish or fresh water to spawn.

The American eel is a **catadromous** fish: it spends most of its life in fresh water and returns to the sea to spawn.

Native to Florida waters: Spotted moray eel (*Gymnothorax moringa*). This is one of many species of strictly marine eels.

Green Turtle (*Chelonia mydas*). Called green because their body fat is greenish, these turtles were swimming in the Caribbean by the millions when Columbus arrived. Adults weigh up to 300 pounds and are the only sea turtles that live exclusively on plants. They feed on turtlegrass, hence the name. Young turtles swim in coastal waters all along the Americas, and adults nest mostly along the Central American coast.

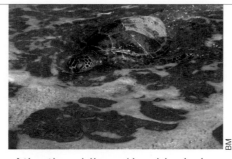

Atlantic ridley (*Lepidochelys kempii*). This is the world's rarest sea turtle, and weighs at most 100 pounds when mature. It feeds on the shellfish, fish, and jellyfish it finds in seagrass beds and estuaries along the Big Bend coast. Along U.S. shores it ranges from Texas to Maine, but it nests almost solely on a 20-mile stretch of beach in the western Gulf of Mexico.

Hawksbill (*Eretmochelys imprecata*). This turtle has a beautiful brown shell decorated with gold flecks. It is an agile, small animal that lives on coral reefs and can climb over rocks to nest on beaches that other turtles cannot reach. It feeds on sponges and probably nests in Florida Bay.

Leatherback (*Dermochelys coriacea*). This is the largest living sea turtle in the world. It may grow to 8 feet long and weigh 1500 pounds. Leatherbacks travel thousands of miles, dive thousands of feet deep, and swim in cold water as far north as Alaska. Their favorite foods are jellyfishes, especially the Portuguese man o' war.

Loggerhead (*Caretta caretta*). So named because its head may be up to ten inches wide, the loggerhead feeds around coral reefs and rocks. It uses its powerful jaws to crush heavy-shelled clams and crabs, including horseshoe crabs, for its food. It weighs around 300 to 400 pounds and is the most common sea turtle in Florida, with about 50,000 nests recorded annually. It is threatened.

FIGURE 21–8

Sea Turtles that Nest on Florida's Shores

The loggerhead turtle is a threatened species. The other four turtles are endangered.

The larvae are neither male nor female, and don't even look like animals. They look like transparent willow leaves, and only their tiny black eyes identify them as living things. For some 15 months, until spring a year after hatching, the larvae hide and feed in the sargassum weeds. Then they swim with ocean currents toward the shore. How they sort themselves out remains a mystery, but the American eels return to America and the European eels to Europe.

When they reach the coast, the young eels continue growing until they are 3 1/2 inches long; now they are called elvers. Finally, some become female and swim up the rivers, while some become male and stay in tidal areas. They become "yellow eels," yellow-green on top, pale beneath. Young ones feed at the mouths of caves and other dusky openings. Older ones develop better night vision and feed only at night.

Archaeologist Michael Wisenbaker marvels at the eels' ability to travel over land: "Eels scale dams, wriggle through aqueducts and navigate sub-

MS

Native to Florida waters: American eel (*Anguilla rostrata*). Thousands of eels inhabit Florida's underwater caves. These were photographed in Morrison Spring In Walton County.

terranean streams to reach their destinations. Sometimes they even crawl over dew-drenched fields—winding up in lakes with no links to the sea." There they may remain for 10, 20, or even 40 years before they return to the ocean to spawn.[11]

FLORIDA AND THE OCEAN

The migrations of fish, sharks, whales, sea turtles, and eels link Florida to the ocean and the rest of the world and demonstrate again that, as in the terrestrial realm, more than one ecosystem is necessary to ensure the survival of wide-ranging creatures. But besides being a hub for marine migrations, Florida plays another, even more crucial, part in supporting life on this planet. It has to do with the health of the ocean itself.

The ocean modulates the climate, absorbing heat in summer and releasing it in winter, absorbing heat in the tropics and releasing it at the poles. The ocean also stores carbon, 20 times more carbon than all of the world's forests and other green plants combined, and this trapping of carbon helps prevent excessive greenhouse warming of the planet.[12]

It is the oceans' phytoplankton that accomplish this feat. They produce a third to a half of the planet's oxygen by way of photosynthesis. In the process, they convert atmospheric carbon dioxide, together with water, into sugars. Some 90 percent of the carbon is recycled through marine food webs to the atmosphere, but 10 percent drops to the ocean floor and does not return to surface currents for a thousand years.[13]

The ocean begins at the shore, and the shore is the edge of the land. How we manage Florida's terrestrial ecosystems, then, affects all of the life in the ocean and on the planet. And in case any doubt remains about the unique and irreplaceable nature of Florida's ecosystems, the next chapter traces their extraordinary history.

PART SEVEN

THE PAST AND THE FUTURE

Florida's rich heritage of natural ecosystems and native species has a history millions of years long, nearly all of it prior to human influence. Today, what becomes of them depends entirely on us.

Mistletoe Cactus (*Rhipsalis baccifera*). A single wild population of this plant grows in the Everglades.

CHAPTER TWENTY-TWO

THE PATH TO THE PRESENT

Florida today has thousands of species of plants, animals, and other living things. How did they all get here? What did it take to produce the astounding diversity of species and living communities that are native to this land? The answer is billions of years of time in which life arose, diversified, and came to occupy the air, land, and water of this region. The great explosion-into-being of the universe itself is thought to have taken place about 14 or so billion years ago. Since then, just to touch on the most momentous of the great events that have led to the present (and to give a very approximate sense of the times involved):

Florida native: Burrmarigold (*Bidens laevis*). This flower grows in freshwater marshes over much of mainland Florida.

14+ billion years ago:	The Universe came into being.
4.5 billion years ago:	Earth was born.
3.8 billion years ago:	Self-replicating molecules had come to exist.
1.3 billion years ago:	The first true cells had arisen.
550 million years ago:	Many-celled plants and animals were present in the ocean and Florida was taking shape beneath the ocean.[1]

Then for more hundreds of millions of years, life evolved in the ocean and on land, while Florida was being formed beneath the sea.

As a body of land above water, Florida came into being between 35 and 25 million years ago, but its hundreds of millions of years beneath the ocean before that are significant. It was during all those years that Florida's massive layers of limestone were laid down, and limestone gives Florida's ecosystems much of the character that they possess today. The time at which Florida emerged is also significant, because it was after the dinosaurs had gone extinct and flowering plants, insects, and mammals were becoming dominant on the land. Figure 22—1 briefly traces Florida's history beneath the sea, and this narrative picks up at 65 million years ago, at a time of dramatic change in Earth's history.

Historians of Earth's long time periods use **mya** to refer to "million years ago." Once the times involved become numbered in the thousands of years, they use **ybp**, meaning "years before the present."

OPPOSITE: Flowering dogwood (*Cornus florida*). Flowering plants were already present in North America before Florida surfaced from under the sea. At the first opportunity, they swarmed all over Florida.

At one time in the past, all of Earth's continents were a single land mass.

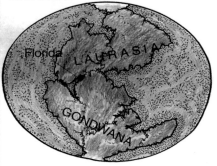

550 mya or before: Florida lies deep in the rift at the corners of several huge land masses, together known as the supercontinent Pangaea (pan-JEE-uh). Later, these land masses will become today's continents.*

FIGURE 22–1

The Geological Origins of Florida

Note: *Many lines of evidence testify to the origin of Florida's bedrock as part of what is now Africa. Fossils in Florida's and Africa's limestones from that time are the same. The slant of the magnetic field in materials laid down at that time is 49 degrees in Florida as in Africa, whereas the slant of North American magnetic fields from the same time is 28 degrees. The rocky sediments deep under Florida's surface materials are of the same age as those of the African mountains. Younger sediments from the Appalachian mountains lie above them.

Then the continents split apart . . .

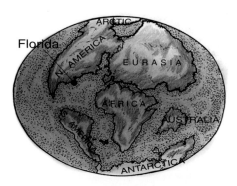

250 to 75 mya: First North America, then South America split away from Africa and migrate apart, leaving the great Atlantic Ocean basin between them. Florida, still under water, moves with North America.

and began moving toward their present positions.

75 to 35 mya: The ocean basins widen and the continents take on further definition.

North America is connected with Asia by a land bridge (just out of view over the north pole), and animals can migrate back and forth across this bridge. However, North and South America are separated by the deep ocean. Their plant and animal populations cannot mingle, but are evolving separately for hundreds of millions of years. Florida remains under water.

Sources: Pojeta and coauthors 1976; Opdyke and coauthors 1987; Dallmayer 1987.

THE BIRTH OF OUR PRESENT WORLD

At about 65 million years ago (mya), a giant meteor slammed into the floor of the Caribbean Sea, making a crater nearly 200 miles across. (The seas have risen since and thick layers of limestone have been deposited on top of it, forming the Yucatan Peninsula of Mexico, but the crater is still there, far below.) The impact splattered molten magma onto what is now Haiti and the Texas Gulf coast, spewed sulfurous ejecta all over the planet, and destroyed the ozone layer in the outer atmosphere. Forest fires ignited and raged across all continents. This terrifying event, the Cretaceous cataclysm, delivered the death blow to all of the dinosaurs that were then roaming the earth, and drastically altered the environment for the next tens of millions of years.

Without the ozone layer to screen out the sun's harsh ultraviolet rays, photosynthesis drastically slowed, and most of the earth's plant life was extinguished. Not only did all of the dinosaurs die; so did 70 percent of all the living things on earth. No land animals remained that weighed more than 50 pounds. The only survivors were those that could escape the killing light: nocturnal animals, burrowing animals, animals in lakes and in the ocean. But even though most plants had died, their seeds remained alive—especially fern spores and the seeds within the fruits of flowering plants.

Geologists identify the time since 65 million years ago as the Cenozoic Era, the era in which we live. Background Box 22–1 identifies the three eras since the beginning of multicellular life on earth, and shows how the last one, the Cenozoic, is divided into periods of significance to Florida history.

BACKGROUND   **22–1**

Time Periods since the First Many-Celled Organisms (550 mya)

Three eras span the last 550 million years—the Paleozoic, Mesozoic, and Cenozoic eras. Details are shown only for the last of the three, "our" era. Florida's terrestrial history begins in the late Oligocene and continues through the Miocene, Pliocene, and Pleistocene epochs.

All spans of time used here are rounded off roughly. For purposes of this treatment, only the durations and approximate starting and ending times are important.

PALEOZOIC ERA (550 to 250 mya)

MESOZOIC ERA (250 to 65 mya)

CENOZOIC ERA (65 mya to the present)

Tertiary Period	Millions of years ago (mya)
Paleocene epoch	65–53
Eocene epoch	53–36
Oligocene epoch	36–23
Miocene epoch	23–5
Pliocene epoch	5–2

Quaternary Period

Pleistocene epoch (2 mya to 10,000 ybp)

Holocene (Recent) epoch (10,000 ybp to the present)

Sources: Hayes 1996; Mojsis and coauthors 1996.

The world changed dramatically during the early Cenozoic Era. Most groups of organisms on land and in the sea made new beginnings and the earth took on a new look. This was no longer the age of gymnosperms and reptiles; it was the age of angiosperms and mammals. By 53 million years ago, the clouds had cleared. Sunlight was again pouring down. The earth grew warm again, plants flourished anew, and their metabolism restored ozone to the atmosphere, recreating conditions in which life could thrive.

Until 35 million years ago, Florida remained under water, but species and communities were developing on exposed land nearby that would migrate into Florida when the seas receded. The stage was set for the widespread dominance of flowering plants and vertebrates that characterize our world today. The coastal plain of Alabama and Georgia supported many trees we would recognize today: birch, alder, oak, elm, sycamore, basswood, chestnut, maple, beech, and hickory. Warmer areas held tupelo, magnolia, and sweetgum; and along the coast were tropical plants such as seagrapes and figs.

Angiosperms are flowering plants, which reproduce by means of seeds contained in closed ovaries.

Gymnosperms (for example, pines and other conifers) reproduce by means of naked seeds contained in cones.

Among trees, the gymnosperms are often called "softwoods" and the angiosperms "hardwoods." In general, the distinction holds true.

Florida native: Sweet pinxter azalea (*Rhododendron canescens*)

At 37 million years ago, the end of the Eocene, the planet was so warm that crocodiles were roaming as far north as Wyoming. A group of animals known as cetaceans were swimming in terrestrial lakes and walking about on land. Later, some of their descendants would evolve to become hoofed animals while others would become today's whales. Roaming the Great Plains were many peculiar-looking creatures related to today's camels, horses, giraffes, elephants, bears, wolves, and many more. Some lines went extinct, other lines branched and prospered. Then another major extinction event occurred and many of the peculiar-looking animals of the Plains went extinct. It may be that the deep ocean circulation changed at about this time. Whatever the reason, massive ice sheets accumulated on top of Antarctica, which by now lay at the south pole. The climate grew much cooler, and the sea level dropped dramatically. This led to the emergence of Florida.

FLORIDA'S DEBUT

At about 35 million years ago, a few of the highest parts of Florida-to-be broke the surface for the very first time. Immediately, the newly exposed high ground began to receive immigrants from nearby terrestrial ecosystems, but soon the sea rose and washed them away again. The sequence repeated many times over. By about 25 million years ago, though, the ocean had receded from the Panhandle's northern highlands and they remained above water. The Florida peninsula, which was lower, grew, then shrank, then grew, but after 23 million years ago it never again disappeared completely beneath the ocean.[2] Figure 22–2 shows that at times Florida was a group of islands, at times a peninsula.

Florida appeared at a time when flowering plants were diversifying greatly. Composites, grasses, and legumes flourished on land, aquatic plants thrived in lakes and streams, and seagrasses spread beneath coastal waters. Herbivores flourished, eating all these plants. Carnivores diversified, finding abundant herbivores to eat. *La Florida*, the Land of the Flowers, was born at an auspicious time to develop immense diversity.

Thanks to the record left behind by Florida's abundant fossils, paleontologists know that fascinating menageries of creatures have been here at different times. Imagine trying to describe a rhinoceros or an alligator to someone who had never seen one. Florida's paleontologists face a similar challenge when they try to convey to us what earlier worlds were like, when cameloids, rhinocerotids, giraffids, and other such fantastical creatures were here in abundance. The very names convey that the animals would have looked strikingly peculiar by today's standards.

Enough fossil finds are available from different times to offer richly detailed scenarios from past epochs. The discoveries described below are separated by at least *several* million years—a long enough time to bring about drastic changes in species and ecosystems. For perspective, keep in

A **composite** is a special kind of flowering plant that belongs to the family Compositae or Asteraceae (daisies, sunflowers) and has many blossoms in a single flowerhead. Composites are the flowers that butterflies like best. Examples appeared on page 265 in photos A, C, and E.

At about 30 mya, only a few islands were above the surface of the water.

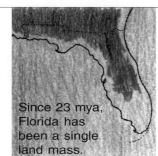
Since 23 mya, Florida has been a single land mass.

At times, Florida has been twice as big as today.

FIGURE 22–2

After 35 mya: Florida Emerges from the Sea

Green areas represent the land mass(es) of Florida at various earlier times. Black lines show the contours of today's Florida.

The exact shapes that Florida assumed from its first emergence to the present will never, of course, be known. However, geologists can easily tell whether sediments were deposited above or below sea level and make maps approximating the dynamic history of shifting shorelines. Some layers accumulated on top of the land during times when it was below sea level, as currents carrying heavy loads of sediments slowed down and dropped them in piles on the ocean floor. Then they were scrubbed off and sometimes completely washed away by erosion when sea level fell again.

Above sea level, the land performed many disappearing acts, as its limestone layers dissolved and were carried to the sea by underground rivers. The land sank down as this dissolution took place, but then in some areas, as its overburden of sediments grew lighter, deeper layers heaved upwards, raising the surface again. Florida's geographical history is pieced together through the use of geological clues—and importantly, from fossils entombed in the rocks.

mind that only *two* million years ago, early hominids were just beginning to learn to use tools in Africa.

FLORIDA'S EARLY NATIVE PLANTS AND ANIMALS

To capture the thrill of discovery of Florida's ancient history, let us pretend that, instead of being paleontologists digging up fossils, we are time travelers who can actually visit sites in Florida millions of years ago when the fossils were first laid down. The entries that follow are written as if in the voices of people who can *see* the ancient plants and animals—still alive. For scholarship's sake, though, references for each entry direct you to reports of the actual fossil finds.

The Late Oligocene (25 to 23 mya). Our time machine first takes us back 25 million years, to a site near Gainesville (the "I-10 site"), close to the very beginning of Florida's terrestrial history. We might write in our journal as follows:

> *Route I-10, 25 mya. We have pitched our tent by a sinkhole in a mesic forest. We spy many animals both in the trees and on the ground. Some look vaguely familiar, as if out of focus. A tribe of beaverlike animals is at work felling trees nearby to dam a stream and we see a hedgehog amble by. Turtles are numerous. Other animals are of recognizable groups— there are rodents and a rabbit that look like those of today; lizards,*

Reminder: A **mesic** forest is one with moist soil.

Florida native: Virginia opossum (*Didelphis marsupialis*). 'Possums have been roaming in Florida since more than 25 million years ago.

snakes, and frogs that seem very different. There's a tortoise that may be a distant relative of today's gopher tortoise. At night, we see flocks of bats of at least five different species, and peccaries (piglike animals with boar tusks). There is an insect-eating animal rather like an anteater, and a most familiar night roamer, the opossum, is already here.

One day we explore a savanna inland from our campsite, and find a thrilling assortment of large mammals. One is a kind of horse, two are like deer, and two others are like goats. We can tell that fires burn frequently across this grassy plain. The plants nearly all grow low to the ground and signs of a recent fire are apparent as blackened twigs of shrubby growth.

On another day we hike to the shore. There, too, we see creatures we can almost name—seven different species of sharks, a sting ray, and a bony fish. Birds of all kinds are numerous, but we cannot get good impressions of them—literally: they have left few and faint fossil traces.

There is danger here—there are large and small carnivores. We know there are sabercats at other sites nearby.

If this campsite is typical, and we hope it is, then we can correctly describe the Florida of this time. Mesic forest predominates, with animals that are adapted to it, and they are so well distributed into the many available habitats that we think they have been evolving here for a long time. We will be back to see how well they persist, but we won't find another informative site until the early Miocene.[3]

The Miocene (23 to 5 mya). Early in the Miocene, communities north and south of the Suwannee River are developing somewhat differently. Perhaps the river presents a barrier at this time. Whatever the reason, some differences have persisted to the present. Many trees found north and west of the Suwannee seem never to have crossed that river to occupy peninsular Florida.

In central Florida, special scrub communities are developing, each on its own "island"—that is, on a high sand hill surrounded by mesic forests that most of its plants and animals cannot cross. Special freshwater communities are developing, too. Many freshwater mollusks of central Florida are endemic, as are many freshwater fishes.[4]

Of our first visit, which takes us to Gilchrist County early in the Miocene, we might write:

Thomas Farm, 18.5 mya: We are camping in a patch of deciduous forest complete with ponds, sinks, and caves, where we can observe a tremendous number of animals. Turtles and frogs are numerous in the ponds, and the caves are full of bats of many species. On the ground and in the trees we recognize many cousins of today's treefrogs, squirrels, turkeys, and deer (or at least "hoofed animals that browse"). Peccaries are still here; so is the opossum. Songbirds are calling in the branches above our campsite.

Our reptile specialist is excited about the variety we find. There are half a dozen snakes, some ground dwellers, others arboreal. One is a racer; one, a tree boa; one, a vine snake. Among the lizards are "a skink, a gecko, a Gila monster, a curly-tailed lizard, an anole, and four types of iguanids." The amphibians are "exceedingly diverse" and most are of the same genera as today—sirens, newts, and salamanders; many frogs and toads.[5]

Florida native: Southern leopard frog (*Rana pipiens sphenocephala*). The protective coloration of this frog reflects millions of years of evolution within its habitat.

From our own and others' observations, we believe that quite a few animals are by now confined (endemic) to the Gulf coast along Florida and Texas. They do not occur in the interior of the continent, probably because the Great Plains are too dry. Among the endemics are three cameloids and a rhinocerotid.

This appears to be a well-established community. All of the animals show adaptations to forest life. Many are browsers, rather than grazers; and most, such as the bush hogs, have short legs that enable them to run through undergrowth.[6]

Other investigators have hiked eastward and report vast savannas that support great herds of grazing animals. Three-toed horses are especially numerous and make up 80 percent of our sample. Other grazers are even-toed: camelids, deerlike ruminants, an oreodont, and a cameloid with a peculiar "slingshot" horn above its nose. This strange beast will soon go extinct—soon, meaning within less than ten million years. A grazing tortoise also thrives on the savanna's herbs and forbs, and little pocket mice are scurrying about in the grasses collecting seeds. Several kinds of kites fly over the grasses searching out small animals.

Preying on all the small herbivores are diverse types of bears, wolves, and cats—some large, some medium, some small—at least ten distinct species. It is hard to convey how thrilled we are to have come upon this scene for the first time. The most astonishing discovery comes at dusk when huge flights of bats emerge from caves deep in the woods. Some of these bats are large nectar feeders; others are small, swiftly-soaring insect feeders; and there are at least three of intermediate size, food habits unknown.[7]

We think these sites are typical of early Miocene Florida. It is a patchwork of mesic forest and savanna, somewhat more tropical than today.[8]

Our time machine next takes us to Alachua County, later in the Miocene. We might write in our journal:

Love Bone Bed, about 10 mya: Again we are near the Gulf of Mexico. A stream runs out here and we have pitched tents on platforms over a freshwater marsh where the water rises and falls in concert with the tides. This is the richest late Miocene vertebrate site in eastern North America. The marsh holds chicken turtles, soft-shell turtles, garfish, and alligators. Every time the tide rises, sharks, whales, and other swimmers move inland in great schools. Along the stream bank a lush forest rings with the songs of numerous birds.

Florida native: Seminole bat (*Lasiurus seminolis*). This animal's ancestors were present in Florida by 20 million years ago.

Vast grasslands have opened up across North America and grazing animals find abundant food. Herbivores, including dozens of species of horses, grow large. We hike inland to explore a grassy savanna and find a stupendous number of species. The samples are so numerous that we have had to resort to a list to convey their extent (see List 22–1).[9]

The ground sloths are from South America and the temptation to speculate on how they got here is too great to resist. We imagine they may have floated in on a great pile of debris washed down a South American river into the Gulf of Mexico. As for the sabercats and bears, they must have come from Eurasia across a land bridge in the vicinity of Alaska or Greenland.

While we are exploring the Love Bone Bed, news arrives from some botanists who are camping in the Panhandle. They stumbled upon a botanically rich site at Alum Bluff on the Apalachicola River, at the north border of Liberty County. They have found several tropical plant assemblages growing along the river including palms, figs, breadfruit, orchids, and camphortrees. The gopherwood will persist at Alum Bluff for at least 13 million years, all the way to the 21st century.

The Pliocene (5 to 2 mya). We return later, again by time machine. The western Great Plains have now become very dry, predominantly grasslands. Many grazers have gone extinct in the middle of the continent, but in Florida, enough watering holes remain on the savannas to support huge herds of large animals. We make these observations in Polk County:

Bone Valley, 5 mya: The climate is now very dry. Florida is huge and semiarid. The sea level is so low that animals can roam to the very edge of the continental shelf and climb some distance down the slope almost to its base where the deep seafloor is today. Browsing animals are present (such as a large browsing horse) but grazing animals greatly outnumber them—there are ten species of grazing horses. Giant mammals include two elephantlike genera. And as if these weren't enough, there are still other large grazers: a large camelid, a medium-sized giraffoid, and a rhinocerotid.

This vast panorama reminds us of the Serengeti Plain of Africa. It has numerous hoofed animals (like the Serengeti's zebras, gazelles, wildebeest, pronghorns, giraffes, and others); and it has cat and dog carnivores (like cheetahs, leopards, lions, jackals, hyenas, and others). We are staggered by its richness and abundance.[10]

Subtropical forests also remain in central Florida. In them we see a large flying squirrel, an early version of the white-tailed deer, and a long-nosed peccary.

This is the last trip we make in our time machine and the last entry in our imaginary journal. For several million years, we can make no more observations, because the seas rise dramatically and remain high from 5 to 2 million years ago, eroding the perimeter of the highlands. When sea level is highest at about 4 million years ago, there may again be a deep channel across north Florida where the Suwannee River now flows, but if so, it soon closes again. Henceforth, Florida remains a single land mass. Our narrative now resumes as regular text.

The Pleistocene (beginning about 2 mya). By the time of the Pleistocene Epoch, the world was transformed. South America became joined to North America (see Figure 22–3), and both in the oceans and on the continents, the effects were profound. The land bridge between North and South America separated the Atlantic and Pacific oceans, forcing global ocean currents to change course. The waters of the Atlantic Ocean once flowed steadily into the Pacific through a deep channel between the two continents, but after about 2.5 million years ago, they were blocked by Panama. As a result, Atlantic and Pacific shrimp, once all of the same species, diverged to become distinct species.

Before the separation of the oceans, a major current from the Pacific Ocean carried heavy loads of nutrients into the Gulf of Mexico through the deep gap between the Americas. After the upheaval of Panama, an Atlantic current washed into the Gulf, circulated around it, and exited at the tip of Florida to flow up the east coast—in short, the Gulf Stream began to circulate as it does today. The warm, shallow Gulf waters became nutrient poor, which is ideal for the growth of coral reefs. A dozen earlier genera of corals went extinct, but 21 new ones evolved, including beautiful staghorn and elkhorn corals. The reefs off the Florida Keys became unlike any others in the world.[11]

As the warm Gulf current flowed north, it encountered cold Arctic air and the cooling created rain. Rain on the poles turned to ice, and polar ice spread over the continents. Repeated ice ages followed.

The Great American Interchange. During each ice age, ocean water became tied up in land ice and the seas fell dramatically. Land areas expanded, and new land routes became available to and from distant places. The isthmus of Panama became a broad land bridge above water. North and South America's plants and animals, which had evolved independently for nearly 300 million years, began crossing that bridge, invading each other's territory, and mingling (see Figure 22—4). New relationships evolved as predators and prey began to mix in new combinations.

Florida's appearance changed dramatically. When it first appeared above sea level some 35 to 25 million years ago, Florida bore predominantly hardwood forests in which patches of pine forest and savanna lay scattered. At 2 million years ago, the pattern was reversed: pine forest and savanna were continuous and the hardwood forests had broken up into patches. A major new, dry-adapted ecosystem was now in evidence in southern Levy County: coastal longleaf pine savanna, the precursor of today's pine grasslands.

Now the wet areas were isolated from each other by dry land. Plants and animals along the Apalachicola River could not enter or cross the surrounding, dry terrain, but remained confined to river bluffs. Today, Florida's oldest endemic species remain in the Apalachicola river area.

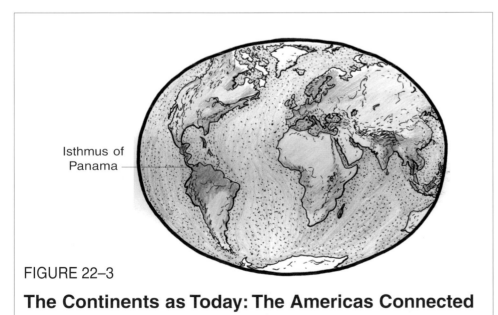

Isthmus of Panama

FIGURE 22–3

The Continents as Today: The Americas Connected

The isthmus of Panama began to rise and seal off the gap between the Americas around 12.5 mya. As the land rose, joining the continents, it separated the oceans, diverted ocean currents, altered the climate, and dramatically changed the world.

Native to Florida waters: Elkhorn coral (*Acropora palmata*). This coral evolved in the shallow seas off Florida within the last ten million years.

The fossil record shows that many major species of animals vanished from North America during the Pleistocene Epoch. In their place were many others, never seen there before. New animals were present from South America, the Far West, and the northern mountains. Again, an abbreviated record has to serve (see List 22–2).

At around 1 million years ago, the climate warmed, the ice sheets retreated, and huge volumes of melt water roared down North American rivers, widening their channels and delivering sediment that built broad deltas at their mouths. On the Mississippi delta, vast wetlands developed, too wide for most plants and animals and even some birds to cross. Florida's connection with the west weakened and many groups split into eastern and western subpopulations.

At least ten more cycles of cooling and warming followed during the Pleistocene's last million years. The climate cooled and warmed about every 100,000 years and sea level fell and rose along with it. Whenever sea level fell, water tables in coastal regions also fell and Florida became very arid, favoring savanna and high pine communities. Whenever sea level rose again, wetter conditions returned. Hardwood forests expanded their ranges, wetlands enlarged, and lakes became numerous. At 18,000 years before the present, when the Wisconsinan ice age was most intensely cold

FIGURE 22–4

The Great American Interchange

Solid lines outline the land as it is today. Blue-green areas show how much larger the continent was 2 million years ago when sea level was lower. White arrows show some of the routes that animals could follow at that time. The land bridge between the Americas was broad, and animals and plants could readily cross between the continents. Routes were also open along the Apalachicola River and between the East and the Far West. (A third route along the Atlantic coast, although a possible route to Florida, did not contribute many species.)

Pine grassland. Fire-dependent pinelands have been present in Florida for more than two million years.

and Florida was dry, the sea level was 300 feet lower, and the peninsula's area was twice as broad as it is today. The whole continental shelf lay exposed. A vast, dry pine savanna flourished over a hundred million acres, decorated with hardwood forests, mostly oak, including turkey oak. Lakes, rivers, and streams were less prominent than now, but thousands of acres of wet flats and seepage bogs were ablaze with wildflowers and carnivorous plants.[12]

Animals were again traipsing into North America from Asia over the land bridge between Siberia and Alaska. Some long-horned bison migrated from Asia into Alaska, then crossed the entire span of the Great Plains and entered Florida to gradually become today's familiar short-horned bison. Herds of giant animals roamed from one water hole to the next: mam-

Florida native: White-eyed vireo (*Vireo griseus*). This little songbird is one of 45 species of birds that depend on Florida's hardwood hammocks for their breeding and nursery grounds.

Seepage bog. By 18,000 ybp, bogs were spread all across the Panhandle.

moths, mastodons, camels, sabertooth cats, dire wolves, giant sloths, armadillos, tapirs, peccaries, bison, giant land tortoises, glyptodonts, horses, giant beavers, capybaras, and short-faced bears.[13] Naturalist Archie Carr, contemplating what Florida must have been like at that time, marveled at the sights we might have seen:

Florida native, now extinct: Mammoth *(Mammuthus colombi)*. Paleontologists discovered this skeleton in the bottom of the Aucilla River in 1968. It is more than 12 feet tall at the shoulder and 16,000 years old by radiocarbon dating.

Florida natives, now extinct. Left: dire wolf (*Canis diris*). This wolf, present in Florida during the Pleistocene, was larger than today's grey wolf. It weighed about 150 pounds. Right: giant armadillo (*Holmesina septentrionalis*). This beast was the size of a large hog and rooted for its food much as armadillos do today.

Right in the middle of Paynes Prairie itself there used to be creatures that would stand your hair on end. Pachyderms vaster than any now alive grazed the tall brakes or pruned the thin-spread trees. There were llamas and camels of half a dozen kinds; and bison and sloths and glyptodonts; bands of ancestral horses; and grazing tortoises as big as the bulls . . . making mammal landscapes that, you can see in even the dim evidence of bones, were the equal of any the world has known.[14]

Figure 22–5, on the next two-page spread, shows what Florida's great savannas might have looked like when those great animals were here. Many remain from that time, and the figure legend identifies only those that have since gone extinct.

THE COMING OF HUMANS

After 15,000 ybp, another warming trend began. The ice sheets began to melt, and at about 13,000 ybp the first bands of nomadic hunters migrated from Alaska down an ice-free strip of land east of the Canadian rockies. Around 12,000 ybp, they arrived in Florida (Figure 22–6).

As on the Plains, the giant animals all died off at about the time the humans arrived (see List 22–3). In Florida's case, however, human hunters are not thought to be the only agent causing the extinctions. In Florida, small mammals as well as big ones went extinct, often just as other similar animals migrated in, suggesting that newcomers were displacing residents. Moreover, some of Florida's extinctions predated human arrival. Perhaps climate change pushed the animals to the brink and human hunters struck the last blow in some cases. Whatever the reason, only the fossils of those giant animals remain.

Like all prior comers, Florida's first people became part of its ecosystems. They selected certain wild plants to eat, they hunted, and they cleared some forests for hunting, sometimes by burning them. Once the large game

FIGURE 22–5

Pleistocene Giant Animals in Florida

At left is a mesic forest with peccaries (1), a giant sloth (2), a glyptodont (3), dwarf rhinos (4), as well as some dire wolves (5) and a sabercat (6). At center and right is a savanna of the type that probably occupied most of Florida during much of the Pleistocene. Roaming the savanna are herds of giraffe-camels (7), horses (8), and bison (9).

animals of the past were extinct, they adapted to eating snails, then oysters, scallops, and sea turtles. By 5,000 ybp, they were making pottery, and by 1,000 ybp they were engaging in corn agriculture.

Meanwhile, new species of insects, birds, and small mammals evolved and filled newly available spaces in woodlands. Coastal wetlands were swamped by rising seas and Florida shrank to half its former area. Cypress

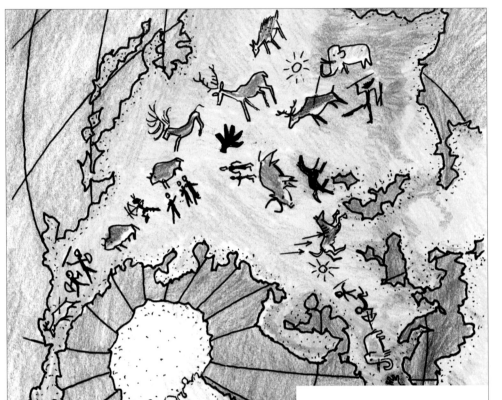

LIST 22–3
Changes in the fossil record since 13,000 ybp

EXTINCT in North America
  Large mammals, 35 genera
  Large birds, 2 (*Gymnogyps*, *Teratornis*)
  Giant tortoise (*Geochelone*)

GONE, E of the Mississippi[a]
  Jack rabbits
  Pronghorn antelopes

GONE FROM FLORIDA[b]
  Bog lemming (*Synaptomus*)
  Porcupine (*Erythizon*)
  Muskrat (*Ondatra*)

SPLIT into two populations E and W of Mississippi River
  Cactuses, 8 genera
  Scrub-jay
  Whip scorpions

REMAINING
  Reptiles: Of 31 species that came in from the Far West, 24 are still here in dry savannas and high pinelands

Notes:
  [a]Eastern North America became too wet for these desert animals.
  [b]Summers became too hot and wet in Florida and these animals moved north.

Source: Adapted from Webb 1990.

FIGURE 22–6

The Arrival of Humans in Florida (13,000–12,000 ybp)

America's first people worked their way across a land bridge from Siberia into Alaska, western Canada, and Washington, following great herds of game animals. They crossed the Great Plains, wiped out 35 genera of mammals and 10 classes of birds within 3,000 years, and left their spearpoints strewn among the skeletal remains. By 12,000 ybp they had arrived in Florida.

Scores of different tribes occupied different parts of Florida. Nearly 10,000 villages, camps, and mounds peppered the state; some 885 such sites are on Eglin Air Force Base alone (in Santa Rosa, Okaloosa, and Walton Counties).

Source: Anon. 1991 (Survey unearths Eglin's past).

Mangrove island. Mangroves began growing along Florida's shores 5,000 years ago or less.

and other wetland species escaped extinction by finding refuge in deep, wet habitats in the interior. They persisted there, as if awaiting the next climatic turn to recolonize coastal lowlands. Seepage bogs dwindled in size and separated into remnants, the Everglades marsh developed, and mangroves began to grow along the coast.

THE EUROPEANS (500 YBP)

When the Europeans arrived, the pace of change quickened. In the early 1500s, the Spanish "discovered" Florida and preempted it as their own. Spanish aggression and diseases decimated the native peoples, and by about 1700, had largely wiped them out or driven them from Florida. (A few tribes returned to Florida later, notably, in the mid to late 1700s, members of the Mikasuki and Creek tribes—the latter renamed Seminole.)[15]

Other Europeans followed the Spanish and established forts, settlements, and citrus plantations along the coasts, on the rivers, and on the highlands. By 1775, St. Augustine, St. Marks, and Pensacola were major Florida towns. Then, in the 1800s, industry came to Florida, and in the 1900s, roads, cars—and some 17 million people.

Today, global gas concentrations are changing, the earth's temperature is rising, more ultraviolet rays are penetrating the atmosphere, and as a result, the earth's biodiversity is diminishing. But those natural ecosystems that remain are the once-on-a-planet, once-in-the-universe products of a history hundreds of millions of years long, and they still hold an enormous and diverse legacy from the past. To the extent that we are willing, we can protect, maintain, and restore them.

Florida native: Ghost orchid (*Dendrophylax lindenii*). Now rare, this orchid grows in wooded swamps in Collier County and in Cuba.

CHAPTER 23

THE NEXT GENERATIONS

The preceding chapters have touched on each of Florida's natural ecosystems and have, we hope, demonstrated that they are beautiful and fascinating. They are also valuable in many ways, and they are worth preserving or, to the extent possible, restoring. This chapter offers a brief look at several aspects of the value of Florida's ecosystems and of their native species—the time it has taken to produce them, the information they contain, and their diversity.

AN INVESTMENT OF TIME

Every plant and animal species native to Florida has an almost unimaginably long evolutionary history. Its first cell-like ancestors took shape in the ancient ocean nearly four billion years ago. Then nearly three and a half billion years went into the evolution of the first multicellular organisms that left detectable fossils behind. Those fossils, which represent the species living on earth some 550 million years ago, are recognizable as the early progenitors of all the modern phyla of living things: sponges, mollusks, corals, chordates, and others of the Paleozoic Era. (For a review, see the the time line in Background Box 22–1.)

Since the start of the Paleozoic, those early representatives of multicellular life on earth have produced many species. Nearly all of those species have since gone extinct, but new species have evolved from them at a slightly greater rate, and some lines have produced offspring generation after generation without a break, all the way to the present. Among the lines that are still going are "our" plants and animals: Florida's native species.

To appreciate the spans of time that have gone into shaping these organisms, imagine the whole history of multicellular life (since the start of the Paleozoic) compressed into a single 24-hour day. Start the clock at midnight, 550 million years ago, and let the next midnight be today, a day in the 21st century. On such a time scale, each million years goes by in 2.62 minutes.

The first fish arose at about 2 o'clock in the morning of this "day," and

Florida native, next generation: Black skimmer nestlings (*Rynchops niger*). The black skimmer is one of thirteen bird species that nest on Florida beaches.

OPPOSITE: Florida flame azalea (*Rhododendron austrinum*) in the Apalachicola Bluffs and Ravines, Liberty County.

379

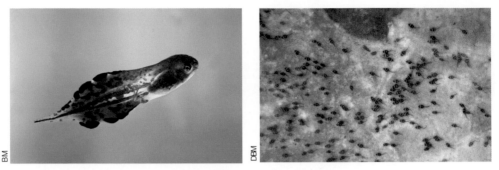

Florida natives, next generation: Pine woods treefrog tadpole (*Hyla femoralis*), left, and river frog tadpoles (*Rana hecksheri*), right. Florida's natural wetland waters swarm with astronomical numbers of these and other frog, salamander, and fish fry.

the first land plants at about 4 AM. By 6 AM the first amphibians and insects were present, and by 8 AM, there were reptiles. By the early afternoon, dinosaurs had come to dominate the earth, and shortly after that, the first mammals were present. Birds first appeared at around 6 o'clock in the evening, and flowering plants arose at about the same time. Then, at about 10:30 in the evening, Florida emerged from the sea. As Florida took shape, diverse plants and animals moved in, began combining into ecosystems, and continued to evolve.

For perspective on these time spans, note that human beings were very late arrivals in this "day" of earth's history. Florida's first, primitive people appeared just *two seconds* before the very end of the day. The first Europeans arrived in *the last tenth of a second.*

In comparison with the long time since the start of the Paleozoic, the 25- to 35-million-year history of Florida as a body of land above water has been relatively short, only one and a half of the twenty-four hours. Yet from another perspective, compared with the few hundred years since Columbus, Florida's history is long and all of it has been invested in tailoring each native species to local lands and waters, as shown next.

A Treasure Trove of Information

An immense amount of information is stored in each of Florida's species. Some of that information is a set of "directions for life," such as how to respire, grow, and repair injuries. Some is specific and provides instructions for adapting to natural, local environments: cypress trees to swamps; mussels to river beds; caterpillars to their host plants; corals to their sites on the continental shelf. Taken all together, the quantity of detail in each species is staggering. Even a single bacterial "cell," which is smaller and simpler than a "true" cell, is built according to a blueprint of some 3,000 genes, each of which in turn is composed of tens of thousands of atoms. A more complex organism, such as a microscopic worm, inherits and passes on much more information—some 19,000 genes. A common mold may have as many as 100,000 genes.

If each "simple" species uses thousands of genes to direct the development of its members, how much more information must the barking treefrog, scrub balm, horseshoe crab, and longleaf pine have? How much

Florida native, next generation: Red maple seeds (*Acer rubrum*). The tree is growing in a swamp in the Waccasassa Bay State Preserve, Levy County.

RB/OAR/NURP

Native to Florida waters: Purple reeffish (*Chromis cyanea*). These fish are swimming by a reef off the Florida Keys.

information is contained in a whole ecosystem, or in the dozens of ecosystems that we have in Florida? To convey how awesome such a quantity is, biologist E. O. Wilson offers this description of a single, tiny ecosystem in soil:

> *A handful of soil and leaf litter . . . contains more order and richness of structure . . . than the entire surface of all the other (lifeless) planets . . . Every species in it is the product of millions of years of history. . . . Each organism is the repository of an immense amount of genetic information.*
>
> *For the sake of illustration, let us use the "bit," the smallest unit of information content used in computer science, to measure the amount of genetic information contained in an organism. . . . A single bacterium possesses about ten million bits of genetic information, a fungus one billion, and an insect from one to ten billion bits according to species. . . .*
>
> *If the amount of information in just one insect—say an ant or a beetle—were to be translated into a code of English words and printed in letters of standard size, the string would stretch over a thousand miles. The life in our one little sample of litter and soil would just about fill all the published editions of the* Encyclopedia Britannica.[1]

In light of the time and information each one represents, then, Florida's natural ecosystems are of great value. One aspect of that value is that natural ecosystems perform services useful to life and society.

ECOSYSTEM SERVICES

Ecosystem "services," meaning natural processes that support life on earth, were referred to in the first chapter, and succeeding chapters pointed out some of them. To bring them back into focus, a brief summary follows.

First, consider what green plants do for life on earth. Green plant-based

KLP/FPS

Florida native: Golden-winged skimmer (*Libellula auripennis*). in Washington Oaks Gardens State Park, Flagler County.

Florida native, next generation: Red mangrove seedling (*Rhizophora mangle*). The parent trees grow in a mangrove swamp on Florida's west coast.

Florida native, next generation: White-tailed deer (*Odocoileus virginiana*). This fawn was seen in woods near the Aucilla River.

ecosystems capture the sun's energy and bind it into fuels and building materials. All living communities and all human societies depend on these ecosystem products. The more green there is, the more total living material there can be. Green plants also consume carbon dioxide and release oxygen, thereby supporting oxygen-breathing animals and oxygen-consuming activities such as the burning of fossil fuels. Healthy plants on land, in surface water, and in the ocean also purify air. They absorb sulfur and nitrogen compounds, ozone, and heavy metals into their tissues, leaving the air clean, breathable by animals, and less laden with tissue-burning acid rain.

Then there is the contribution of green plants to the health of the planet. Oxygen released by the plants on earth forms ozone in the outer atmosphere, and ozone screens out wavelengths of sunlight that, if they penetrated freely to the earth's surface, would destroy all exposed life. Moreover, green plants consume carbon dioxide (which holds heat in the atmosphere), thereby helping to keep the global temperature within limits suitable for life. At the same time, green plants and their organic residues in ecosystems hold huge volumes of carbon out of the atmosphere and thereby prevent overheating of the globe. Among carbon-holding systems, the ocean's communities of algae, the earth's soil communities, and the plant populations of wetlands and forests are especially important.

Florida native, next generation: Salamander eggs. Probably the eggs of an Apalachicola dusky salamander (*Desmognathus apalachicolae*), these develop best in muck formed by seepage in a steephead ravine near the Apalachicola River.

Earlier chapters have already enumerated many other ecosystem services: water purification, soil conservation, erosion control, flood abatement, recycling of nutrients and minerals, aeration of soil, disposal of wastes. Finally (and this has been the focus of this book), natural ecosystems provide habitat: nurseries, feeding and breeding grounds, and shelter that permit native species populations to conduct all phases of their life cycles unaided by us.[2]

Florida native, next generation in the making: Okaloosa darter (*Etheostoma okaloosae*). These fish were spawning in Toms Creek, Okaloosa County.

Florida native rearing next generation: Perdido Key beach mouse (*Peromyscus polionotus trisyllepsis*). This mother was tending her tiny, pink newborns on a barrier island beach.

Florida natives: Green treefrog (*Hyla cinerea*) and string-lily (*Crinum americanum*) in Washington Oaks Gardens State Park, Flagler County.

SPECIES DIVERSITY

Species diversity greatly enhances ecosystem value. The greater the number of species in an ecosystem, the greater the system's productivity and resilience. In a grassland, for example, a mixture of different grasses can produce a greater mass of living material than a crop of any one grass, and species-rich grasslands can withstand and recover from stresses such as drought much more successfully than can low-diversity systems.[3]

Because species diversity has value, the loss of just one species can weaken a whole ecosystem. Species diversity is expressed most fully in large, ancient, undisturbed ecosystems like Florida's original pine grasslands, rocklands, and scrub, whose component plant and animal populations have evolved and interacted for millions of years. In contrast, less biodiversity is seen in small, isolated patches of habitat and in modified communities such as parks and farms, which are of more recent origin and are usually much less complex than natural communities.[4]

To appreciate the diversity of Florida's species, consider the photos that have appeared in this book. Nearly all were of plants and animals—two of the five kingdoms of living things. Of some 4,000 possible plants, about 150 appeared on these pages. Of more than 1,000 native freshwater fish, about 75 were shown. Of some 30,000 terrestrial invertebrates native to Florida, about 130 photos appeared, mostly of insects and spiders. The vertebrates were favored, of course, because they are so well loved and familiar; even so, we showed only 150 of a possible 700 (see List 23-1). Three entire kingdoms of living things, the bacteria, protists, and fungi, were almost completely left out.

We mentioned more species than we showed: we squeezed as many names as we could into the lists of species that appeared in most chapters. But even all the mentions add up to less than five percent of Florida's total species. Imagine, then, the state's total species diversity. Estimate a value for the eons of evolutionary time each species represents. Add credit for the information each species contains. The conclusion is clear: they are worth preserving and, to the extent possible, restoring. What has to happen to make this possible?

What will it take to ensure that the priceless biodiversity of this corner of the continent will be preserved? Look at the photographs arrayed on the

LIST 23-1
Native Florida plants and animals shown in this book (approximate numbers)

PLANTS	150
ANIMALS	
Invertebrates	130
Vertebrates	150
Amphibians ~30	
Reptiles ~35	
Birds ~50	
Mammals ~25	

Florida native, next generation: Wiregrass (*Aristida stricta* var. *beyrichiana*) in bloom. A recent, growing-season fire has produced a haze of golden seedheads, presaging ecosystem renewal in this pine grassland.

Florida native, guarding next generation: Swallow-tailed kite (*Elanoides forficatus*). The nest is on Plantation Key in Monroe County.

pages of this chapter. Each is of a plant or animal in the context that best supports it—a newborn deer in its woodland bed, a clutch of salamander eggs in the muck where they develop best, a pair of darters spawning in the aquatic garden where they live. They can thrive because they have the habitats they need.

Florida's species are adapted to local conditions, and Florida is the best place to preserve them. To ask a species to survive in some other setting is to ask the impossible. For us, to take on the care of a species outside of its natural habitat would be costly, burdensome, energy-intensive, time consuming, and unlikely to succeed. Small sets of specimens can at best be maintained elsewhere for short times under controlled conditions in greenhouses and zoos, but they cannot perpetuate themselves without help except under natural conditions. The best site in which to conserve each species is its own natural habitat.

Beyond the value of diversity, beyond the utilitarian value of ecosystem services, and beyond their beauty and fascination, the other creatures, plants as well as animals, simply have the right to be here. They have traversed the same, long, evolutionary time span, and the same bumpy road of life and death through myriad generations, as we have. They share with us the eloquent language of DNA; they are our kin. They need us to pay attention to their plight in the wake of our hurtling journey into the future. And indeed, if we don't pay attention and safeguard them against the repercussions of our own pursuits, we will learn too late that we needed them, too, more than we knew.

Green heron (*Butorides striatus*) stalking prey by a Florida stream.

385

REFERENCE NOTES

CHAPTER 1—NATURAL ECOSYSTEMS AND NATIVE SPECIES

1. Discussion of the term *natural* inspired by Clewell 2000.

2. *Guide* 1990, iv; emphases ours.

3. Species names for plants are from Wunderlin and Hanson 2003; for freshwater fish from Eddy and Underhill 1978, for marine fish from Anon. c. 1994 (*Fishing Lines*), for amphibians and reptiles from Collins and Taggart 2002, and for birds, Bruun and Zim 2001.

4. Fish count from Florida Biodiversity Task Force 1993, 3.

5. Florida diversity from Gatewood and Hardin 1985; deep ocean species from Broad 1997 and Russell 1998.

6. Vascular plants from Wunderlin and Hansen 2003, 1.

CHAPTER 2—THE CONTEXT: CLIMATE, LAND, SOIL, WATER

1. Heath and Conover 1981, 43; Winsberg 1990, 110; Anon. 1991 (Rainy rescue).

2. Florida rainfall amounts from Fernald and Patton 1984, 19.

3. Number of tropical storms from Livingston (E. H.) 1989.

4. Wolfe, Reidenauer, and Means 1988, 33.

5. Lightning frequency from Waters 1988 (Nature notes); Frank 1982; Ray 1992.

6. Frank 1982.

7. Brown, Stone, and Carlisle 1990.

8. Sea level has risen more than eight inches since about 1910, compared with about one inch per century for the prior 3,000 years; Klingener 1999. It may currently be rising at about 15 inches per century; Nerem 1995. An 8- to 12-inch rise is expected within the next 50 years for northeast Florida; Patterson 2000. Along Florida's Gulf coast, sea level rise is causing tidal flooding and forest deaths; Williams, Ewel, Stumpf, Putz, and Workman 1999.

9. Fernald and Purdum 1992, 37.

CHAPTER 3—HIGH PINE GRASSLANDS

1. Platt 1999.

2. Wells and Shunk 1931; Bridges and Orzell 1989.

3. Komarek 1965.

4. Platt and Schwartz 1990, based on data in early land survey records reviewed by Schwartz and Travis 1995.

5. Komarek 1964; Clewell 1981/1996, 76–79; Fowells 1965.

6. Snedaker and Lugo 1972, 120; Andre F. Clewell, note to Ellie Whitney, December 1995.

7. Walter. R. Tshinkel, letter to Ellie Whitney, 25 January 1998.

8. Begley 1995.

9. Need for snags from Wuerthner 1995; snakes from Means 1996.

10. Minno and Minno 1993; Weigl, Steele, Sherman, Ha, and Sharpe 1989, 48.

11. Weigl, Steele, Sherman, Ha, and Sharpe 1989.

12. Leveton 1995.

13. Svingen 1995; Hooper, Robinson, and Jackson 1990.

14. Information on the arboreal ant from Hahn and Tschinkel 1997.

15. Hooper, Robinson, and Jackson 1990.

16. MacArthur and MacArthur 1961; Anon. 1993 (Some like it old).

17. McKibben 1996.

18. Fossil records from Means 1996; interdependence from Clewell, 1981/1996 160–164.

19. Means 1996; Clewell 1981/1996, 150, 157–158.

20. Norse 1990.

21. Hooper, Robinson, and Jackson 1990.

CHAPTER 4—FLATWOODS AND PRAIRIES

1. Significance from DeSelm and Murdock 1993; present status from Platt 1999.

2. Glitzenstein, Streng, and Platt 1995.

3. Root adaptations from Stout and Marion 1993.

4. Researchers who sampled the ground cover at seven different times during the year counted 147 species of forbs in a frequently burned longleaf pine flatwoods. Platt et al 1988b as cited by Stout and Marion 1993.

5. Stout and Marion 1993. The ant species are *Crematogaster lineolata*, *Aphaenogaster traetae*, *Aphaenogaster fulva*, and *Conomyrma insana*.

6. Earthworm species worldwide from Margulis and Schwartz 1990, 216; Florida earthworms from Reynolds 1994; calculations from Darwin 1881; and Hickman, Roberts, and Hickman 1984, 343.

7. Brower 1997, 82–83.

8. Hutto 1995, 117.

9. That the turkey nests on dry ground burned a year earlier is from Boyles-Sprenkel 1991.

10. Rat statistics from Means 1985 (Cotton rat).

Chapter 5—Interior Scrub

1. Myers 1990, 155.

2. Myers 1990, 173; Snedaker and Lugo 1971, 102.

3. Myers 1990, 167–172.

4. Burns 1992; Tebo 1988 (Florida's desert islands); Fitzpatrick 1992 (Vanishing Florida scrub).

5. Stap 1994.

6. Root grafting from Snedaker and Lugo 1971, 162.

7. Plant chemicals from Deyrup and Eisner 1993; rosemary competition from Stap 1994.

8. Eisner 1997.

9. Eisner 1997.

10. Deyrup 1989; Deyrup and Eisner 1993.

11. Deyrup and Eisner 1993.

12. Deyrup and Eisner 1996.

13. Deyrup 1994, 254–256.

14. Deyrup and Menges 1997.

15. Layne 1992.

16. Wolfe, Reidenauer, and Means 1988, 41, 51.

17. Snedaker and Lugo 1972, 101–102.

18. Fitzpatrick 1994.

Chapter 6—Temperate Hardwood Hammocks

1. Gap succession from Clewell 1981/1996, 193-194; scatter-hoarding from Platt and Hermann 1986.

2. Duffy and Meier 1992.

3. Will Whitcomb, conversation with D. Bruce Means, c. 1975.

4. Florida Biodiversity Task Force 1993, 2.

5. Mitchell 1963.

6. Carr 1994, 165.

7. Description of, and succession in, oak groves from Carr 1994, 170; size of oak from Canfield 1998; dimensions of oaks from Godfrey 1988, 313; and Laessle and Monk 1961.

8. Carr 1994, 178-179.

9. Carr 1994, 180.

10. Carr 1994, 181-182.

11. Begley 1995.

12. Foresters say that "Any simplification of a natural system [erodes] the system's capacity to resist and recover from disturbances." Perry and Amaranthus 1997, 32.

Chapter 7—Rocklands and Terrestrial Caves

1. Roger L. Hammer, note to Ellie Whitney, c. March 2003; Snyder, Herndon, and Robertson 1990, 248; Platt 1999. 2. Hammer, note, 2003; Doren, Platt and Whiteaker 1993.

3. Snyder, Herndon, and Robertson 1990.

4. Snyder, Herndon, and Robertson 1990.

5. Toops 1998, 53; Anon. 1990 (Wild Florida, vol. 2).

6. Hurchalla 1998.

7. Endemism from Snyder, Herndon, and Robertson 1990, 236; Anon. 1998 (Key Largo Hammocks State Botanical Site); Sawicki 1997; Toops 1998, 56.

8. Snyder, Herndon, and Robertson 1990, 248.

9. Toops 1998, 57–59.

10. Sawicki 1997.

11. Cerulean 1992.

12. Huffstodt 1998.

13. Wisenbaker 1989.

Chapter 8—Wetlands

1. Gilbert, Tobe, Cantrell, Sweeley, and Cooper 1995, 12.

2. Willoughby 1897.

3. Statistic from Loftin 1992.

4. Grow 1984.

5. Clewell 1990.

6. Grow 1984.

7. Douglas 1994.

Chapter 9—Seepage Wetlands

1. Walker and Peet 1983 published the record for flowering plant species in a seepage bog in the Green Swamp of North Carolina: 47 species per square meter. Numbers of species in north Florida seepage bogs may be even greater.

2. Species numbers from Hermann 1995, 78.

3. Information on the moth from T. E. Miller, letter to Ellie Whitney, 7 April 1998. The moth is a member of the genus *Exyra*.

4. Information on the maggot from Miller, letter, 1998; other inhabitants from Carr 1994, 150–151; Hermann 1995, 83; information on aquatic insects from Fish and Hall 1978; and on dividing the resource from Folkerts 1982.

5. Wolfe, Reidenauer, and Means 1988, 154.

6. A couple of these giant cedars still stand along the upper Escambia River. Ward and Clewell 1989, 19, 29.

7. Ward and Clewell 1989, 36–38.

CHAPTER 10—MARSHES

1. Barada 1983.
2. Kushlan 1990, 325.
3. Kushlan 1990, 327.
4. *Guide* 1990, 40.
5. Kushlan 1990, 336–337.
6. Toops 1998, 21.
7. Means 1990 (Temporary ponds).
8. The published world record is for the flats in the Green Swamp of North Carolina; Walker and Peet 1992. Records for Florida flats remain to be published.
9. There are 20 crayfish species, and one has 3 variants, so some say 22 altogether. Hobbs 1942, 19–21.
10. Dwarf cypress from Mitsch and Gosselink 1986, 321–322.
11. Tilt angle from Kushlan 1990, 329–330: 3 cm per km over the entire original 10,000 sq km of the Everglades.
12. Urquhart 1994.
13. Hinrichsen 1995.
14. Urquhart 1994.
15. Hurchalla 1998.
16. Toops 1998, 27.
17. Toops 1998, 44.
18. Sharing system from Toops 1998, 31–41.

CHAPTER 11—FLOWING-WATER SWAMPS

1. Lickey and Walker 2002.
2. Cypress root adaptations from Clewell 1981/1996, 132.
3. Clewell 1981/1996, 131–132.
4. Disease resistance from Ward and Clewell 1989, 38.
5. Clewell 1981/1996, 137.
6. Andre F. Clewell, letter to Ellie Whitney, 14 July, 2000.
7. Toops 1998, 44, 46.
8. Strahler 1964.
9. Clewell 1991 (Physical environment) and Clewell 1991 (Vegetational mosaic).
10. Leitman, Sohm, and Franklin 1983, A44–A45.
11. Barry, Garlo, and Wood 1996.
12. That distribution of tree species changes along the river is from Leitman, Sohm, and Franklin 1983.
13. Clewell 1991 (Vegetational mosaic), 54–55.
14. Statistic from Harris, Sullivan, and Badger 1984.
15. Southall 1986.
16. Animal count from Suwannee River Task Force 1989, 9.

CHAPTER 12—LAKES AND PONDS

1. Griffith, Canfield, Horsburgh, and Omernik 1997, 5.
2. Marjory Stoneman Douglas, quoted in Huffstodt 1992.
3. Huffstodt 1991.
4. Griffith, Canfield, Horsburgh, and Omernik 1997, 7.
5. Statistics from Heath and Conover 1981.
6. Robert K. Godfrey, comment to D. Bruce Means 1987.
7. Brenner, Binford, and Deevey 1990, 370.
8. Karr and Chu 1999.
9. Brenner, Binford, and Deevey 1990, 380–381.
10. Franz 1992.

CHAPTER 13—ALLUVIAL, BLACKWATER, AND SEEPAGE STREAMS

1. Muir 1916, 100–101, 111.
2. Heath and Conover 1981, 109; Clewell, 1991 (Physical environment), 18.
3. River sizes from Heath and Conover 1981, 120, fig 149.
4. Nordlie 1990.
5. Endemic snails from Nordlie 1990.
6. Stolzenburg 1992.
7. O'Brien and Box 1999.
8. James D. Williams, comment to Ellie Whitney, June 2003.
9. Small stream fish from Bass (Gray) 1991 (Okaloosa darter); Bass 1993 (Harlequin darter); Abdul-Samaad 1996 (Blackmouth shiner).
10. Anon. 1986 (Upper Suwannee River).
11. Southall 1986; Anon. 1986 (Upper Suwannee River); Spivey 1992.
12. Southall 1986.
13. Sampat 1996.
14. Harris, Sullivan, and Badger 1984.
16. Holtz 1986; Karr 1994; Karr and Chu 1999.

CHAPTER 14—AQUATIC CAVES, SINKS, SPRINGS, SPRING RUNS

1. Fernald and Patton 1984, 39.
2. Wolfe, Reidenauer, and Means 1988, 172.
3. Means 1985 (Georgia blind salamander).
4. Stamm 1991.
5. Florida Springs Task Force 2000; Scott, Means, Means, and Meegan 2002.

CHAPTER 15—BEACH-DUNE SYSTEMS

1. Doyle, Sharma, Hine, Pilkey, and coauthors 1984, 17.
2. Clewell 1981/1996, 181.
3. Plants on foredunes from Clewell 1981/1996, 188–190; Holzer 1986; Gore 1986. Rapid growth from Johnson and Barbour 1990.
4. Johnson and Barbour 1990.
5. Number of sand gnat species from Tunstall 1998.
6. Cox 1988; Dunne 1995.
7. Emmel 2000.
8. Ball 1992; Stevens 1990.
9. Bates 1986.

Chapter 16—Tidal Marshes

1. Palmer 1975.
2. Coultas and Hsieh 1997, 18–19.
3. Clewell 1976, 25.
4. Plant mortality due to salt wrack from Clewell 1997, 97.
5. Clewell 1997, 101–103; Odum 1993, 237.
6. Rey and McCoy 1997, 200.
7. Montague and Wiegert 1990, 497, 503.
8. Salt marsh snake from Moler 1992.
9. Montague and Wiegert 1990, 504.
10. Coultas and Hsieh 1997, 14.
11. Clewell 1997, 89.

Chapter 17—Mangrove Swamps

1. Toops 1998, 72.
2. Odum, McIvor, and Smith, 47; isopod from Estevez 1978.
3. Toops 1998, 83.
4. Donaldson and Ratterman 1996.

Chapter 18—Estuarine Waters and Seafloors

1. Livingston (R. J.) 1990, 550.
2. A. Rudloe, Unpublished manuscript, 2000.
3. Gore 1992, 152.
4. Livingston (R. J.) 1990, 563–564.
5. Livingston (R. J.) 1990, 566.
6. Livingston (R. J.) 1990, 581.
7. Florida Biodiversity Task Force 1993, 3.
8. Florida Biodiversity Task Force 1993, 3.

Chapter 19—Submarine Meadows

1. Iverson and Bittaker 1986; Zieman and Zieman 1989, 1.
2. Clewell 1976, 5.
3. Zieman and Zieman 1989, 24–25; Whaley 1990.
4. Holdfasts from Zieman and Zieman 1989, 43; seagrass structure from Clewell 1976, 7.
5. A. Rudloe, Unpublished manuscript, 2000.
6. Yokel 1984; Yokel 1985; Roger L. Hammer, letter to Ellie Whitney, November 2003.
7. Woolfenden and Schreiber 1973.
8. Importance of seagrass beds from Jaap and Hallock 1990.
9. Reiger 1990.

Chapter 20—Sponge, Rock, and Reef Communities

1. Jaap and Hallock 1990; Reed, Gore, Scotto, and Wilson 1982.
2. Jaap and Hallock 1990.
3. Voss 1988, 30, 32.

4. Wolkomir 1995; Kaufman 1977; Brawley and Adey 1981; Jaap and Hallock 1990.
5. Jaap and Hallock 1990, 603; Den Hartog 1977.
6. Eldredge 1992.

Chapter 21—The Gulf and the Ocean

1. Estimate of ocean species from Norman 1998.
2. Angel 1993.
3. Gore 1992, 52.
4. Gore 1992, 139–140.
5. Salt dome ecosystem from Anon. 1997 (Gulf worms); description and number of salt domes from Gore 1992, 65-67.
6. Anon. 1997 (Gulf worms).
7. Anon. 1997 (Gulf worms).
8. Right whale information from Neuhauser 1993.
9. Lohmann 1992.
10. Horton 1987, 48–52.
11. Wisenbaker 1997.
12. Weber 1993, 41–43.
13. Weber 1993, 42–43.

Chapter 22—The Path to the Present

1. Time spans are from Harland, Smith, and Wilcock 1964; Gould 1996 (adapted).
2. S. David Webb, letter to Ellie Whitney, 22 December 1998.
3. Patton 1969.
4. Webb 1990, 90, 95, 100.
5. Meylan 1984.
6. Pratt 1990.
7. S. David Webb, letter to Ellie Whitney, 22 December 1998.
8. Webb 1981; Meylan 1984.
9. Webb, MacFadden, and Baskin 1981; Hulbert 1982.
10. Webb and Crissinger 1983.
11. Ross 1996.
12. Turkey oak was probably already present at 18,000 BC. Andre C. Clewell, letter to Ellie Whitney, 10 July 2000.
13. Osborne and Tarling 1996; Wisenbaker 1988; Wisenbaker 1989.
14. Carr 1994, 19.
15. Hiaasen 1993.

Chapter 23—The Next Generations

1. Wilson 1983.
2. The value of nature's services was estimated by Daily, ed., 1997.
3. Yoon 1994.
4. Roush 1982; May 1993; Raven, Berg, and Johnson 1993, 344.

BIBLIOGRAPHY

Abdul-Samaad, M. 1996. Delicate balance (Blackmouth shiner). *Florida Wildlife* 50 (July-August): 32.

Abrahamson, W. G., and D. C. Hartnett. 1990. Pine flatwoods and dry prairies. In Myers and Ewel 1990 (q.v.), 103-149.

Angel, M. V. 1993. Biodiversity of the pelagic ocean. *Conservation Biology* 7:4 (December): 760-772.

Anon. 1986. Upper Suwannee River. *ENFO* (September): 9-10.

Anon. 1990. Wild Florida, vol. 1: The Florida Scrub, a poster with text on back. Tallahassee: Florida Game and Fresh Water Fish Commission, Nongame Wildlife Education Section.

Anon. 1990. Wild Florida, vol. 2: Tropical Hardwood Hammock, a poster with text on back. Tallahassee: Florida Game and Fresh Water Fish Commission, Nongame Wildlife Education Section.

Anon. 1991. Longleaf pine communities vanishing. *Skimmer* 7:2 (Summer), 1, 7, 8.

Anon. 1991. Rainy rescue. *Tallahassee Democrat*, 12 January.

Anon. 1991. Survey unearths Eglin's past. *Tallahassee Democrat*, 25 August.

Anon. 1993. Some like it old. *Tallahassee Democrat*, 4 October.

Anon. c. 1994. *Fishing Lines*. Tallahassee: Department of Environmental Protection.

Anon. 1997. Gulf worms. *Scientific American* (November): 24.

Anon. 1998. Key Largo Hammocks State Botanical Site. *Florida Living Magazine* (October): 62.

Ashton, R. E., ed. 1992. *Rare and Endangered Biota of Florida*. Vol. 1, *Mammals*, ed. S. R. Humphrey. Gainesville: University Press of Florida.

——. 1992. *Rare and Endangered Biota of Florida*. Vol. 3, *Amphibians and Reptiles*, ed. P. E. Moler. Gainesville: University Press of Florida.

——. 1994. *Rare and Endangered Biota of Florida*. Vol. 4, *Invertebrates*, ed. M. Deyrup and R. Franz. Gainesville: University Press of Florida.

——. 1996. *Rare and Endangered Biota of Florida*. Vol. 5: *Birds*, ed. J. A. Rodgers, Jr., H. W. Kale, II, and H. T. Smith. Gainesville: University Press of Florida.

Ashton, R. E., and P. Ashton. 2004. *The Gopher Tortoise*. Sarasota: Pineapple Press.

Ball, S. 1992. Florida—last stand for monarchs? *Florida Wildlife* 46 (January-February): 46.

Barada, B. 1983. The human impact on Florida's weather. *ENFO* (October): 1-10.

Barry, W. J., A. S. Garlo, and C. A. Wood. 1996. Duplicating the mound-and-pool microtopography of forested wetlands. *Restoration and Management Notes* (Summer): 15-21.

Bass, D. G., Jr. 1991. Riverine fishes of Florida. In Livingston 1991 (q.v.), 65-83.

Bass, Gray. 1991. Delicate balance (Okaloosa darter). *Florida Wildlife* 45 (May-June): 29.

——. 1993. Delicate balance (Harlequin darter). *Florida Wildlife* 47 (March-April): 9.

Bates, S. K. 1986. Migratory butterflies flutter by Florida. *Skimmer* 2:2 (Summer), 1, 7.

Begley, S. 1995. Why trees need birds. *National Wildlife* (August/September): 42-45.

Bennington, S. 2003. Caribbean conga. *Blue Planet* (Summer): 8-10.

Boyles-Sprenkel, C. 1991. Multiply your turkeys. *Florida Wildlife* 45 (January-February): 38-39.

Brawley, S. H., and W. H. Adey. 1981. The effect of micrograzers on algal community structure in a coral reef microcosm. *Marine Biology* 61: 167-177.

Brenner, M., M. W. Binford, and E. S. Deevey. 1990. Lakes. In Myers and Ewel 1990 (q.v.), 364-391.

Bridges, E. L., and S. L. Orzell. 1989. Longleaf pine communities of the west Gulf coastal plain. *Natural Areas Journal* 9: 246-263.

Broad, W. J. 1997. *The Universe Below*. New York: Simon and Schuster. Excerpted in *Sky* (June): 100-105.

Brouillet, L., and R. D. Whetstone. 1993. Climate and physiography. In *Flora of North America, North of Mexico*, ed. Flora of North America Editorial Committee: 15-46. Oxford: Oxford University Press.

Brower, K. 1997. *Our National Forests*. Washington: National Geographic Society.

Brown, R. B., E. L. Stone, and V. W. Carlisle. 1990. Soils. In Myers and Ewel 1990 (q.v.), 35-69.

Burns, B. 1992. Partnership protects rare scrub. *Nature Conservancy Florida Chapter News* (Spring): 1.

Canfield, C. 1998. Highlands Hammock State Park. *Florida Living Magazine* (September): 66-67.

Carr, A. 1994. *A Naturalist in Florida*. New Haven, Conn.: Yale University Press.

Cerulean, S. 1992. Delicate balance (Key Largo woodrat). *Florida Wildlife* 46 (May-June): 40.

———. 1994. Swallow-tailed kites. *Florida Wildlife* 48 (July-August): 7-9.

Clewell, A. F. 1971. The vegetation of the Apalachicola National Forest (unpublished manuscript). Prepared under contract number 38-2249, USDA Forest Service, Atlanta and submitted to the office of the Forest Supervisor, Tallahassee.

———. 1976. Coastal wetlands (unpublished manuscript).

———. 1985. *Guide to the Vascular Plants of the Florida Panhandle*. Tallahassee: Florida State University Press.

———. 1981/1996. *Natural Setting and Vegetation of Panhandle Florida*. U.S. Army Corps of Engineers, Mobile District, Report No. COESAM/PDEI-86/001.

———. 1990. Cited by B. Orth. 1990. Clewell's world. *Builder/Developer* (January): 10-11.

———. 1991. Florida rivers (The physical environment). In Livingston 1991 (q.v.), 17-30.

———. 1991. Florida rivers (The vegetational mosaic). In Livingston 1991 (q.v.), 47-63.

———. 1997. Vegetation. In Coultas and Hsieh 1997 (q.v.), 77-109.

———. 2000. Restoring for natural authenticity. *Ecological Restoration* 18 (Winter): 216-217.

Collins, J. T., and T. W. Taggart. 2002. *Standard Common and Current Scientific Names for North American Amphibians, Turtles, Reptiles, & Crocodilians*. 5th ed. Hays, Kansas: Center for North American Herpetology.

Conover, A. 1998. To reproduce, mussels go fishing. *Smithsonian* 29 (January), 64-71.

Coultas, C. L., and Y.-P. Hsieh, eds. 1997. *Ecology and Management of Tidal Marshes*. Delray Beach, Fla.: St. Lucie Press.

Cox, J. 1988. The influence of forest size on transient and resident bird species occupying maritime hammocks of northeastern Florida. *Florida Field Naturalist* (May): 25-34.

Cressler, A. 1993. The caves of Dade County, Florida. *Florida Speleologist* 30 (Summer-Fall): 44-52.

Croker, T. C., Jr., and W. D. Boyer. 1975. *Regenerating Longleaf Pine Naturally*. Forestry Service, Research Paper SO-105. New Orleans, La.: USDA Southern Forestry Experiment Station.

Daily, G. C., ed. 1997. *Nature's Services*. Washington: Island Press.

Dallmayer, R. D. 1987. 40Ar/39Ar age of detrital muscovite within lower Ordovician sandstone in the coastal plain basement of Florida. *Geology* 15: 998-1001.

Darwin, C. 1881. *The Formation of Vegetable Mould, through the Actions of Worms, with Observations on Their Habits*. London: Murray.

Dawes, C. J., D. Hanisak, and W. J. Kenworth, 1995. Seagrass biodiversity in the Indian River Lagoon. *Bulletin of Marine Science* 57: 59-66.

Den Hartog, J. 1977. The marginal tentacles of *Rhodactis sanctithomae* (Corallimorpharia) and the sweeper tentacles of *Montastraea cavernosa* (Scleractinea). *Proceedings of the Third International Coral Reef Symposium*, Miami, Florida 1: 463-469.

Derr, M. 1989. *Some Kind of Paradise*. New York: Morrow.

DeSelm, H., and N. Murdock. 1993. Grass-dominated communities. In *Biodiversity of the Southeastern United States: Upland Terrestrial Communities*, ed. W. Martin, S. Boyce, and A. Echternacht: 87-141. New York: Wiley.

Deyrup, M. 1989. Arthropods endemic to Florida scrub. *Florida Scientist* 52: 254-270.

———. 1994. Florida scrub millipede. In Ashton 1994 (q.v.), 254-256.

Deyrup, M., and T. Eisner. 1993. Last stand in the sand. *Natural History* (December): 42-47.

———. 1996. Photosynthesis beneath the sand in the land of the pygmy mole cricket. *Pacific Discovery* (Winter): 44-45.

Deyrup, M., and E. S. Menges. 1997. Pollination ecology of the rare scrub mint *Dicerandra frutescens* (Lamiaceae). *Florida Scientist* 60:3 (Summer).

Donaldson, C., and L. Ratterman. 1996. Quotables to remember. *Palmetto* (Summer): 7, 12.

Doren, R. F., W. J. Platt, and L. D. Whiteaker. 1993. Density and size structure of slash pine stands in the everglades region of south Florida. *Forest Ecology and Management* 59: 295-311.

Douglas, M. S. 1947. *The Everglades* (50th anniversary edition 1997). Sarasota: Pineapple Press.

———. 1994. Cited by B. O'Donnell. 1994. The Everglades. *Florida Wildlife* 48 (July-August): 2-6.

Doyle, L. R., D. C. Sharma, A. C. Hine, O. H. Pilkey, Jr., W. J. Neal, and O. H. Pilkey, Sr. 1984. *Living with the West Florida Shore*. Durham, N.C.: Duke University Press.

Doyle, R. 1997. Plants at risk in the U.S. *Scientific American* (August): 26.

Duffy, D. C., and A. J. Meier. 1992. Do Appalachian herbaceous understories ever recover from clearcutting? *Conservation Biology* (June): 196-201.

Dunkle, S. W. 1991. Florida's dragonflies. *Florida Wildlife* 45 (July-August): 38-40.

Dunne, P. 1995. The top ten birding spots. *Nature Conservancy* (May/June): 16-23.

Eddy, S., and J. C. Underhill. 1978. *How to Know the Freshwater Fishes*. 3d ed. Boston: McGraw-Hill.

Eisner, T. 1997. Tribute to a mint plant. *Wings* (Fall): 3-6.

Eldredge, M. 1992. Management of Florida Keys must involve entire ecosystem. *Marine Conservation News* (Winter): 10.

Emmel, T. C. 2000. Fabulous Florida butterflies. *Florida Wildlife* 54 (May-June): 3-5.

Engstrom, R. T. 1982. In Thirty-fourth winter bird population study, ed. R. L. Boyd and C. L. Cink, 31. *American Birds* 36: 28-49.

Estevez, E. D. 1978. Ecology of *Sphaeroma terebrans Bate* (an unpublished Ph.D. dissertation). Broward City, Fla.: University of South Florida.

Evered, D. S. 1993. Report from the field. *Barrier Island Trust Newsletter* (June): 1-4.

Ewel, K. C. 1990. Swamps. In Myers and Ewel 1990 (q.v.), 281-323.

Fernald, E. A., and D. J. Patton, eds. 1984. *Water Resources Atlas of Florida*. Tallahassee: Florida State University Institute of Science and Public Affairs.

Fernald, E. A., and E. D. Purdum, eds. 1992. *Atlas of Florida*. Gainesville: University Press of Florida.

FICUS. 2002. Ecosystem description: Pine flatwoods and dry prairies. http://www.ficus.-usf.edu/docs/fl_ecosystem/upland/up-flat.htm. 8 February.

Fish, D., and D. W. Hall. 1978. Succession and stratification of aquatic insects inhabiting the leaves of the insectivorous pitcher plant, *Sarracenia purpurea. American Midland Naturalist* 99(7): 172-183.

Fitzpatrick, J. W. 1992. Vanishing Florida scrub. *Florida Naturalist* (Spring): 8-9.

———. 1994. Cited by D. Stap. 1994. Along a ridge in Florida. *Smithsonian* (September): 36-45.

Flack, S., and R. Chipley. 1996. *Troubled Waters*. Arlington, Va.: The Nature Conservancy.

Florida Biodiversity Task Force. 1993. Conserving Florida's biological diversity: A report to Governor Lawton Chiles. Tallahassee: State of Florida, Office of the Governor. Photocopy.

Florida Springs Task Force. 2000. *Florida's Springs*. Prepared for D. B. Struhs, secretary, Department of Environmental Protection and the citizens of the state of Florida (November).

Folkerts, G. W. 1982. The Gulf coast pitcher plant bogs. *American Scientist* 70 (May-June): 259-267.

Fowells, H. A. 1965. *Silvics of Forest Trees of the United States*. USDA Forest Service Handbook #271. Washington: Government Printing Office.

Frank, N. L. 1982. Tropical storms and hurricanes. *ENFO* (September): 1-8.

Franz, S. 1992. Swarming mayflies. *Florida Wildlife* 48 (May-June): 6-7.

Garland, J. L. 1995. Potential extent of bacterial biodiversity in the Indian River Lagoon. *Bulletin of Marine Science* 57: 79-83.

Gatewood, S., and D. Hardin. 1985. La Florida. *Nature Conservancy News* (September-October): 6-12.

Gilbert, K. M., J. D. Tobe, R. W. Cantrell, M. E. Sweeley, and J. R. Cooper. 1995. *The Florida Wetlands Delineation Manual*. Tallahassee: Florida Department of Environmental Protection and the South Florida, St. Johns River, Suwannee River, Southwest Florida, and Northwest Florida Water Management Districts.

Glitzenstein, J. S., D. R. Streng, and W. J. Platt. 1995. *Evaluating the Effects of Season of Burn on Vegetation in Longleaf Pine Savannas*. Tallahassee: Florida Game and Fresh Water Fish Commission.

Godfrey, R. K. 1988. *Trees, Shrubs, and Woody Vines of Northern Florida and Adjacent Georgia and Alabama*. Athens: University of Georgia Press.

Gore, J. 1994. The diversity of Florida's bats. *Skimmer* 10:2 (Summer), 1-2.

Gore, R. H. 1986. Barrier islands. *Florida Wildlife* 40 (January/February): 18-22.

———. 1992. *The Gulf of Mexico*. Sarasota, Fla.: Pineapple Press.

Gould, S. J., ed. 1996. *The Historical Atlas of Earth*. New York: Henry Holt.

Griffith, G., D. E. Canfield, Jr., C. A. Horsburgh, and J. M. Omernik. 1997. *Lake Regions of Florida*. Corvallis, Ore.: U.S. Environmental Protection Agency.

Grimes, B. H. 1989. *Species Profiles* (Atlantic marsh fiddler). Raleigh, N.C.: U.S. Fish and Wildlife Service, North Carolina Cooperative Fishery Research Unit and North Carolina State University.

Grow, G. 1984. What are wetlands good for? *ENFO* (February): 3, 5.

Guide to the Natural Communities of Florida. 1990. Tallahassee: Florida Natural Areas Inventory and Department of Natural Resources.

Hahn, D. A., and W. R. Tschinkel. 1997. Settlement and distribution of colony-founding queens of the arboreal ant, *Crematogaster ashmeadi*, in longleaf pine forest. *Insectes Sociaux* 44: 323-336.

Hall, M. O., and N. J. Eiseman. 1981. The seagrass epiphytes of the Indian River, Florida, I. Species list with descriptions and seasonal occurrences. *Botanica Marina* 24: 139-146.

Harland, W. B., A. G. Smith, and B. Wilcock, eds. 1964. The Phanerozoic time-scale. *Quarterly Journal of the Geologic Society of London* 120S (supplement).

Harris, L. D., and R. D. Wallace. 1984. Breeding bird species in Florida forest fragments. *Proceedings of the Annual Conference of Southeastern Associations of Fish and Wildlife Agencies* 38: 87-96.

Harris, L. D., with R. Sullivan and L. Badger. 1984. *Bottomland Hardwoods*. IFAS document 8-20M-84. Gainesville: IFAS.

Haug, E. A., B. A. Millsap, and M. S. Martell. 1993. *Speotyto cunicularia*, Burrowing owl. In *The Birds of North America*, ed. A. Poole and F. Gill, pages. Philadelphia: Academy of Natural Sciences and Washington: American Ornithologists' Union.

Hayes, J. M. 1996. The earliest memories of life on earth. *Nature* 384 (7 November): 21-22.

Heath, R. C., and C. S. Conover. 1981. *Hydrologic Almanac of Florida*. Open-File Report 81-1107. Tallahassee: Florida Department of Environmental Regulation.

Hermann, S. M. 1995. *Status and Management of Florida's Carnivorous Plant Communities*. Nongame Wildlife Program Project Report GFC084-033 (December). Tallahassee: Florida Game and Fresh Water Fish Commission.

Hiaasen, C. 1993. The Miccosukee: The Seminoles and the Glades. *Tallahassee Democrat*, 4 April.

Hickman, C. P., L. S. Roberts, and F. M. Hickman. 1984. *Integrated Principles of Zoology*. 7th ed. St. Louis: Mosby.

Hinrichsen, D. 1995. Waterworld. *Amicus Journal* (Summer): 23-27.

Hobbs, H. H., Jr. 1942. The crayfishes of Florida. *University of Florida Biological Series* 3:2 (November).

Holtz, S. 1986. Bringing back a beautiful landscape. *Restoration and Management Notes* (Winter): 56-61.

Holzer, R. F. 1986. Ecological Communities of the Big Bend (an unpublished masters thesis). Tallahassee: Florida State University.

Hooper, R. G., A. F. Robinson, Jr., and J. A. Jackson. 1990. *The Red-Cockaded Woodpecker*. Atlanta: U.S. Department of Agriculture, Forest Service, Southeastern Area, State and Private Forestry.

Horton, T. 1987. The passion of eels. In *Bay Country*, 48-52. New York: Ticknor and Fields.

Hubbard, M. D., and C. S. Gidden. 1997. Terrestrial vertebrates of Florida's Gulf coast tidal marshes. In Coultas and Hsieh 1997 (q.v.), 331-338.

Huffstodt, J. 1991. Lake Okeechobee. *Florida Wildlife* 45 (March-April): 10-12.

———. 1998. Saving the Florida Keys. *Florida Wildlife* 52 (May-June): 16-19.

Humphreys, J., S. Franz, and B. Seaman. 1993. *Florida's Estuaries*. Sea Grant Extension Bulletin #23. Gainesville: Florida Sea Grant College Program and Tallahassee: Department of Community Affairs.

Hurchalla, M. 1998. Crossing the Everglades. *Palmetto* (Summer): 16-19.

Hutto, J. 1995. *Illumination in the Flatwoods*. New York: Lyons and Burford.

Iverson, R. L., and H. F. Bittaker. 1986. Seagrass distribution and abundance in eastern Gulf of Mexico coastal waters. *Estuarine Coastal and Shelf Science* 22: 577-602.

Jaap, W. C., and P. Hallock. 1990. Coral reefs. In Myers and Ewel 1990 (q.v.), 574-616.

Johnson, A. F., and M. G. Barbour. 1990. Dunes and maritime forests. In Myers and Ewel 1990 (q.v.), 429-480.

Johnson, C. 1996. Spring into summer. *Florida Water* (Spring/Summer): 2-5.

Jones, W. K., J. H. Cason, and R. Bjorklund. 1990. *A Literature-Based Review of the Physical, Sedimentary, and Water Quality Aspects of the Pensacola Bay System*. Water Resources Special Report 90-3 (May). Pensacola: Northwest Florida Water Management District.

Kangas, P. 1994. Dwarf mangrove forests and savannas of south Florida. In *Proceedings of the North American Conference on Barrens and Savannas*, ed. J. S. Fralish, R. C. Anderson, J. E. Ebinger, and R. Szafoni, 375. Normal, Ill.: Illinois State University, October 15-16. Environmental Protection Agency, Great Lakes National Program Office.

Kaplan, E. H. 1982. *A Field Guide to Coral Reefs: Caribbean and Florida*. Boston: Houghton Mifflin.

Karr, J. 1994. Marine and estuarine bioassessment, a presentation made at the Florida Surface Water Quality Conference, Tallahassee, 23 September.

Karr, J. R., and E. W. Chu. 1999. *Better Biological Monitoring*. Covelo, Calif.: Island Press.

Kaufman, L. 1977. The three spot damsel fish. *Proceedings of the Third International Coral Reef Symposium*, Miami, Florida 1: 559-564.

Klingener, N. 1999. Rise in ocean levels poses real threat, Keys group told. *Miami Herald,* 28 May.

Komarek, E. V. 1964. The natural history of lightning. *Proceedings of the Tall Timbers Fire Ecology Conference* 3: 139-183.

———. 1965. Fire ecology. *Proceedings of the Tall Timbers Fire Ecology Conference* 4: 169-220.

Kruczynski, W. L., and B. F. Ruth. 1997. Fishes and invertebrates. In Coultas and Hsieh 1997 (q.v.), 131-173.

Kushlan, J. A. 1990. Freshwater marshes. In Myers and Ewel 1990 (q.v.), 324-363.

Laessle, A. M., and C. D. Monk. 1961. Some live oak forests of northeastern Florida. *Quarterly Journal of the Florida Academy of Sciences* 24(1): 39-55.

Layne, J. N. 1992. Florida mouse. In Ashton 1992 (*Mammals*) (q.v.), 250-264.

Leitman, H. M., J. E. Sohm, and M. A. Franklin. 1983. *Wetland Hydrology and Tree Distribution of the Apalachicola River Flood Plain*. Water-Supply Paper 2196. Tallahassee: U.S. Geological Survey.

Leveton, D. 1995. Delicate balance (Fox squirrel). *Florida Wildlife* 49 (November/December): 15.

Lewis, R. R., III, and E. D. Estevez. 1988. *Ecology of Tampa Bay*. U.S. Fish and Wildlife Services Biological Report 85(7.18). Springfield, Va.: National Technical Information Service.

Lickey, E. B., and G. L. Walker. 2002. Population genetic structure of baldcypress and pondcypress. *Southeastern Naturalist* 1 (2): 131-148.

Livingston, E. H. 1989. Overview of stormwater management. *ENFO* (Feb): 1-6.

Livingston, R. J. 1983. *Resource Atlas of the Apalachicola Estuary*. Sea Grant Project No. T/P-1. Tallahassee: Florida Sea Grant College Program.

————. 1990. Inshore marine habitats. In Myers and Ewel 1990 (q.v.), 549-573.

————, ed. 1991. *The Rivers of Florida*. New York: Springer Verlag.

Loftin, J. P. 1992. Rediscovering the value of wetlands. *Florida Water* (Fall): 16-23.

Lohmann, K. J. 1992. How sea turtles navigate. *Scientific American* (January): 100-106.

Lovejoy, T. E. 1986. Species leave the ark one by one. In *The Preservation of Species*, ed. B. G. Norton, 13-27. Princeton, N.J.: Princeton University Press.

Lyons, W. G., D. G. Barber, S. M. Foster, F. S. Kennedy, Jr., and G. R. Milano. 1981. The spiny lobster, *Panulurus argus*, in the middle and upper Florida Keys. *Florida Marine Research Publication* 38 (February).

MacArthur, R., and J. MacArthur. 1961. On bird species diversity. *Ecology* 42: 594-598.

Margulis, L., and K. V. Schwartz. 1990. *Five Kingdoms*. New York, Freeman.

May, R. M. 1993. The end of biological history? (a review of E. O. Wilson's book *The Diversity of Life*, 1992). *Scientific American* (March): 146-149.

McCook, A. 2002. You snooze, you lose. *Scientific American*, April: 31.

McKibben, B. 1996. What good is a forest? *Audubon* 98 (May): 54-63.

Means, D. B. 1977. Aspects of the significance to terrestrial vertebrates of the Apalachicola River drainage basin. *Florida Marine Research Publication 26*, 37-67. Apalachicola: Florida Marine Research Institute.

————. 1985. Delicate balance (Georgia blind salamander). *Florida Wildlife* 39 (November-December): 37.

————. 1985. The cotton rat. *ENFO* (February): 6-7.

————. 1990. Temporary ponds. *Florida Wildlife* 44 (November-December): 12-16.

————. 1991. Florida's steepheads. *Florida Wildlife* 45 (May-June): 25-28.

————. 1994. Longleaf pine forests. *Florida Wildlife* 48 (September-October): 2-6.

————. 1994. Temperate hardwood hammocks. *Florida Wildlife* 48 (November-December): 20-23.

————. 1996. Longleaf pine forest, going, going, . . . In *Eastern Old-Growth Forests*, ed. M. B. Davis, 210-229. Washington: Island Press.

————. 2000. Southeastern U.S. coastal plain habitats of the Plethodontidae. In *Biology of Plethodontid Salamanders*, ed. R. C. Bruce, R. G. Jaeger, and L. D. Houck, 287-302. New York: Kluwer Academic/Plenum Publishers.

Means, D. B., and P. E. Moler. 1979. The pine barrens treefrog. In *Proceedings of the Rare and Endangered Wildlife Symposium, Technical Bulletin WL-4,* ed. R. R. Odum and J. L. Landers, Athens, Georgia, August 3-4, 77-83. Georgia Department of Natural Resources.

Meylan, P. A. 1984. A history of fossil amphibians and reptiles in Florida. *Plaster Jacket* 44 (February): 5-29.

Mikkelsen, P. M., P. S. Mikkelsen, and D. J. Karlen. 1995. Molluscan biodiversity in the Indian River Lagoon. *Bulletin of Marine Science* 57: 94-127.

Minno, M., and M. Minno. 1993. Interdependence. *Palmetto* (Summer): 6-7.

Mitchell, R. S. 1963. Phytogeography and floristic survey of a relic area in the Marianna Lowlands. *American Midland Naturalist* 69: 328-366.

Mitsch, W., and J. Gosselink. 1986. *Wetlands*. 2d ed. New York: Van Nostrand-Reinhold.

Mojsis, S. J., G. Arrhenius, K. D. McKeegan, T. M. Harrison, A. P. Nutman, and C. R. L. Friend. 1996. Evidence for life on earth before 3,800 million years ago. *Nature* 384 (7 November). 55-59.

Moler, P. 1992. Salt marsh survivors. *Skimmer* 8:2 (Summer), 3, 7.

Montague, C. L., and R. G. Wiegert. 1990. Salt marshes. In Myers and Ewel 1990 (q.v.), 481-516.

Muir, J. [1916.] *Thousand-Mile Walk to the Gulf*, reprinted 1981. Boston: Houghton Mifflin.

Muller, J. W., E. D. Hardin, D. R. Jackson, S. E. Gatewood, and N. Claire. 1989. *Summary Report on the Vascular Plants, Animals and Plant Communities Endemic to Florida*. Technical Report No. 7 (June). Tallahassee: Florida Game and Fresh Water Fish Commission Nongame Wildlife Program.

Myers, R. L. 1990. Scrub and high pine. In Myers and Ewel 1990 (q.v.), 150-193.

Myers, R. L., and J. J. Ewel, eds. 1990. *Ecosystems of Florida*. Orlando: University of Central Florida Press.

Nauman, C. E. 1986. The increasing rarity of Florida's ferns. *Florida Naturalist* (Winter): 2-4.

Nerem, R. S. 1995. Global mean sea level variations from TOPEX/POSEIDON altimeter data. *Science* 268: 708-710.

Neuhauser, H. 1993. Florida waters critical to right whale survival. *Florida Naturalist* (Fall): 4-7.

Nordlie, F. G. 1990. Rivers and springs. In Myers and Ewel 1990 (q.v.), 392-425.

Norman, M. E. 1998. The year of the oceans. *Earth Island Journal* (Spring): 23.

Norse, E. A. 1990. What good are ancient forests? *Amicus Journal* (Winter): 42-45.

Norton, B. G., ed. 1986. *The Preservation of Species*. Princeton, N.J.: Princeton University Press.

O'Brien, C. A., and J. B. Box. 1999. Reproductive biology and juvenile recruitment of the shinyrayed pocketbook, *Lampsilis subangulata* (Bivalvia: Unionidae) in the Gulf coastal plain. *American Midland Naturalist* 142: 129-140.

Odum, E. P. 1993. *Ecology and Our Endangered Life Support Systems*. 2d ed. Sunderland, Mass.: Sinauer Associates.

Odum, W. E., and C. C. McIvor. 1990. Mangroves. In Myers and Ewel 1990 (q.v.), 517-548.

Odum, W. E., C. C. McIvor, and T. J. Smith, III. 1982. *The Ecology of the Mangroves of South Florida*. Report number FWS/OBS 81/24. Washington: U.S. Fish and Wildlife Service, Office of Biological Services.

Opdyke, N. D., D. S. Jones, B. F. MacFadden, D. L. Smith, P. A. Mueller, and R. D. Shuster. 1987. Florida as an exotic terrane. *Geology* 15: 900-903.

Osborne, R., and D. Tarling. 1996. Human origins and migration. *Historical Atlas of the Earth*, 156-157. New York: Holt.

Osorio, R. 1991. Three pine rock land shrubs. *Palmetto* (Fall): 8-9.

Palumbi, S. 2000. Evolution, synthesized. *Harvard Magazine* (March-April): 26-30.

Palmer, J. D. 1975. Biological clocks of the tidal zone. *Scientific American* (February): 70-79.

Patterson, S. 2000. Rising waters, trouble ahead. *Florida Times-Union* (Jacksonville), 6 March.

Patton, T. H. 1969. An Oligocene land vertebrate fauna from Florida. *Journal of Paleontology* 43:2 (March): 543-546.

Perry, D. A., and M. P. Amaranthus. 1987. Disturbance, recovery, and stability. In *Creating a Forestry for the 21st Century*, ed. K. A. Kohm and J. E. Franklin, 31-56. Washington, D.C.: Island Press.

Platt, W. J. 1999. Southeastern pine savannas. In *Savannas, Barrens, and Rock Outcrop Plant Communities of North America*, ed. R. C. Anderson, J. S. Fralish, and J. M. Baskin, 23-51. Cambridge, England: Cambridge University Press.

Platt, W. J., and S. M. Hermann. 1986. Relationships between dispersal syndrome and characteristics of populations of trees in a mixed-species forest. In *Frugivores and Seed Dispersal*, ed. A. Estrada, T. H. Fleming, C. Vasques-Yanes, and R. Dirzo, 309-321. The Hague, Netherlands: Junk (publisher).

Platt, W. J., and M. W. Schwartz. 1990. Temperate hardwood forests. In Myers and Ewel 1990 (q.v.), 194-229.

Pojeta, J., Jr., J. Kriz, and J. M. Berdan. 1976. Silurian-Devonian pelecypods and Palaeozoic stratigraphy of subsurface rocks in Florida and Georgia and related Silurian pelecypods from Bolivia and Turkey. *U.S. Geological Survey Professional Papers* No. 879: 1-39.

Pratt, A. E. 1990. Taphonomy of the large vertebrate fauna from the Thomas Farm locality (Miocene, Hemingfordian), Gilchrist County, Florida. *Bulletin of the Florida Museum of Natural History, Biological Sciences* 35(2): 35-130.

Puri, H. S., and R. O. Vernon. 1964. Summary of the geology of Florida and a guidebook to the classic exposures. *Florida Geological Survey Special Publication* No. 5.

Raven, P. H. 1994. Defining biodiversity. *Nature Conservancy* (January-February): 10-15.

Raven, P. H., L. R. Berg, and G. B. Johnson. 1993. *Environment*. Fort Worth, Texas: Saunders.

Ray, P. S. 1992. The sunshine state's other claim to weather fame. *Tallahassee Democrat*, 1 December.

Reed, J. K., R. Gore, L. Scotto, and K. Wilson. 1982. Community composition, structure, areal, and trophic relationships of decapods associated with shallow- and deep-water *Oculina varicosa* coral reefs. *Bulletin of Marine Science* 32: 761-786.

Reiger, G. 1990. Symbols of the marsh. *Audubon* (July): 52-58.

Repenning, R. W., and R. F. Labisky. 1985. Effects of even-age timber management on bird communities of the longleaf pine forest in northern Florida. *Journal of Wildlife Management* 49: 1088-1098.

Rey, J. R., and E. D. McCoy. 1997. Terrestrial arthropods. In Coultas and Hsieh 1997 (q.v.), 175-208.

Reynolds, J. W. 1994. Earthworms of Florida. *Megadrilogica* 5(12): 125-141.

Robbins, C. S., B. Bruun, and H. S. Zim. 1983. *Birds of North America*. Racine, Wisc.: Western Publishing.

———. 2001. *Birds of North America*. Rev. ed. New York: St. Martin's Press.

Rodriquez, J. P., and W. M. Roberts. 1997. Drawing entitled "Endangered species hot spots in the U.S.," adapted by A. Bombay in *EDF Letter* (April): 5.

Ross, J. F. 1996. A few miles of land arose from the sea—and the world changed. *Smithsonian* (December): 112-121.

Roush, J. 1982. On saving diversity. *Nature Conservancy News* (January-February): 4-10.

Rupert, F. R. 1998. Florida caves map (unpublished).

Rupert, F., and S. Spencer. 1988. *Geology of Wakulla County, Florida*. Florida Geological Survey Bulletin No. 60. Tallahassee: Florida Geological Survey.

Russell, D. 1998. Deep blues. *Amicus Journal* (Winter): 25-29.

Salmon, M. 1988. Ecology and behavior of fiddler crabs. *Florida Naturalist* (Summer): 5-7.

Sampat, P. 1996. The River Ganges' long decline. *World Watch* (July/August): 25-32.

Sanibel Seashells, Inc., www.seashells.com.

Sawicki, R. 1997. Tropical flyways and the white-crowned pigeons. *Florida Naturalist* (Summer): 6-7.

Schwartz, M. W., and S. M. Hermann. 1993. *The Population Ecology of Torreya Taxifolia*. Nongame Wildlife Program Project HG89-030. Tallahassee: Florida Game and Fresh Water Fish Commission.

Schwartz, M. W., and J. Travis. 1995. The distribution and character of natural habitats in pre-settlement northern Florida. *Public Land Survey Records Project Report* (December). Tallahassee: Florida Game and Fresh Water Fish Commission Nongame Wildlife Program.

Scott, T. M., G. H. Means, R. C. Means, and R. P. Meegan. 2002. *First Magnitude Springs of Florida, Open File Report No. 85*. Tallahassee: Florida Geological Survey

Skeate, S. T. 1987. Interactions between birds and fruits in a northern Florida hammock community. *Ecology* 68(2): 297-309.

Smith, M. L. 1997. Scientists discover new life forms during Caribbean expeditions. 1997. *Marine Conservation News* (Autumn): 4.

Snedaker, S. C., and A. E. Lugo. 1972. Ecology of the Ocala National Forest. Ocala, Fla.: U.S. Forest Service (unpublished manuscript).

Snyder. J. R., A. Herndon, and W. B. Robertson, Jr. 1990. South Florida rockland. In Myers and Ewel 1990 (q.v.), 230-277.

Southall, P. D. 1986. Living resources of the Suwannee. *ENFO* (September): 13-14.

Spivey, T. 1992. The Suwannee River. *Florida Water* (Fall): 12-14.

Stamm, D. 1991. Springs of passage. *Florida Wildlife* 45 (July-August): 19-23.

Stap, D. 1994. Along a ridge in Florida. *Smithsonian* (September): 36-45.

Stevens, W. K. 1990. Monarchs' migration. *New York Times*, 4 December.

Stiling, P. 1989. Delicate balance (Purse-web spider). *Florida Wildlife* 43 (Jul-August): 9.

Stolzenburg, W. 1992. The mussels' message. *Nature Conservancy Magazine* (November/December): 16-23.

Stout, I. J., and W. R. Marion. 1993. Pine flatwoods and xeric pine forests of the southern (lower) coastal plain. In *Biodiversity of the Southeastern United States*. Vol. 1. *Lowland Terrestrial Communities*, ed. W. H. Martin, S. G. Boyce, and A. C. Echternacht, 373-446. New York: Wiley.

Strahl, S. D. 1997. Florida's Everglades. *Florida Naturalist* (Spring): 6-7.

Strahler, A. N. 1964. Quantitative geomorphology of drainage basins and channel networks. In *Handbook of Applied Hydrology*, ed. Ven te Chow, 4-39 to 4-76. New York: McGraw-Hill.

Stys, B. 1997. *Ecology of the Florida Sandhill Crane*. Nongame Wildlife Technical Report No. 15 (July). Tallahassee: Florida Game and Fresh Water Fish Commission.

Suwannee River Task Force. 1989. Report pursuant to Governor Bob Martinez's executive order 88-246. Issues and Draft Recommendations. Tallahassee: State of Florida, Office of the Governor, 21 July. Photocopy.

Svingen, K. 1995. Group aims to rescue woodpecker. *Tallahassee Democrat*, 27 November.

Tebo, M. 1988. Florida's desert islands. *Skimmer* 4:1 (Winter), 1-2.

Toops, C. 1998. *The Florida Everglades*. Rev. ed. Stillwater, Minn.: Voyageur Press.

Tousignant, M. 1995. Wildlife crusader comes to the rescue of the much-maligned bat. *Tallahassee Democrat*, 25 September.

Tunstall, J. 1998. Tiny swamp, beach dwellers pack big bite. *Tallahassee Democrat*, 12 April.

Urquhart, J. C. 1994. Everglades National Park. In *Our Inviting Eastern Parklands*, 112-157. Washington: National Geographic Society.

USGSLIST (World Wide Web): 13 March 1998.

Virnstein, R. W. 1995. Anomalous diversity of some seagrass-associated fauna in the Indian River Lagoon, Florida. *Bulletin of Marine Science* 57: 75-78.

Voss, G. L. 1988. *Coral Reefs of Florida*. Sarasota, Fla.: Pineapple Press.

Wahlenberg, W. G. 1946. *Longleaf Pine*. Washington: Charles Lathrop Pack Forestry Foundation.

Walker, J., and R. K. Peet. 1983. Composition and species diversity of pine-wiregrass savannas of the Green Swamp, North Carolina. *Vegetatio* 55: 163-179.

Ward, D. B., and A. F. Clewell. 1989. Atlantic White Cedar (*Chamaecyparis thyoides*) in the Southern states. *Florida Scientist* 52(1): 8-47.

Ware, S., C. Frost, and P. Doerr. 1993. Southern mixed hardwood forest: The former longleaf pine forest. In *Biodiversity of the Southeastern United States: Upland Terrestrial Communities*, ed. W. Martin, S. Boyce, and A. Echternacht: 447-493. New York: Wiley.

Waters, J., Jr. 1986. Nature notes. *Florida Wildlife* 40 (May-June): 47.

————.. 1988. Nature notes. *Florida Wildlife* 42 (September-October): 47.

Webb, S. D. 1981. The Thomas Farm fossil vertebrate site. *Plaster Jacket* 37 (July): 6-25.

————. 1990. Historical biogeography. In Myers and Ewel 1990 (q.v.), 70-100.

Webb, S. D., and D. B. Crissinger. 1983. Stratigraphy and vertebrate paleontology of the central and southern phosphate districts of Florida. Geology Society of America, Southeast Section Meeting, *Field Trip Guidebook*: 28-72.

Webb, S. D., B. J. MacFadden, and J. A. Baskin. 1981. Geology and paleontology of the Love Bone Bed from the late Miocene of Florida. *American Journal of Science* 281(5): (May), 513-544.

Weber, P. 1993. Safeguarding oceans. In *State of the World* 1994, ed. L. R. Brown, 41-60. New York: Norton.

Weigl, P. D., M. A. Steele, L. J. Sherman, J. C. Ha, and T. L. Sharpe. 1989. Ecology of the fox squirrel (*Sciurus niger*) in North Carolina. *Bulletin of Tall Timbers Research Station* 24: i-91.

Wells, B. W., and I. V. Shunk. 1931. The vegetation and habitat factors of the coarser sands of the North Carolina coastal plain. *Ecological Monographs* 1: 465-520.

Whaley, R. 1990. Causes of seagrass decline in the Tampa Bay area. *Palmetto* (Spring): 7-10.

Williams, K., K. C. Ewel, R. P. Stumpf, F. E. Putz, and T. W. Workman. 1999. Sea-level rise and coastal forest retreat on the west coast of Florida. *Ecology* 80: 2045-2061.

Willoughby, H. 1897. Cited by Lt. J. Huffstodt. 1997. Across the Everglades by canoe. *Florida Wildlife* 51 (November-December): 25-27.

Wilson, E. O. 1983. An introduction. *Nature Conservancy News* (November/December): 344-346.

Winsberg, M. D. 1990. *Florida Weather*. Orlando, Fla.: University of Central Florida Press.

Winston, J. E. 1995. Ectoproct diversity of the Indian River coastal Lagoon. *Bulletin of Marine Science* 57: 84-93.

Wisenbaker, M. 1988. Florida's true natives. *Florida Naturalist* (Spring): 2-5.

———. 1989. Sinkholes. *Florida Naturalist* (Winter): 3-6.

———. 1991. Floridana. *Florida Wildlife* 45 (September-October): 49.

———. 1994. Florida's aquatic cave animals. *Florida Wildlife* 48 (January-February): 14-17.

———. 1997. A well-traveled fish. *Florida Wildlife* 51 (March-April): 27-29.

Wolfe, S. H., J. A. Reidenauer, and D. B. Means. 1988. *An Ecological Characterization of the Florida Panhandle*. Biological Report 88(12). Washington: U.S. Fish and Wildlife Service and New Orleans, La.: Minerals Management Service.

Wolkomir, R. 1995. Seeking gifts from the sea, Sanibel-style. *Smithsonian* (August): 60-69.

Woolfenden, G. E., and R. W. Schreiber. 1973. The common birds of the saline habitats of the eastern Gulf of Mexico. In *A Summary of Knowledge of the Eastern Gulf of Mexico*, ed. J. I. Jones, R. E. Ring, M. L. Rinkel, and R. E. Smith, Section 3J. St. Petersburg: State University System of Florida Institute of Oceanography.

Worthington, L. V. 1980. Sargasso Sea. In *McGraw-Hill Encyclopedia of Ocean and Atmospheric Sciences*, ed. S. P. Parker, 397. New York: McGraw-Hill.

Wuerthner, G. 1995. Why healthy forests need dead trees. *Earth Island Journal* (Fall): 22.

Wunderlin, R. P., and B. F. Hanson. 2003. *Guide to the Vascular Plants of Florida*. 2d. ed. Gainesville: University Press of Florida.

Yoon, C. K. 1994. Do ecosystems need biodiversity? In Drugs from bugs, 28-29. *Garbage* (Summer): 22-29.

Zieman, J. C. 1982. *The Ecology of the Seagrasses of South Florida*. Fish and Wildlife Service Biological Report 82/25. Washington, D.C.: National Coastal Ecosystems Team, Office of Biological Services, U.S. Department of the Interior.

Zieman, J. C., and R. T. Zieman. 1989. *The Ecology of the Seagrass Meadows of the West Coast of Florida*. Fish and Wildlife Service Biological Report 85 (7.25). Springfield, Va.: National Technical Information Service.

APPENDIX

ECOSYSTEMS

This book uses the ecosystem classification scheme employed by the Florida Natural Areas Inventory (FNAI), as presented in the *Guide to the Natural Communities of Florida* (Tallahassee: Florida Natural Areas Inventory and Department of Natural Resources, 1990). FNAI recognizes 69 ecosystems, or natural communities, in Florida. They fall naturally into six categories: interior uplands, wetlands, and waters; and coastal uplands, wetlands, and waters. This appendix shows how FNAI's natural communities are distributed into this book's chapters. (Chapters 1, 2, and 8 are introductory and Chapters 22 and 23 are closing chapters, and do not deal with specific ecosystems. Chapter 21 covers the pelagic waters and seafloors around Florida, which are not included in FNAI's scheme.)

Part and Chapter	FNAI Community	FNAI Category
INTERIOR UPLANDS		
3 High Pine Grasslands	Sandhill	Xeric uplands
	Upland Pine Forest	Mesic uplands
4 Flatwoods and Prairies	Dry Prairie	Mesic flatlands
	Mesic Flatwoods	Mesic flatlands
	Scrubby Flatwoods	Mesic flatlands
5 Interior Scrub	Scrub	Xeric uplands
6 Hardwood Hammocks	Bluff	Mesic uplands
	Prairie Hammock	Mesic flatlands
	Slope Forest	Xeric uplands
	Upland Glade	Xeric uplands
	Upland Hardwood Forest	Xeric uplands
	Upland Mixed Forest	Xeric uplands
	Xeric Hammock	Xeric uplands
7 Rocklands, Terrestrial Caves	Pine Rocklands	Rocklands
	Rockland Hammock	Rocklands
	Sinkhole	Rocklands
	Terrestrial Cave	Subterranean
INTERIOR WETLANDS		
9 Seepage Wetlands	Baygall	Seepage wetland
	Seepage Slope	Seepage wetland

399

Part and Chapter	FNAI Community	FNAI Category
INTERIOR WETLANDS (continued)		
10 Marshes	Bog	Basin wetland
	Depression Marsh	Basin wetland
	Marl Prairie	Wet flatlands
	Slough	Floodplain wetland
	Swale	Floodplain wetland
	Wet Flatwoods	Wet flatlands
	Wet Prairie	Wet flatlands
11 Flowing-Water Swamps	Bottomland Forest	Floodplain wetland
	Floodplain Forest	Floodplain wetland
	Floodplain Swamp	Floodplain wetland
	Freshwater Tidal Swamp	Floodplain wetland
	Hydric Hammock	Wet flatlands
	Strand Swamp	Floodplain wetland
INTERIOR WATERS		
12 Lakes and Ponds	Basin Marsh	Basin wetland
	Basin Swamp	Basin wetland
	Clastic Upland Lake	Lacustrine
	Coastal Dune Lake	Lacustrine
	Dome Swamp	Basin wetland
	Flatwoods/Prairie Lake	Lacustrine
	Marsh Lake	Lacustrine
	River Floodplain Lake	Lacustrine
	Sandhill Upland Lake	Lacustrine
	Swamp Lake	Lacustrine
13 Streams	Alluvial Stream	Riverine
	Blackwater Stream	Riverine
	Floodplain Marsh	Floodplain wetland
	Seepage Stream	Riverine
14 Caves, Sinks, Springs	Aquatic Cave	Subterranean
	Coastal Rockland Lake	Lacustrine
	Sinkhole Lake	Lacustrine
	Spring-Run Stream	Riverine
COASTAL UPLANDS		
15 Beach-Dune Systems	Beach/Dune	Coastal
	Coastal Berm	Coastal
	Coastal Grasslands	Coastal
	Coastal Rock Barren	Coastal
	Coastal Strand	Coastal
	Maritime Hammock	Coastal
	Shell Mound	Coastal

Florida native: Gray fox (*Urocyon cinereo argenteus*).

Florida native: Ogre-faced spider (*Deinopis spinosa*). The spider puts out a scent to attract moths while holding a web with its long legs. Then when a moth flies by, it streches its legs out and slaps the streched web over its prey.

Florida native: Copes gray treefrog (*Hyla chrysoscelis*).

PHOTOGRAPHY CREDITS

AB/FKNMS = Commander Alan Bunn for FKNMS (q.v.)

ABS = Allen Blake Sheldon

AC = Andrew Chapman, Green Water Labs/BCI Engineers & Scientists

ALA = Alabama Cooperative Fish and Wildlife Research Unit

ARP/NOAA = Dr. Anthony R. Picciolo for NOAA (q.v.)

BAR/TNC = Barry A. Rice for The Nature Conservancy

BC = Bruce Colin Photography

BM = Barry Mansell

BP = Bill Petty, Amateur mycologist, Crawfordville, Fla.

CC/FPS = Carrie Canfield for FPS (q.v.)

CC/NOAA = C. Clark for NOAA (q.v.)

CJ = Chris Johnson, www.FloridaLeatherbacks.com

CWH/NOAA = Cmdr William Harrigan for NOAA (q.v.)

DA = Doug Alderson

DA/FWM = Doug Alderson, Associate Editor, Florida Wildlife Magazine

DB/FPS = Dana Bryan for FPS (q.v.)

DBM = D. Bruce Means, Coastal Plains Institute and Land Conservancy

DEP/AQP = Department of Environmental Protection Aquatic Preserves

DN/GSML = David Norris for the Gulf Specimen Marine Laboratory

ELA = Emi and Larry Allen

EW = Ellie Whitney

FFWCC = Florida Fish and Wildlife Conservation Commission

FHF = Finley-Holiday Film Corporation, www.finley-holiday.com

FKNMS = Florida Keys National Marine Sanctuary

FLAUSA = Visit Florida

FMNH = Florida Museum of Natural History, Exhibits and Public Programs Department

FPS = Florida Park Service (www.FloridaStateParks.org)

GB = Giff Beaton, www.giffbeaton.com

GB/FFWCC = Gray Bass for FFWCC (q.v.)

GN = Gil Nelson, Ph.D., Botanist/Author/Photographer

GSML = Gulf Specimen Marine Laboratory

GW/OAR/NURP = G. Wenz for OAR, NURP (q.v.)

HD/FKNMS = Heather Dine for FKNMS (q.v.)

HLJ/USGS = Howard L. Jelks for the U.S. Geological Survey

IAW/FPS = Ivy Aletheia Wilson for FPS (q.v.)

IM/OAR/NURP = Ian MacDonald for OAR/NURP (q.v.)

JCN/FPS = J. C. Norton for FPS (q.v.)

JD/FFWCC = Jay DeLong for FFWCC (q.v.)

JH = Joe Halusky

JH/RG = Jerry Hawthorne of River Graphics

JP/IQM = Justin Peach for Image Quest Marine

JR = Jeff Ripple, Photographer, PO Box 142613, Gainesville, FL 32614

JS/RP = John Sullivan, Ribbit Photography

JT = Jerry Turner, 28 Pebble Pointe, Thomasville, GA 31792

JV = James Valentine, Photographer

JV/FPS = James Valentine for FPS (q.v.)

JW/FPS = Jerry White for FPS (q.v.)

KC/FPS = Kathleen Carr for FPS (q.v.)

KD/BPS = Kerry Dressler of Bio-Photo Services, Inc.

KLP/FPS = www.KeniLeePhotos.com for FPS (q.v.)

KME = Kevin M. Enge

KWM/USFWS = Kenneth W. McCain for USFWS (q.v.)

LB = Larry Busby, Ranger, Waccasassa Bay State Preserve

LF = Lois Fletcher, Executive Director, Gilchrist County Chamber

LH = Lisa K. Horth, Ph.D, Biologist, University of Virginia

LL/FPS = Larry Lipsky for FPS (q.v.)

LZ/FKNMS = Larry Zetwoch for FKNMS (q.v.)

MAB = Mikhail A. Blikshteyn, www.CHEREPASHKA.com

MFH = Museum of Florida History, Tallahassee

MJA = Matthew Aresco, Biologist, Florida State University

MM = Mary Matthews, Photographer, Tallahassee, Florida

MS = Michael Spelman

MW = Michael Wisenbaker, Outdoor Photographer, Tallahassee, Florida

NB = Nancy Bissett, Green Horizon Land Trust

NB/USGS = Noah Burkhalter for the U.S. Geological Survey

NOAA = National Oceanographic and Atmospheric Administration

NURP = National Undersea Research Program

OAR = Oceanographic and Atmospheric Research, NOAA (q.v.)

PB/IQM = Peter Batson for Image Quest Marine

PEM = Paul E. Moler, FFWCC (q.v.)

PG/FKNMS = Paige Gill for FKNMS (q.v.)

PG/NOAA = Paige Gill for NOAA (q.v.)

PKS = Pam Sikes, Photographer, P.O. Box 997 Folkston, GA, stpsikes@alltel.net

PP/IQM = Peter Parks for Image Quest Marine

PS = Peter Stiling

RB/OAR/NURP = R. Bray for OAR, NURP (q.v.)

RC/OAR/NURP = R. Cooper for OAR, NURP (q.v.)

REA = Ray E. Ashton, Jr., Ashton Biodiversity Research and Preservation Institute, Inc.

RLH = Roger L. Hammer

RP/NOAA = Ron Phillips for NOAA (q.v.)

RPP = Rick Poley Photography, 12410 Green Oak Ln, Dade City, FL 33525

RS/IQM = Roger Steene for Image Quest Marine

RS/OAR/NURP = R. Steneck for OAR, NURP (q.v.)

RW = Russ Whitney

SC = Stephen Coleman, Nature Photographer, sdcme1@juno.com

SFWMD = South Florida Water Management District

SMS = Steven M. Sikes

SSC = Seashells.com

TCH = Terrence Hitt, Nature Coast Expeditions, P.O. Box 218, Cedar Key, FL 32625

TE = Thomas Eisner, Naturalist, Ithaca, New York

TKG/FKNMS = Thomas K. Gibson for FKNMS (q.v.)

TN/NOAA/NURP = T. Niesen, NOAA, NURP (q.v.)

USFWS = United States Fish and Wildlife Service

WBSP = Waccasassa Bay State Preserve

WCM = William C. Maxey

WR = William Rossiter, Cetacean Society Intl., www.csiwhalesalive.org

WRT = Walter R. Tschinkel

WS/KP = Wes Skiles, Karst Productions

Index to Species and Other Categories

Following are the approved common names of the plants and animals that were mentioned in this book, together with small samplings of algae, bacteria, fungi, lichens, and protists. For each, the scientific name is given where available. (Scientific names for species were explained in Chapter 1, Box 1–2,

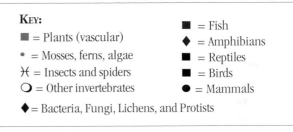

"Categories of Living Things.") An example is the first entry below, Acadian flycatcher: the genus name is *Empidonax* and the species name is *Empidonax virescens*. The authorities used here for common and scientific names are listed in the Reference Notes for Chapter 1 (note 3).

Categories Other than Species. In a few instances, categories smaller than species are given. Species may be subdivided into varieties (for plants) or subspecies or races (for animals). An example of a plant, Chapman's rhododendron, is the endemic variety, *Rhododendron minus* var. *chapmanii*. An example of an animal, the Pugnose minnow, is the Florida pugnose minnow, *Opsopoeodus emiliae peninsularis*. When organisms are not identified precisely to the species level or lower, the genus or a higher category is given. Box 1-2 on pages 13-14 explained these categories.

Finding Species and Categories by Common Name. Common names change. For some of the organisms whose common names have changed since they were described in the reference we used, the old name is included with a cross reference to the new name. Should it prove difficult to locate a species by a familiar name (such as "Alligator" or "Magnolia," try prefacing the name with "American," "Common," "Eastern," "Florida," "Northern," "Southeastern," or "Southern." The alligator's approved common name is American alligator; the magnolia is the Southern magnolia.

Non-Native Plants and Animals. For animals, all non-native species are designated, simply, "alien." For plants, those that, according to the Exotic Pest Plant Council, have proven invasive and harmful to ecosystems are designated "invasive exotic."

Art and Definitions. Numbers in *italics* refer to photos, drawings, or figures. Numbers in **boldface** refer to definitions.

○Apple snail, *Pomacea paludosus*, 241, *241*

●Armadillo (nine-banded), *Dasypus novemcinctus* (alien), 49, 99

■Arrowfeather threeawn, *Aristida purpurascens*, 42

■Arrowhead, *Sagittaria* species, 157, 158, 209, 273

■Ashe's magnolia. *See* Bigleaf magnolia

■Aster (family Asteraceae), 164. *See also* Flaxleaf, Perennial saltmarsh, *and* Whitetop aster

■Atamasco lily, *Zephyranthes atamasca*, i, *187*

■Atlantic bonito, *Sarda sarda*, 349

■Atlantic croaker, *Micropogon undulatus*, *312*

■Atlantic pupfish. *See* Sheepshead minnow

■Atlantic ridley, *Lepidochelys kempii*, 280, 297, *357*

■Atlantic sharpnose shark, *Rhizoprionodon terraenovae*, *312*

●Atlantic spotted dolphin, *Stenella frontalis*, 350

■Atlantic St. John's-wort, *Hypericum reductum*, 133

■Atlantic white cedar, *Chamaecyparis thyoides*, 128, 141, 147, *151*, 151–153

■Awl-leaf arrowhead, *Sagittaria subulata*, 273

B

■Bachman's sparrow, *Aimophila aestivalis*, 48, 51, 61

■Bachman's warbler, *Vermivora bachmanii*, 113

◆Bacteria, **13**, 208, **208**

■Bald-cypress. *See* Cypress

■Bald eagle, *Haliaeetus leucocephalus*, 61, 135, 160, 168, 191, 306, 329

■Baldrush. *See* Shortbeak beaksedge

■Balloonfish, *Diodon holocanthus*, *349*. *See also* Spiny pufferfish

✕Banana spider, *Heteropoda venatoria*, 63

■Bandana-of-the-everglades, *Canna flaccida*, *155*

■Banded sunfish, *Enneacanthus obesus*, 211, 224

■Banded topminnow, *Fundulus cingulatus*, 224

■Banded water snake, *Nerodia fasciata*, 159, 204, 278

■Baneberry. *See* White baneberry

■Bannerfin shiner, *Notropis leedsi*, 224

■Barbara's buttons, *Marshallia* species, 98

■Barbour's map turtle, *Graptemys barbouri* (endemic), 190, *192*, 220

■Bark anole, *Anolis distichus* (alien), 113, 297

◆Barking treefrog, *Hyla gratiosa*, 50, 81, 162, *212*

■Barn owl, *Tyto alba*, 90

■Barracuda. *See* Great barracuda

■Barred owl, *Strix varia*, 51, 91, *91*, 99, 168, 191, 193, 211

✕Bartram's hairstreak, *Strymon acis*, 112

○Basket sponge, *335*

■Bass, *Micropterus*, 135, 170, 210, *211*, 220, 222. *See also* Largemouth, Shoal, Spotted, Striped, Suwannee bass *and* Red drum

■Basswood, *Tilia americana*, 87

●Bats (order Chiroptera), 119–120. *See also* Seminole *and* Yellow bat

■Bay, 46, 149, **150**, 172. *See also* Loblolly, Red, Silk, *and* Swamp bay *and* Sweetbay

■Bay cedar, *Suriana maritima*, 257, 258

■Bayleaf capertree, *Capparis flexuosa*, 110

■Beach morning-glory, *Ipomoea imperati*, *252*, 253, 257. *See also* Railroad vine

●Beach mouse, *Peromyscus polionotus*, 259, *260*, *383*

■Beach peanut, *Okenia hypogaea*, *253*

○Beaded periwinkle, *Tectarius muricatus*, *315*

■Beakrush. *See* Beaksedge

■Beaksedge, *Rhynchospora* species, 73, 141, 164. *See also* Shortbeak *and* Tracy's beaksedge

●Beaver, *Castor canadensis*, 224, 235

■Bedstraw St. John's-wort, *Hypericum galioides*, 188

■Belted kingfisher, *Ceryle alcyon*, 160, 278

●Big brown bat, *Eptesicus fuscus*, 120, *120*

■Big cordgrass, *Spartina cynosuroides*, 273

■Big floatingheart, *Nymphoides aquatica*, 158, *173*, 209

■Big-head sea robin, *Prionotus tribulus*, 294

■Bigleaf magnolia, *Magnolia macrophylla ashei* (endemic), *94*, 96

✕Big Pine Key conehead, *Belocephalus micanopy*, 112

■Big threeawn, *Aristida condensata*, 42

◆Bird-voiced treefrog, *Hyla avivoca*, 190

●Bison (extinct), *Bison antiquus*, 63–64. *See also* American bison

○Bivalve (phylum Mollusca), 316

■Black-and-white warbler, *Mniotilta varia*, *150*

■Blackbanded darter, *Percina nigrofasciata*, 219, 225

■Blackbanded sunfish, *Enneacanthus chaetodon*, *215*

■Blackbead, *Pithecellobium* species, 110

■Black-billed cuckoo, *Coccyzus erythrophthalmus*, 261

✕Black carpenter ant, *Camponotus pennsylvanicus*, *44*

■Black crappie, *Pomoxis nigromaculatus*, 211, 224, 228. *See also* Speckled perch

■Black-crowned night-heron, *Nycticorax nycticorax*, 281

■Black drum, *Pogonias cromis*, *312*

■Blackgum, *Nyssa sylvatica*, **178**. *See also* Swamp tupelo

■Black ironwood, *Krugiodendron ferreum*, 188

■Black mangrove, *Avicennia germinans*, 258, 270, *285*, 286–287, *288*, *289*, 291, 292, 293

■Blackmouth shiner, *Notropis melanostomus* (endemic), 220

■Black mullet. *See* Striped mullet

■Black-necked stilt, *Himantopus mexicanus*, *281*

■Black needlerush. *See* Needle rush

■Black racer. *See* Southern black racer

●Black rat, *Rattus rattus* (alien), 297

■Black rush. *See* Needle rush

■Black skimmer, *Rynchops niger*, 258, *259*, 280, *379*

■Blacksnakeroot, *Sanicula* species, 98

■Blackspotted topminnow, *Fundulus olivaceus*, 219, *225*

■Black swamp snake, *Seminatrix pygaea*, 159, 190, 297

■Blacktail redhorse, *Moxostoma poecilurum*, 219

■Blacktail shiner, *Notropis venustus*, 219, 224, *225*

■Blacktip shark, *Carcharhinus limbatus*, *330*

■Black titi, *Cliftonia monophylla*, 147, *148*, **148**

■Blacktorch, *Erithalis fruticosa*, 110

■Black tupelo. *See* Swamp tupelo

■Black vulture, *Coragyps atratus*, 61, 296

■Black walnut, *Juglans nigra*, 87, 98

✕Black widow spider, *Latrodectus mactans*, 97

■Black willow, *Salix nigra*, 133, 185

■Bladderwort, *Utricularia* species, 141, 142, 158, 167, 209. *See also* Eastern purple *and* Zigzag bladderwort

●Blainville's beaked whale, *Mesoplodon densirostris*, 350

■Blanketflower. *See* Firewheel

■ Flagfin shiner, *Pteronotropis signipinnis*, 219, *219*

■ Flagfish. *See* American-flag fish

○ Flamingo tongue snail, *Cyphoma gibbosum*, *315*

♦ Flatwoods salamander, *Ambystoma cingulatum*, *61*, 162

■ Flaxleaf aster, *Ionactis linariifolia*, 42

■ Flier, *Centrarchus macropterus*, 224, 228

■ Florida anisetree, *Illicium floridanum*, *94*, *95*, 97

⋊ Florida atala, *Eumaeus atala*, *106*, 112

■ Florida beargrass, *Nolina atopocarpa* (endemic), *53*

● Florida black bear, *Ursus americanus*, 14, 44, 63, 81, 151, *152*, 153, 297

■ Florida bluestem, *Andropogon floridanus*, 106

♦ Florida bog frog, *Rana okaloosae*, *149*

■ Florida butterfly orchid, *Encyclia tampensis*, *105*

■ Florida calamint, *Calamintha dentata*, 42

○ Florida cave amphipod, *Crangonyx grandimanus*, 236, *236*

♦ Florida chorus frog, *Pseudacris nigrita verrucosa* (endemic), 162

■ Florida clover ash, *Tetrazygia bicolor*, 106

■ Florida cooter, *Pseudemys floridana*, 224, 243, 278

■ Florida dropseed, *Sporobolus floridanus*, 42

■ Florida flame azalea, *Rhododendron austrinum*, 98, *378*

■ Florida gar, *Lepisosteus platyrhincus*, 210, 293, 294

■ Florida grape, *Vitis cinerea* var. *floridana*, 63

■ Florida greeneyes, *Berlandiera subacaulis* (endemic), 42

■ Florida green water snake, *Nerodia floridana*, 159, 297

⋊ Florida harvester ant, *Pogonmyrmex badius*, 75

● Florida Key deer, *Odocoileus virginianus clavium* (endemic subspecies), 113, 114, 297

■ Florida Keys mole skink, *Eumeces egregius egregius* (endemic), 113

■ Florida Keys noseburn, *Tragia saxicola* (endemic), 106

⋊ Florida leafwing, *Anaea floridalis*, 112

♦ Florida leopard frog, *Rana sphenocephala sphenocephala*, 159

■ Florida maple, *Acer saccharum floridanum*, 87, *94*

● Florida mouse, *Podomys floridanus* (endemic), *49*, 75, 79–80

● Florida panther, *Felis concolor coryi* (endemic subspecies), 44, 63, 113, *181*, 297

■ Florida poisontree, *Metopium toxiferum*, 106, 110

■ Florida pompano, *Trachinotus carolinus*, *312*

■ Florida pondweed, *Potamogeton floridanus* (endemic), 158

■ Florida pugnose minnow, *Opsopoeodus emiliae peninsularis*, 210

⋊ Florida purplewing, *Eunica tatila* (endemic), *107*

■ Florida redbelly turtle, *Pseudemys nelsonii*, 179, 297

■ Florida rosemary, *Ceratiola ericoides*, 73–74, *74*, **74**. *See also* Apalachicola false rosemary, False rosemary

■ Florida royal palm, *Roystonea regia*, *109*

■ Florida sandhill crane, *Grus canadensis pratensis*, 61, *62*, 65, 160, 168

■ Florida scrub-jay, *Aphelocoma coerulescens* (endemic), 75, 79, *79*

■ Florida scrub lizard, *Sceloporus woodi* (endemic), 75, *78*

○ Florida scrub millipede, *Floridobolus penneri* (endemic), 76, *76*

■ Florida softshell, *Apalone ferox*, 243, 297

○ Florida tree snail, *Liguus fasciatus*, 112, 113, *113*

■ Florida willow, *Salix floridana*, *16*, 153

■ Florida worm lizard, *Rhineura floridana* (endemic), 75, *78*

■ Flounder, *Paralichthys* species, 242, 325, 310. *See also* Gulf flounder, Lined sole

■ Flowering dogwood, *Cornus florida*, 87, *95*, 101, *360*

■ Flycatcher, *Myiarchus* species, *193*

■ Flying fish (family Exocoetidae), 349

● Flying squirrel (southern), *Glaucomys volans*, 44, 46, 99

■ Flyr's nemesis, *Brickellia cordifolia*, 98

■ Forb, **36**

⋊ Four-spotted pennant, *Brachymesia gravida*, *159*

● Fox squirrel, *Sciurus niger*, 45–46, *45*, 63, 65, 91, 297

■ Fragrant ladiestresses, *Spiranthes odorata*, 156

• Fragrant maidenhair, *Adiantum melanoleucum*, 116

● Fraser's dolphin, *Lagenodelphis hosei*, 350

■ Freshwater flounder. *See* Lined sole

■ Frog's bit, *Limnobium spongia*, 209

♦ Fungi (kingdom Fungi), **13**, **40**, 45–46, 74, 291

G

■ Gadwall, *Anas strepera*, 193, 211

■ Gallberry, *Ilex glabra*, 55, 58, 133, 147

■ Gar, *Lepisosteus* species, 170, 235, 242, 293. *See also* Florida, Shortnose, *and* Spotted gar

○ Gastropod (phylum Mollusca), **315**, *315*

■ Gayfeather, *Liatris* species, 42. *See also* Godfrey's gayfeather

♦ Georgia blind salamander, *Haideotriton wallacei* (endemic), *236*, 236–237

● Gervais' beaked whale, *Mesoplodon europaeus*, 350

○ Ghost crab, *Ocypode quadrata*, 259, *259*

■ Ghost orchid, *Dendrophylax lindenii*, *377*

● Giant armadillo, *Holmesina septentrionalis* (extinct), *373*

♦ Giant toad. *See* Cane toad

■ Gizzard shad, *Dorosoma cepedianum*, 211, 224, 241

■ Glasswort, 258, 285. *See also* Annual glasswort, Perennial glasswort

■ Glossy crayfish snake, *Regina rigida rigida*, 190

■ Glossy ibis, *Plegadis falcinellus*, 204

■ Goat's rue, *Tephrosia virginiana*, 42

■ Gobies (family Gobiidae), 334

■ Godfrey's gayfeather, *Liatris provincialis* (endemic), *265*

■ Goldenclub, *Orontium aquaticum*, 158, 223

○ Golden crab, *Chaceon fenneri*, *349*

● Golden mouse, *Ochrotomys nuttalli*, *193*

• Golden polypody, *Phlebodium aureum*, 111

■ Goldenrod, *Solidago* species, 36, *263*, 264. *See also* Bush, Sweet, *and* Woody goldenrod

■ Golden shiner, *Notemigonus crysoleucas*, 210

⋊ Golden silk spider, *Nephila clavipes*, 14

■ Golden stars, *Bloomeria crocea*, 98

■ Golden topminnow, *Fundulus chrysotus*, *171*, 211

⋊ Golden-winged skimmer, *Libellula auripennis*, *381*

■ Goldspotted killifish, *Floridichthys carpio*, 293

■ Goliath grouper, *Epinephelus itajara*, 293, *312*

♦ Gopher frog, *Rana capito*, *49*, 50, 81, 202. *See also* Dusky gopher frog

■ Gopher snake. *See* Eastern indigo snake

■ Gopher tortoise, *Gopherus polyphemus*, 37, 45, 48–50, *49*, 61, 72, 75, 79, 80, 259

■ Gopherwood, *Torreya taxifolia*, 16, *95*, *96*, 368

■ Grass, **36**

■Marlin, *Tetrapturus* species, 349, 352. *See also* White marlin

■Marsh fimbry, *Fimbristylis spadicea*, 273

■Marshhay cordgrass, *Spartina patens*, 257, 272

■Marsh killifish, *Fundulus confluentus*, 293

■Marshpennywort, *Hydrocotyle* species, 209

●Marsh rabbit, *Sylvilagus palustris*, 151, *152*, 153, 181, 281, 297

●Marsh rice rat, *Oryzomys palustris natator* (endemic), 281, 297

■Marsh wren, *Cistothorus palustris*, 160, 278, 281

■Mayapple, *Podophyllum peltatum*, 98

■Meadowparsnip, *Thaspium* species, 98

■Meadowrue. *See* Rue anemone

○Medusa worm, *Eteone heteropoda*, 292

■Melaleuca. *See* Punktree

●Melon-headed whale, *Peponocephala electra*, 350

■Menhaden. *See* Yellowfin menhaden

■Merlin, *Falco columbarius*, 114

■Mexican primrosewillow, *Ludwigia octovalvis*, *130*

• Microalgae (division Thallophyta), *208*, **208**, **271**, **321**

■Milkbark, *Drypetes diversifolia*, 110

■Milkpea, *Galactia* species, 42, 73, 258

■Milkweed, *Asclepias* species, 164, 188, *262*, 263

■Milkwort, *Polygala* species, 164

■Mimic glass lizard, *Ophisaurus mimicus*, *146*

●Minke whale, *Balaenoptera acutorostrata*, 350

■Mississippi kite, *Ictinia mississippiensis*, 90, 191, 193

■Mistletoe. *See* Mahogany *and* Oak mistletoe

■Mistletoe cactus, *Rhipsalis baccifera*, *359*

■Mockernut hickory, *Carya alba*, *94*, 96, 100

■Mock pennyroyal, *Hedeoma graveolens* (endemic), *52*, 59

■Mohr's threeawn, *Aristida mohrii*, 42

■Mojarra (family Gerreidae), *241*

○Mole crab, *Emerita talpoida*, 268, *268*

◆Mole salamander, *Ambystoma talpoideum*, 61, 81, 162

■Mole skink, *Eumeces egregius* (endemic), 50. *See also* Bluetail *and* Florida Keys mole skink

■Mollies, *Poecilia* species, 173

○Mollusks (phylum Mollusca), 313–318

✕Monarch, *Danaus plexippus*, 262–264, *263*, *264*

○Moon jellyfish, *Aurelia aurita*, *338*

○Moon snail. *See* Shark eye

• Mosquito fern, *Azolla caroliniana*, 209

■Mosquitofish, *Gambusia* species, *171*, 173, 204, 224, 293

✕Mosquito hawk. *See* Royal river cruiser

✕Moth, specific to Feay's palafox, *Shinia gloriosa*, *83*

■Mountain laurel, *Kalmia latifolia*, *94*, 95

■Mountain spurge, *Pachysandra procumbens*, 98

■Mourning dove, *Zenaida macroura*, 51, 90, 296

○Mud crab, *Scylla serrata*, *293*

■Mudfish, 228. *See also* Bowfin

■Mud snake, *Farancia abacura*, 159, 190, *221*

■Mud sunfish, *Acantharchus pomotis*, 222, 224

■Muhly, *Muhlenbergia* species, 158, 164, 257

■Mule-ear oncidium, *Trichocentrum undulatum*, *110*

■Mullet (family Mullidae), 215, 230, 241, 242, 293, 296, 310. *See also* Striped mullet

■Mummichog, *Fundulus heteroclitus*, 278

■Muscadine, *Vitis rotundifolia*, 109, 151

◆Mushroom(s) (kingdom Fungi), *15*, *90*. *See also* Fungi

○Mussels (class Bivalvia), 221–222, *223*. *See also* Blue mussel, Shinyrayed pocketbook

■Mutton snapper, *Lutjanus analis*, *294*

◆Mycorrhizae (kingdom Fungi), **40**

■Myrsine, *Rapanea punctata*, 106

■Myrtle dahoon, *Ilex cassine* var. *myrtifolia*, 141, 147, 151

■Myrtle oak, *Quercus myrtifolia*, 69, 73, 256

N

■Narrowleaf blue-eyed grass, *Sisyrinchium angustifolium*, *58*

■Narrowleaf yellowtops, *Flaveria linearis*, *254*

◆Narrowmouth toad (eastern), *Gastrophryne carolinensis*, 50, 162, *192*.

■Needlefish (many different genera), 242, 293, 294, 310

■Needle palm, *Rhapidophyllum hystrix*, *94*, 95

■Needle rush, *Juncus roemerianus*, 270, 271, 272, 273, *274*, 278, 294

• Netted chain fern, *Woodwardia areolata*, 151

■Neverwet. *See* Goldenclub

■New Jersey tea, *Ceanothus americanus*, 98

◆Newts, *Notophthalmus* species, 162. *See also* Striped newt

■Northern bobwhite, *Colinus virginianus*, 48, 50, 51, *52*, 61, 81, 90, 112

■Northern cardinal, *Cardinalis cardinalis*, 61, 81, 90, 114, 160, 193

◆Northern cricket frog, *Acris crepitans*, *192*

■Northern flicker, *Colaptes auratus*, 51, 90

■Northern harrier, *Circus cyaneus*, 160, 281

■Northern mockingbird, *Mimus polyglottos*, *85*, 90, 263

■Northern needleleaf, *Tillandsia balbisiana*, *180*

■Northern parula, *Parula americana*, 90, 100, 193

●Northern right whale, *Eubalaena glacialis*, 350, *353*

○Nudibranch (phylum Mollusca), *304*. *See also* Fire *and* Lettuce nudibranch

O

■Oakleaf hydrangea, *Hydrangea quercifolia*, *97*

■Oak mistletoe, *Phoradendron leucarpum*, 100, 101

◆Oak toad, *Bufo quercicus*, 50, 81, 112, 162

■Oblongleaf twinflower, *Dyschoriste oblongifolia*, 106

■Octoberflower, *Polygonella polygama*, 73

○Octopus (family Octopodidae), 329

■Odorless bayberry, *Myrica inodora*, 147, 151

■Ogeechee tupelo, *Nyssa ogeche*, **178**, *179*

✕Ogre-faced spider, *Deinopis spinosa*, *401*

■Ohoopee shiner. *See* Bannerfin shiner

■Okaloosa darter, *Etheostoma okaloosae* (endemic), 220, 225, *383*

■Okefenokee pygmy sunfish, *Elassoma okefenokee*, 211

○Oligochaetes (phylum Annelida), **313**

◆One-toed amphiuma, *Amphiuma pholeter*, 189, 190, *191*, 225

◆Onion-stalk lepiota, *Lepiota cepaestipes*, *90*

■Orchids (family Orchidaceae), 141. *See also* Florida butterfly, Ghost, Rosebud, Waterspider false rein-, *and* Wormvine orchid

◆Ornate chorus frog, *Pseudacris ornata*, 50, *61*, 162

■Ornate diamondback terrapin, *Malaclemys terrapin macrospilota* (endemic), *297*

⋈Scarab beetle (family Scarabaeidae), 112

■Scareweed, *Baptisia simplicifolia* (endemic), 59

■Scarlet calamint, *Calamintha coccinea*, 73

■Scorpion fish, *Scorpaenopsis* species, 334

■Scrub balm, *Dicerandra frutescens* (endemic), 74, *75*, 77–78

■Scrub hickory, *Carya floridana* (endemic), 73

■Scrub-jay, *Aphelocoma* species, 376. *See also* Florida scrub-jay

◆Scrub lichen, *Cladina evansii* and *Cladonia perforata*, 72

■Scrub oak, *Quercus inopina* (endemic), **67**. *See also* Scrub oaks

■Scrub oaks (*Quercus* species), **67**, 69, 71, 79, 254, 256, 257. *See also* Oak scrub, in General Index

■Scrub palmetto, *Sabal etonia* (endemic), 73

❍Scrub snail, *Veronicella floridana* (slug), 77

■Scrub wild olive, *Osmanthus megacarpus* (endemic), 73

⋈Scrub wolf spider, *Geolycosa* species, 75

❍Sea anemone (phylum Coelenterata), 329, 338. *See also* Pink-tipped anemone

■Seacoast marshelder, *Iva imbricata*, 257

■Seagrape, *Coccoloba uvifera*, 110, 254, 257, 258, 363

■Seagrasses, *9*, *299*, 321–331, *320*

■Seahorse, *Syngnathus* species, 329

■Seaoats, *Uniola paniculata*, *5*, *249*, 252–253, 257, 259, *260*

■Seapurslane, *Sesuvium* species, 257, 258

■Sea robin, *Prionotus* species, 334

■Seashore dropseed, *Sporobolus virginicus*, 258

■Seaside sparrow, *Ammodramus maritimus*, 278, 280, 281

❍Sea squirt (phylum Chordata), *327*

■Sea torchwood, *Amyris elemifera*, 110

■Sea turtles (order Testudinata), 354–357, *357*. *See also* Atlantic ridley, Green, Hawksbill, Leatherback, *and* Loggerhead

• Seaweed. *See* Macroalgae

■Sedge, **157**

●Sei whale, *Balaenoptera borealis*, 350

●Seminole bat, *Lasiurus seminolus*, 100, 120, *367*

■Seminole killifish, *Fundulus seminolis*, *171*, 211

■Semipalmated plover, *Charadrius semipalmatus*, 295

■Shad, *Alosa* species, 218. *See also* Gizzard *and* Threadfin shad

❍Shark eye, *Polinices duplicatus*, *315*

■Sharp-shinned hawk, *Accipiter striatus*, 261, 296

■Sheepshead, *Archosargus probatocephalus*, 241, 277, 294, *312*

■Sheepshead minnow, *Cyprinodon variegatus*, 205, 293

■Shiny blueberry, *Vaccinium myrsinites*, 58, 73

❍Shinyrayed pocketbook, *Lampsilis subangulata*, *222*

■Shoal bass, *Micropterus cataractae* (endemic), 220

■Shoalweed, *Halodule wrightii*, *322*, 323

• Shoestring fern, *Vittaria lineata*, 111

■Shoregrass. *See* Keygrass

■Shortbeak beaksedge, *Rhynchospora nitens*, 158

■Short-billed dowitcher, *Limnodromus scoloplaceus*, 269

●Short-finned pilot whale, *Globicephala macrorhynchus*, 350

■Shortleaf pine, *Pinus echinata*, 12, 87

■Shortnose gar, *Lepisosteus platostomus*, 210

■Shortnose sturgeon, *Acipenser brevirostris* (endemic), 220

■Short-tailed hawk, *Buteo brachyurus*, 61, 114

●Short-tailed shrew, *Blarina carolinensis*, 62, 113, 297

■Short-tailed snake, *Stilosoma extenuatum*, 50, 75, *78*

❍Shrimp, *Penaeus* species, 329. *See also* Brown, Cave, and Pink shrimp *and* Red cleaning shrimp

■Silk bay, *Persea borbonia* var. *humilis* (endemic), 73

■Silky camellia, *Stewartia malacodendron*, *95*

■Silverbell, *Halesia* species, 87, 188

■Silver bluestem, *Bothriochloa laguroides* (alien), 106

●Silver-haired bat, *Lasionycteris noctivagans*, 120

■Silver perch, *Bairdiella chrysoura*, 294

■Simpson's applecactus, *Hartisia simpsonii* (endemic), 110

■Six-lined racerunner, *Cnemidophorus sexlineatus*, 50, 50, 81, 259

■Skate, *Raja* and other species, 334. *See also* Clearnose skate

●Skunk (family Mustelidae), 260. *See also* Spotted *and* Striped skunk

■Skyflower, *Hydrolea corymbosa*, *209*

■Slash pine, *Pinus elliottii*, 11, *12*, 34, 38, *39*, 56, *66*, 105, *107*, 140, 141, 147, 150, 151, 204, 254, 256, 270

■Slenderleaf clammyweed, *Polanisia tenuifolia*, 43

■Slimspike threeawn, *Aristida longespica*, 43

◆Slimy salamander, *Plethodon grobmani*, 81, 118, 121, 190

■Slippery elm, *Ulmus rubra*, 133

■Smallfruit varnishleaf, *Dodonaea elaeagnoides*, 106

■Smartweed, *Polygonum* species, 223. *See also* Dotted smartweed

■Smoothbark St. John's-wort, *Hypericum lissophloeus* (endemic), 133

■Smooth Solomon's seal, *Polygonatum biflorum*, 98

■Snail kite, *Rostrhamus sociabilis*, *169*

❍Snails (class Gastropoda), 221, 240, 241. *See also* Apple, Cave, Coffee bean, Flamingo tongue, Florida tree, Periwinkle, and Scrub snail *and* Shark eye

■Snake bird. *See* Anhinga

■Snapper (family Lutjanidae), 293, 294, 298, 329, 334. *See also* Gray *and* Yellowtail snapper

■Snook, *Centropomus* species, 241, 293, 294

■Snowberry, *Chiococca alba*, 109, 257

■Snowy egret, *Egretta thula*, 159, *281*, 281, 295, *296*

■Snowy plover, *Charadrius alexandrinus*, 258

■Soapberry, *Sapindus saponaria*, 110

❍Soft coral (order Alcyonacea), *320*, 338–345, *341*

■Sooty tern, *Sterna fuscata*, 258

■Sora, *Porzana carolina*, 281, 295

●Southeastern bat, *Myotis austroriparius*, 120

■Southeastern crowned snake, *Tantilla coronata*, 50

■Southeastern five-lined skink, *Eumeces inexpectatus*, 50, 81, 112, 149, 192

■Southeastern sunflower, *Helianthus agrestis*, 157

■Southern arrowwood, *Viburnum dentatum*, 87

■Southern black racer, *Coluber constrictor priapus*, 50, 81, 112, 149

■Southern cattail, *Typha domingensis*, 273

◆Southern chorus frog, *Pseudacris nigrita*, 190

◆Southern dusky salamander, *Desmognathus auriculatus*, 190, *191*

■Southern fogfruit, *Phyla stoechadifolia*, 164

■Southern hognose snake, *Heterodon simus*, 50, 259

◆Southern leopard frog, *Rana sphenocephala utricularius*, 162, 190, 278, *367*

• Southern lip fern, *Cheilanthes microphylla*, 116

■Southern magnolia, *Magnolia grandiflora*, 6, 36, 80, 87–91, *89*, *95*,

T

■ Tadpole madtom, *Noturus gyrinus*, 211
■ Taillight shiner, *Notropis maculatus*, 210
■ Tall thistle, *Cirsium altissimum*, 98
■ Tall threeawn, *Aristida patula*, 257
■ Tamarind, *Tamarindus indicus* (alien), 109, 110
■ Tapegrass, *Vallisneria americana*, 240
■ Tarflower, *Bejaria racemosa*, 58
■ Tarpon, *Megalops atlanticus*, 241, *279*, 293
■ Tarpon snook, *Centropomus pectinatus*, *294*
■ Teal, *Anas* species, 211
■ Tennessee leafcup, *Polymnia laevigata*, 98
■ Thickleaf waterwillow, *Justicia crassifolia* (endemic), *16*
■ Threadfin shad, *Dorosoma petenense*, 210, 211, 224
■ Threadleaf sundew, *Drosera filiformis*, *142*
■ Threeawn. *See* Arrowfeather, Big, Mohr's, Slimspike, Tall, *and* Woollysheath threeawn
♦ Three-lined salamander, *Eurycea guttolineata*, 190, *191*
■ Ticktrefoil, *Desmodium* species, 43
♦ Tiger salamander (eastern), *Ambystoma tigrinum*, 50, *135*, 162
⌘ Tiger swallowtail, *Pterouras glaucus*, *65*, 265
■ Titi, *57*, 141, **148**, 151, 260. *See also* Black *and* Swamp titi
■ Toadfish, *Opsanus* species, 334, 337. *See also* Gulf toadfish
■ Toothachegrass, *Ctenium aromaticum*, 140
■ Topminnow, 170, *171. See also* Killifish, Mosquitofish
■ Torreya. *See* Gopherwood
■ Tracy's beaksedge, *Rhynchospora tracyi*, 158
■ Tracy's sundew, *Drosera tracei*, *142*
■ Trailing arbutus, *Epigaea repens*, 94
⌘ Trap-door spider, *Cyclocosmia truncata*, *189*
■ Tread softly, *Cnidoscolus stimulosis*, 73, 253
■ Triangle cactus, *Acanthocereus tetragonus*, 110
■ Tricolored heron, *Egretta tricolor*, 159, *231*, 281, *295*
■ Tropical royalblue waterlily, *Nymphaea elegans*, *209*
● True's beaked whale, *Mesoplodon mirus*, 350
■ Trumpet pitcherplant. *See* Yellow pitcherplant
■ Tuberous grasspink, *Calopogon tuberosus*, 58, 164
○ Tube worms, deep-sea, 351. *See also* Oligochaetes, Polychaetes, *and* Worm shell
■ Tufted titmouse, *Baeolophus bicolor*, 51, 90
■ Tuliptree, *Liriodendron tulipifera*, 46, 87, *95*, 133, 153
■ Tuna, *Thunnus*, 349, 352. *See also* Yellowfin tuna
■ Tupelo, *Nyssa* species, *57*, 125, *141*, 157, 161, 175, **178**, 178–179, 363. *See also* Ogeechee, Swamp, *and* Water tupelo
■ Turkey oak, *Quercus laevis*, *38*, 43, 371
♦ Turkey-tail mushroom, *Trametes versicolor*, *15*
■ Turkey vulture, *Cathartes aura*, 14, 61, 81, 296
○ Turkey wing, *Arca zebra*, *317*
■ Turtlegrass, *Thalassia testudinum*, *322*, 322–323, 324, 329, 357
♦ Two-lined salamander, *Eurycea cirrigera*, 121, 190, *191*, 225
♦ Two-toed amphiuma, *Amphiuma means*, 159

U

■ Umbrellasedge, *Fuirena* species, 256
♦ Upland chorus frog, *Pseudacris triseriata feriarum*, *134*

V

■ Variable witchgrass, *Dicanthelium commutatum*, 188
■ Varnishleaf, *Dodonaea viscosa*, 257
○ Vase sponge, *Ircinia campana*, *335*
■ Veery, *Catharus fuscescens*, 193
■ Velvetseed, *Guettarda* species, 110
• Venus'-hair fern, *Adiantum capillus-veneris*, 116
⌘ Viceroy, *Limenitis archippus*, *263*, 263
○ Vinegaroon, *Mastigoproctus giganteus*, 75
• Virginia chain fern, *Woodwardia virginica*, 151
■ Virginia creeper, *Parthenocissus quinquefolia*, 101, 109
■ Virginia iris, *Iris virginica*, *133*
■ Virginia live oak. *See* Live oak
● Virginia opossum, *Didelphis virginiana*, 44, 63, 81, 99, 113, 114, 151, 153, 160, 297, *366*
■ Virginia rail, *Rallus limicola*, 281, 295
■ Virginia saltmarsh mallow, *Kosteletzkya virginica*, *267*
■ Virginia willow, *Itea virginica*, *95*
■ Vulture. *See* Black *and* Turkey vulture

W

● Wagner's mastiff bat, *Eumops glaucinus*, 120
■ Wahoo, *Acanthocybium solandri*, 349
■ Wakerobin, *Trillium* species, 98
⌘ Walking stick (order Phasmatodea), *60*, 112
■ Wand loosestrife, *Lythrum lineare*, 273
■ Warmouth, *Lepomis gulosus*, 211, 219, 228
■ Water hickory, *Carya aquatica*, 188
■ Water locust, *Gleditsia aquatica*, 188
■ Watermeal, *Wolffia* species, 209
■ Water moccasin. *See* Cottonmouth
■ Waternymph, *Najas* species, 158
■ Water oak, *Quercus nigra*, *37*, 87, 187, *265*
■ Water primrose. *See* Creeping primrosewillow
■ Watershield, *Brasenia schreberi*, 209
■ Waterspider false reinorchid, *Habenaria repens*, *204*, 209
■ Waterthyme, *Hydrilla verticillata* (invasive exotic), 11
■ Water tupelo, *Nyssa aquatica*, *174*, **178**, 186, 187
■ Wax myrtle, *Myrica cerifera*, 58, 87, 106, *132*, 133, 147, 151, 186, 254
■ Weed shiner, *Notropis texanus*, 219, 224
○ Wentletrap, *Epitonium* species, *315*
■ West Indian mahogany, *Swietenia mahagoni*, 110, 111
● West Indian manatee, *Trichecus manatus*, 243, *244*, 280, 297, 309, 349, 350
■ West Indian tufted airplant, *Guzmania monostachia*, *180*
● Whales (order Cetacea), 349, 350. *See also* Northern right whale
■ White baneberry, *Actaea pachypoda*, 94
■ White basswood, *Tilia americana* var. *heterophylla*, 98
■ White birds-in-a-nest, *Macbridea alba* (endemic), *58*
■ White-breasted nuthatch, *Sitta carolinensis*, 51, 90
■ White catfish, *Ictalurus catus*, 211, 241
■ White-crowned pigeon, *Columba leucocephala*, 114, 296
■ White-eyed vireo, *Vireo griseus*, 90, 193, *372*
■ White ibis, *Eudocimus albus*, 159, 168, 173, 277
■ White indigoberry, *Randia aculeata*, 98, 110

■White mangrove, *Laguncularia racemosa*, 258, 286

■White marlin, *Tetrapterus albidus*, *351*

■White oak, *Quercus alba*, 87, *95*, 98

✶White peacock, *Anartia jatrophae*, *168*

○White shrimp, *Penaeus setiferus*, 279, 310

■White stopper, *Eugenia axillaris*, 110

●White-tailed deer, *Odocoileus virginianus*, 63, 80, 81, 82, 181, 281, 297, *382*

■White-tailed kite, *Elanus leucurus*, 61

■White thoroughwort, *Eupatorium album*, 43

■Whitetop aster, *Sericocarpus* species, 43

■Whitetop pitcherplant, *Sarracenia leucophylla*, *138*, *139*

■Wicky, *Kalmia hirsuta*, 58

■Wild azalea. *See* Sweet pinxter azalea

■Wild blue phlox, *Phlox divaricata*, 98

■Wild coffee, *Psychotria* species, 110, 257

■Wild columbine, *Aquilegia canadensis*, *97*, 98

■Wild ginger, *Asarum arifolium*, 94, *96*

■Wild grape, *Vitis* species, 147, 151, 188

■Wild hydrangea, *Hydrangea arborescens*, 94

■Wild lime, *Zanthoxylum fagara*, 112

■Wild olive, *Osmanthus americanus*, 87, 96, 256

■Wild rice, *Zizania aquatica*, 230

■Wild turkey, *Meleagris gallopavo*, 61–63, 65, 91, 99, 112, 181, 211

■Willet, *Catoptrophorus semipalmatus*, 258, 261

■Willow, *Salix* species, 172, 179. *See also* Black *and* Florida willow

■Willow bustic, *Sideroxylon salicifolium*, 106, 110

■Wilson's plover, *Charadrius wilsonia*, 258

■Winged elm, *Ulmus alata*, 87

■Winged loosestrife, *Lythrum alatum*, *137*

■Winged sumac, *Rhus copallinum*, 106

■Wiregrass, *Aristida stricta* var. *beyrichiana*, *2*, 34, 35, 36, 40, 41–42, *41*, *50*, *52*, *54*, *56*, 61, 90, *94*, 96, 106, 107, 140, 141, 161, *384*

◆Witches' butter fungus, *Tremella mesenterica*, 90

■Witchhazel, *Hamamelis virginiana*, 87, 188

○Wood-boring isopod, *Sphaeroma terebrans*, 292

■Wood duck, *Aix sponsa*, 44, 48, 51, 99, 192, 193, 211, *233*

■Woodpeckers, 44, 100, 191–192

■Wood stork, *Mycteria americana*, *160*, 173, 181, 280, 293

■Wood thrush, *Hylocichla mustelina*, 90, *101*, 102

○Woodville cave crayfish, *Procambarus orcinus*, *235*, 237

■Woody goldenrod, *Chrysoma pauciflosculosa*, 254

■Woollysheath threeawn, *Aristida lanosa*, 43

○Worm shell, *Vermicularia knorri*, *315*

■Wormvine orchid, *Vanilla barbellata*, *19*

■Wrasse, *Bodianus* species, 334

Y

■Yaupon, *Ilex vomitoria*, 73

●Yellow bat, *Lasiurus intermedius*, 100, 120

■Yellow-bellied sapsucker, *Sphyrapicus varius*, 44, 51, 261

■Yellow-billed cuckoo, *Coccyzus americanus*, 90, 261, 296

■Yellow bullhead, *Ictalurus natalis*, 211

■Yellow-crowned night-heron, *Nycticorax violacea*, 193, *235*

■Yelloweyed grass, *Xyris* species, 141, 203

■Yellowfin menhaden, *Brevoortia smithi*, *330*

■Yellowfin mojarra, *Gerres cinereus*, 293

■Yellowfin tuna, *Thunnus albacares*, *348*

■Yellow indiangrass, *Sorghastrum nutans*, 43

■Yellow milkwort, *Polygala rugelii* (endemic), 59

■Yellow pitcherplant, *Sarracenia flava*, *143*

■Yellow poplar. *See* Tuliptree

■Yellowtail snapper, *Ocyurus chrysurus*, *344*

■Yellow-throated vireo, *Vireo flavifrons*, 90

■Yellow-throated warbler, *Dendroica dominica*, 61, 90

■Yellow warbler, *Dendroica petechia*, 296

○Zebra arc. *See* Turkey wing

■Zigzag bladderwort, *Utricularia subulata*, *135*

■Zigzag silkgrass, *Pityopsis flexuosa* (endemic), 43

○Zooplankton, **199**, **268**, 304–306, *305*

•Zooxanthellae (division Thallophyta), **339**, 383

General Index

Page numbers for definitions are given in bold type (**123**). Pages for photos and drawings are in italics (*123*). Page numbers on which map locations are shown are in italics with the word *map* (*123map*).

For species and for broader categories such as "Lichens," "Forbs," "Corals," and the like, please see the Index to Species and Other Categories on pages 403–417.

St. Johns River (continued)
215, *217map*, 220, 239, *302map*, 303, 308, 310

St. Joseph Bay, *302map*, 325

St. Lucie County, 8

St. Lucie Inlet, 294

St. Lucie River, 352

St. Marks National Wildlife Refuge, 15, 64, *86map*, 103, 283

St. Marks River/Spring, *217map*, 240, 273, 308

St. Mary's River, *136map*, *217map*, 227

St. Vincent Island, 68, *249*, *251map*

Stalactites/stalagmites, *116*

Steephead
  ravines, 92–97, *93*, *94–95*
  streams, **216**, 217, 224–226

Steinhatchee River/Rise, 136, *217map*, 241, 308

Stock Island, 120

Storms, 9, 20, 21, 23–24, **23**, *23map*, *23*

Strands, **179**, **215**

Strand swamps, *vi*, 175, 179–181

Streams, 215–231, **215**, *217map*
  bed/channel, 181–182, *182*
  order, 181–182, **182**, 218
  see also Flood/floodwater, Floodplains

Submersed plants, **127**, 198, 209

Subsidence, **115**

Subspecies (taxonomic term), 14, 403

Succulent (plants), **272**, 285

Sumter County, 136, 155

Surface fire, **35**, **147**

Surficial aquifer, **139**, *140*

Sutton's Lake, *129*, 179

Suwannee County, 228, 233, 241

Suwannee River/floodplain, 215, *217map*, 239, 243, 308, 366
  biodiversity in, 193, 220
  elements of, 227–230, *227*, *228*, *229*
  see also Bays/estuaries

Swales (wetlands), **126**, 161, 164–165, **165**, 254. *See also* Interdune swale

Swamps, **126**, *174*, 175–194, **269**

Swamp lakes/blackwater lakes, 201, 202–203, **202**, *203*

Sweetwater Creek, 93

T

Tampa Bay, *302map*, 308, 309, 316, 325, 326

Tardily deciduous hardwoods, **88**

Tarpon Springs, *302map*

Tate's Hell, *136map*

Taxonomic categories/diversity, 13–14, 347

Taylor County, 64, 136, 241

Taylor Slough, *170*

Temperate hardwoods/hammocks, *84*, 85–103, **85**

Temporary ponds, **161**, *161*

Ten Thousand Islands, *284*, 285, *302map*, 316, 318

Thomas County (Georgia), 28, 37

Thomas Farm (archaeological) site, 366–367

Threatened species, *17map*

Tidal
  marshes, *2*, *8*, *245*, 266–267, *266*, 269, 270–283, *282*
  swamps, 188, 269
  wetlands, 175, **189**, 269–283, **269**
  wrack, 272–273, *274*. *See also* Seawrack

Tides (spring, neap), **270**

Tiger Creek Preserve, *86map*

Timberholes, 334

Tip-up mounds, *24*, *34*, 89–90

Titi swamps, **148**, *148*

Tolomato River, *226*

Toms Creek, 383

Topography, *27map*, 27–29, **29**

Topsail Hill State Park, 205

Torreya State Park, 97, 222, 226

Tortugas/Tortugas Sanctuary, 326, **328**, 334, 340

Tosohatchee State Reserve, 64

Trail Ridge, 93

Transpiration, 155

Tree islands (Everglades), *108map*, 109–110, *110*

Tropical
  depressions, 23
  hardwoods/hammocks, **85**, *86map*, 107, 110
  storms, **23**. *See also* Storms
  see also Coral reefs

Troy Spring, 240

Tupelo swamps, **178**

Turtle Mound, *86map*

U

Understory, **34**

United Nations World Heritage Site, 169

Upland, *27map*, 31, 33
  glades, 17
  pine, **33**
  plains, *28map*

Upland (continued)
  species, **133**

V

Variety (taxonomic term), 14, 403

Vermetid reefs, *302map*, 318

Volusia Blue Spring, 241

Volusia County, 86, 241

W

Waccasassa Bay State Preserve, 28, 70, 380

Waccasassa River, *217map*, *226*, 308

Wacissa River/Spring, *217map*, 240

Wade Longleaf Forest, *37*

Wakulla County, 64, 86, 102, 117, 125, 129, 149, 179, 233, 241, 265, 283

Wakulla River/Spring, *217map*, *233*, 234–235, *234*, *235*, 239, 241

Walton County, 28, 139, 205, 220, 225, 347, 358

Washington Oaks Gardens State Park, 381, 383

Water Resources Atlas of Florida, 233

Water table, 233, 234

Weather, 19–24, **19**

Weeds, 4

Weeki Wachee River/Spring, 240, 308

Wekiva River, 217

Welaka Research and Education Center, 64

Welaka Reserve, *86map*

Western Red Hills, *92map*

Wet flats/flatwoods/prairies, **126**, 161, 162–164, *163*

Wetlands, **123**, 125–137, *126*, *127map*, *136map*, 193–194. *See also* Plants (wetland types)

Wet sinks. *See* Sinks/sinkholes

Wilson, Edward O., 381

Withlacoochee Bay, *302map*

Withlacoochee River
  (from Green Swamp), 136, *217map*, 308
  (Suwannee tributary), *217map*, 228

Woodville Karst Plain, *27map*

Worm reefs, 301, *302map*, 318

Wright Hammock, 109

X

Xeric (forests/hammocks/soils), **25**

Xeric oak hammocks, 98–100, *98*

Y

ybp, **361**

Yellow River, *217map*, 219, 306

Yucatan Channel, *350map*

OTHER TITLES FROM PINEAPPLE PRESS

If you enjoyed reading this book, here are some other Pineapple Press titles you might enjoy as well. To request our complete catalog or to place an order, write to Pineapple Press, P.O. Box 3889, Sarasota, Florida 34230, or call 1-800-PINEAPL (746-3275). Or visit our website at www.pineapplepress.com.

Common Coastal Birds of Florida and the Caribbean by David W. Nellis. Covers 72 of the most common birds that inhabit the coastal areas of Florida and the islands to the south. Discusses each bird's own ecological niche, manifested by its nesting, feeding, roosting, and migration habits. More than 250 photographs. (hb or pb)

The Everglades: River of Grass, 50th Anniversary Edition by Marjory Stoneman Douglas. This is the treasured classic of nature writing that captured attention all over the world and launched the fight to save the Everglades. This Anniversary Edition offers an update by Cyril Zaneski, environmental writer for the *Miami Herald*, on the events affecting the Glades since 1987. (hb)

The Ferns of Florida by Gil Nelson. Treats Florida's amazing variety of ferns. Includes color plates with more than 200 images, notes on each species' growth form and habit, general remarks about its botanical and common names, unique characteristics, garden use, and history in Florida. (hb or pb)

The Florida Water Story: From Raindrops to the Sea by Peggy Sias Lantz and Wendy A. Hale. Introduces readers 10 to 14 years old to Florida's water systems and describes and illustrates many of the plants and animals that depend on the lakes, springs, sinkholes, rivers, the Everglades, coastal beaches, mangrove forests, marshes, coral reefs, the Gulf of Mexico, and the Atlantic Ocean. (hb)

Florida's Birds: A Handbook and Reference by Herbert W. Kale II and David S. Maehr. Illustrated by Karl Karalus. This guide to identification, enjoyment, and protection of Florida's varied and beautiful population of birds identifies and discusses more than 325 species, with information on distinguishing marks, habitat, season, and distribution. Full-color illustrations throughout. (hb or pb)

The Gopher Tortoise: A Life History by Patricia Sawyer Ashton and Ray E. Ashton Jr. With hundreds of color photos and easy text, this book offers children and adults a first-rate explanation of the critical role this fascinating tortoise plays in shaping upland Florida and the Southeast. (hb or pb)

Guide to the Lake Okeechobee Area by Bill and Carol Gregware. The first comprehensive guidebook to this area includes a 110-mile hike/bike tour on top of the Herbert Hoover Dike encircling the lake. (pb)

The Gulf of Mexico by Robert H. Gore. A synopsis of the history, geology, geography, oceanography, biology, ecology, and economics of this great body of water. The only book of its kind. (hb)

Key Biscayne: A History of Miami's Tropical Island and the Cape Florida Lighthouse by Joan Gill Blank. This is the engaging history of the southernmost barrier island in the United States and the Cape Florida Lighthouse, which has stood at Key Biscayne's southern tip for 170 years. (hb or pb)

A Land Remembered by Patrick Smith. This well-loved, best-selling novel tells the story of three generations of the MacIveys, a Florida family battling the hardships of the frontier, and how they rise from a dirt-poor cracker life to the wealth and standing of real estate tycoons. Also available in a student edition and a teacher's manual. (hb or pb)

A Natural History Atlas to the Cays of the U.S. Virgin Islands by Arthur E. Dammann and David W. Nellis. A survey of the more than 60 small islands lying off the main U.S. Virgin Islands, with full-color photographs and descriptions of each cay and its plant and animal life (over 150 species pictured). (pb)

Paynes Prairie—The Great Savanna: A History and Guide 2nd Edition by Lars Andersen. Offers the sweeping history of the shallow-bowl basin just south of Gainesville, but now adds a guide to outdoor activities that can be enjoyed in the state preserve today, including maps. (pb)